PARTIAL DIFFERENTIAL EQUATIONS OF EVOLUTION

MATHEMATICS AND ITS APPLICATIONS
Series Editor: G. M. BELL,
Emeritus Professor of Mathematics, King's College, University of London

STATISTICS, OPERATIONAL RESEARCH AND COMPUTATIONAL MATHEMATICS Section
Editor: B. W. Conolly,
Emeritus Professor of Mathematics (Operational Research), Queen Mary College, University of London

Mathematics and its applications are now awe-inspiring in their scope, variety and depth. Not only is there rapid growth in pure mathematics and its applications to the traditional fields of the physical sciences, engineering and statistics, but new fields of application are emerging in biology, ecology and social organization. The user of mathematics must assimilate subtle new techniques and also learn to handle the great power of the computer efficiently and economically.

The need for clear, concise and authoritative texts is thus greater than ever and our series endeavours to supply this need. It aims to be comprehensive and yet flexible. Works surveying recent research will introduce new areas and up-to-date mathematical methods. Undergraduate texts on established topics will stimulate student interest by including applications relevant at the present day. The series will also include selected volumes of lecture notes which will enable certain important topics to be presented earlier than would otherwise be possible.

In all these ways it is hoped to render a valuable service to those who learn, teach, develop and use mathematics.

Mathematics and its Applications
Series Editor: G. M. BELL,
Emeritus Professor of Mathematics, King's College, University of London

Anderson, I.	**Combinatorial Designs: Construction Methods**
Artmann, B.	**Concept of Number: From Quarternions to Monads and Topological Fields**
Arczewski, K. & Pietrucha, J.	**Mathematical Modelling in Discrete Mechanical Systems**
Arczewski, K. & Pietrucha, J.	**Mathematical Modelling in Continuous Mechanical Systems**
Bainov, D. D. & Konstantinov, M.	**The Averaging Method and its Applications**
Baker, A. C. & Porteous, H. L.	**Linear Algebra and Differential Equations**
Balcerzyk, S. & Jösefiak, T.	**Commutative Rings**
Balcerzyk, S. & Jösefiak, T.	**Commutative Noetherian and Krull Rings**
Baldock, G. R. & Bridgeman, T.	**Mathematical Theory of Wave Motion**

Series continued at back of book

PARTIAL DIFFERENTIAL EQUATIONS OF EVOLUTION

JAROSLAV BARTÁK, Ph. D., Associate Professor,
Faculty of Education, Charles University,
LEOPOLD HERRMANN, Ph. D., Mathematical Institute,
Czechoslovak Academy of Sciences,
VLADIMÍR LOVICAR, Mathematical Institute,
Czechoslovak Academy of Sciences,
OTTO VEJVODA, Ph. D., Associate Professor,
Faculty of Sciences, Charles University

Translator
LEOPOLD HERRMANN

Translation Editor
Dr E. H. TWIZELL B.Sc., M.A., Ph.D.
Department of Mathematics and Statistics
Brunel University, Uxbridge

ELLIS HORWOOD
NEW YORK LONDON TORONTO SYDNEY TOKYO SINGAPORE

English Edition first published in 1991
in coedition between
ELLIS HORWOOD LIMITED
Market Cross House, Cooper Street,
Chichester, West Sussex, PO19 1EB, England

A division of
Simon & Schuster International Group
A Paramount Communications Company

and
SNTL – Publishers of Technical Literature
Prague, Czechoslovakia

Printed in Czechoslovakia

British Library Cataloguing in Publication Data

Partial differential equations of evolution.
1. Evolution equations
I. Barták, Jaroslav II. Series
515.7
ISBN 0-13-651449-9

Library of Congress CIP data available

Contents

Preface

The present textbook is devoted to partial differential equations of evolution, that is to equations containing partial derivatives with respect to the *time* variable. There are systematically presented all four standard types of linear evolution equations of mathematical physics, namely first order equations, the wave (and telegraph) equation (as a representative of second order hyperbolic equations), the heat equation (as a representative of second order parabolic equations) and the (fourth order) beam and plate equations.

The properties of these equations and also the methods of their treatment show wide differences. Except for numerical analysis all fundamental methods of solution are introduced: the method of characteristics (typical for hyperbolic and first order equations), the method of Laplace integral transformation (most suitably adapted for parabolic equations) and the separation-of-variables method together with the Fourier method (equally well applicable to both hyperbolic and parabolic second order equations as well as to fourth order equations). It should be stressed that the theory of differential equations of evolution has grown to a very extensive and rather complex area and for this reason it is natural that an introductory and elementary book on the subject, such as this, can but touch upon some selected aspects and necessarily neglects many others.

A mathematical approach to solving a certain problem arising in engineering, physical, biological, social sciences, and so forth, consists, in principle, of three stages:

(1) the creation of a mathematical model,
(2) its mathematical treatment,
(3) the interpretation of results.

Throughout the text attention is paid to all three stages. The discussion is concerned with models given by means of partial differential equations of evolution. We attempt to indicate at least briefly how the equations to be studied stem from the theory of vibrations, heat conduction, hydrodynamics, etc. and, on the other hand, how the obtained mathematical results may be interpreted in terms of the original problem. Of course, due to the mathematical character of the work the second stage receives the most attention.

The first chapter provides a brief introduction to the theory of partial differential equations of evolution. We recall some notation, basic concepts and auxiliary results. The second chapter (by J. Barták and O. Vejvoda) presents partial differential equations of the first order and systems of such equations. The wave and telegraph equations are examined in the third chapter (by V. Lovicar). The fourth chapter (by L. Herrmann) deals with the heat (diffusion) equation. Eventually, the aim of Chapter 5 (by L. Herrmann) is to illustrate the use of the Fourier method for solving equations of the fourth order. Chapters 2 to 4 may be read independently to a great extent. A large number of solved examples and exercises with hints and answers are included.

An understanding of the material requires only modest prerequisites; it is dependent actually only upon a knowledge of Calculus, elements of the theory of ordinary differential equations, linear algebra, special functions, integral transforms and (when dealing with the Fourier method) upon a rudimentary knowledge of functional analysis concerning Hilbert space theory. So, the text may be followed by a reader with a mere background in engineering curriculum. The relevant results may be found in various standard textbooks.

In conclusion we wish to express our gratitude to our colleagues for many valuable suggestions and helpful comments.

Jaroslav Barták
Leopold Herrmann
Vladimír Lovicar
Otto Vejvoda

Prague,
September 1989

CHAPTER 1

Preliminaries

1.1 PARTIAL DIFFERENTIAL EQUATIONS OF EVOLUTION

After a brief survey of conventions and notation this introductory section provides basic terminology and concepts of the theory of partial differential equations. We define the classes of evolutionary, linear, nonlinear, homogeneous, inhomogeneous, elliptic, parabolic and hyperbolic partial differential equations, respectively. The notions of initial and boundary conditions will be clarified as well as those of well-posedness and stability.

1.1.1 Mathematical models

To get useful pieces of information on phenomena and processes arising in systems (be they physical, chemical, biological, economic or other) one develops various types of models of these systems. Of particular importance are mathematical models which enable us to predict the evolution of such systems and to control them effectively.

By a mathematical model of a system we mean, loosely speaking, its description by means of mathematical quantities (such as functions, operators, etc.) which satisfy certain relations expressing, for instance, the laws of evolution of the system under study. As an illustration let us consider the motion of a particle within the framework of classical mechanics. According to Newton's second law the motion is described by a vector function x satisfying the ordinary differential equation

$$m\frac{d^2x(t)}{dt^2} = F(t),$$

where $x(t)$ defines the space position vector at the time t of a particle of mass m which is acted upon by the force $F(t)$ at the time t. On the basis of this equation of motion we can determine by purely mathematical means the evolution of many mechanical systems (from a falling particle to planetary orbits). However, we must put up with the fact that the reality is rather complex in most cases and for that reason the mathematical model can take into account a few of its significant traits and neglect many other aspects. Hence, the solution characterizes only the main qualitative features of the system and, possibly, only for a limited time interval.

A great variety of phenomena and processes can be described by integrating partial differential equations. We shall deal with such equations which describe the behaviour of a system as time evolves (the evolution of processes) and which are therefore called partial differential equations of *evolution* or *evolutionary* partial differential equations. These equations stand in contrast to the so-called stationary partial differential equations which describe time-independent phenomena (steady states) and which will not be taken up in this text.

In the following chapters we shall present a number of examples of concrete processes described by evolutionary equations to be examined. Moreover, we shall often restate the results of mathematical treatment in terms of these processes.

1.1.2 Basic concepts and notation

We denote by $\mathbf{N}$ (respectively, $\mathbf{R}$) the set of positive integers (respectively, real numbers). The symbol $\mathbf{R}^+$ will denote the interval $[0, \infty)$.

The linear space of ordered n-tuples ($n \in \mathbf{N}$) of real numbers is denoted by $\mathbf{R}^n$. The elements of the space $\mathbf{R}^n$ (called points or vectors) are denoted by $(x_1, \ldots, x_n)$ or simply x. In the case $n = 2$ we write commonly (x, y) or (x, t) instead of (x_1, x_2). Similarly in the case $n = 3$ often (x, y, z) or (x, y, t) is written instead of (x_1, x_2, x_3). The space $\mathbf{R}^n$ becomes n-dimensional Euclidean space upon defining the (Euclidean) norm[1]

$$\|x\| = \left(\sum_{j=1}^{n} x_j^2 \right)^{\frac{1}{2}}, \quad x = (x_1, \ldots, x_n).$$

If $x = (x_1, \ldots, x_n)$ and $y = (y_1, \ldots, y_n)$ are two vectors from $\mathbf{R}^n$, then $x \cdot y$ denotes their scalar product defined by[2]

$$x \cdot y = \sum_{j=1}^{n} x_j y_j.$$

If x is a unit vector in $\mathbf{R}^n$, that is, $\|x\| = 1$, then $x \cdot y$ is called the projection of y on x.[3]

If $x = (x_1, x_2, x_3)$ and $y = (y_1, y_2, y_3)$ are two vectors from $\mathbf{R}^3$, then $x \times y$ denotes their vector product defined by

$$x \times y = (x_2 y_3 - x_3 y_2,\ x_3 y_1 - x_1 y_3,\ x_1 y_2 - x_2 y_1).$$

♣ By a region (in $\mathbf{R}^n$) we mean an open connected subset of $\mathbf{R}^n$. The closure (respectively, the boundary) of a region $\Omega \subset \mathbf{R}^n$ is denoted by $\bar{\Omega}$ (respectively, $\partial\Omega$).

The boundaries of regions on which partial differential equations are investigated play a very important role. For this reason, there exist different classifications of

---

[1] Other words used are length, magnitude, and so forth.

[2] An equivalent definition: $x \cdot y = \|x\| \, \|y\| \cos\theta$ where θ is the angle between the directions of the two vectors.

[3] Some authors define the projection of y on x as the *vector* $(x \cdot y)x$.

regions and their boundaries.[4] To provide precise definitions lies beyond the scope and intent of this book. [We refer the reader to specialist literature, for example Adams (1975).] In the sequel when formulating the assumptions we content ourselves with a (rather intuitive) saying that "the region is regular" (as, for example, the interior of a rectangle, parallelepiped, sphere, etc.) without worrying about the exact meaning of this term.[5]

A particularly important property enjoyed by bounded regular regions is that for "almost every" point of the boundary there is defined an *outward normal (unit) vector* ν.[6]

♡ By a function (of one or more real independent variables) we commonly mean a mapping from a subset of $\mathbf{R}^n$ (usually a region Ω or a domain $\tilde{\Omega}$, $\Omega \subset \tilde{\Omega} \subset \bar{\Omega}$) into the set $\mathbf{R}$ of real numbers. On occasion (when we work with integral transforms, for example) we use complex-valued functions; throughout the book the symbol i is reserved to denote the imaginary unit.

The partial derivative of a function u with respect to a variable s is denoted by

$$\frac{\partial u}{\partial s} \quad \text{or} \quad u_s$$

and similarly the alternate notation is used for partial derivatives of higher orders, as for instance,

$$\frac{\partial^2 u}{\partial s^2} = u_{ss}, \qquad \frac{\partial^2 u}{\partial s_1 \partial s_2} = u_{s_1 s_2}, \qquad \frac{\partial^4 u}{\partial s^4} = u_{ssss}$$

(we deal only with functions for which the order of differentiation is immaterial). The derivative of a function of one variable is denoted by

$$\frac{du}{ds} \quad \text{or} \quad u'$$

and similarly for higher order derivatives.

In addition to partial derivatives we frequently use *directional derivatives*. If u is a function of n variables and α is a unit vector in $\mathbf{R}^n$ then $\partial u(x)/\partial\alpha$ stands for the derivative of u at x in the direction α which is defined as

$$\frac{\partial u(x)}{\partial \alpha} = \lim_{\tau \to 0+} \frac{u(x+\tau\alpha) - u(x)}{\tau}$$

whenever this limit exists.

---

[4]A typical example: a region Ω is said to have the cone property if there exists a fixed cone such that each point of Ω can be considered a vertex of this cone if the latter is suitably placed in Ω.

[5]For most of our further applications "Ω is regular" is satisfied if Ω has the cone property.

[6]For how to compute ν see § 1.2.1.

♠ Let $\Omega \subset \mathbb{R}^n$ be a region and $k \in \mathbb{N} \cup \{0\}$. By the symbol $C^k(\Omega)$ we denote the set (here and hereafter, in fact, the linear space) of functions having continuous partial derivatives up to order k on Ω. (As usual, the partial derivative of u of order 0 is the function u itself). Equivalently, $C^k(\Omega)$ denotes the set of continuously differentiable functions up to order k on Ω.

By the symbol $C^\infty(\Omega)$ we denote the set of functions having continuous partial derivatives of all orders on Ω (the set of infinitely differentiable functions on Ω). We let $C_0^\infty(\Omega)$ be the set of functions from $C^\infty(\Omega)$ vanishing outside a set K, where $K \subset \Omega$ is bounded and closed (and different, in general, for different functions).

By the symbol $C^k(\bar{\Omega})$ we denote the set of functions from $C^k(\Omega)$ whose partial derivatives up to order k can be *extended by continuity up to the boundary* $\partial\Omega$ of the set Ω. Analogously we define the sets $C^k(\tilde{\Omega})$, where $\Omega \subset \tilde{\Omega} \subset \bar{\Omega}$.

The set $C^0(\Omega)$ (that is, the set of continuous functions on Ω) is denoted simply by $C(\Omega)$ and similarly we write $C(\bar{\Omega})$ instead of $C^0(\bar{\Omega})$.

A function $u \in C^k(\bar{\Omega})$ is said to have continuous partial derivatives (to be continuously differentiable) up to order k on $\bar{\Omega}$, while the values of partial derivatives at a point x of the boundary $\partial\Omega$ are defined as

$$\frac{\partial^l u(x)}{\partial x_1^{l_1} \cdots \partial x_n^{l_n}} = \lim_{\substack{x' \to x \\ x' \in \Omega}} \frac{\partial^l u(x')}{\partial x_1^{l_1} \cdots \partial x_n^{l_n}},$$

$l_1, \ldots, l_n \in \mathbb{N} \cup \{0\}$, $\sum_{j=1}^n l_j = l$, $l = 0, \ldots, k$. The definition of $C^k(\bar{\Omega})$ says nothing other that this limit is well-defined (that is, it exists and is independent of the choice of the approximating points x'.)[7]

The analogous definition applies to directional derivatives $\partial u(x)/\partial\alpha$ at a point x of $\partial\Omega$ in any direction α. (Incidentally, for α "pointing inward Ω" the definition is in agreement with that given above.) Of particular importance in the theory of partial differential equations is the case $\alpha = \nu$, where ν is the unit outward normal vector to $\partial\Omega$ at x; we speak of $\partial u(x)/\partial\nu$ defined at (not necessarily all) points of $\partial\Omega$ as the *outward normal derivative on* $\partial\Omega$.[8]

◇ In particular, let Ω be an interval in $\mathbb{R}$, say $\Omega = (a, b)$. For simplicity we write $C^k(a, b)$ instead of $C^k((a, b))$. The symbols $C^k[a, b]$, $C^k[a, b)$ and $C^k(a, b]$ are defined analogously. A function u belongs to $C^k[a, b]$ (a, b finite) if and only if u belongs to $C^k(a, b)$ and there exist one-sided derivatives up to order k from the right at a and from the left at b, respectively.

A function u is said to be *piecewise continuous* on (a, b) (a, b finite) if there exist a finite number of points $a = x_0 < x_1 < \cdots < x_p < x_{p+1} = b$ such that u [restricted to (x_{j-1}, x_j)] belongs to $C[x_{j-1}, x_j]$, $j = 1, \ldots, p+1$. The points $x_1, \ldots, x_p$ are possible points of jump discontinuity of the function u.

[7]Another equivalent definition: $u \in C^k(\bar{\Omega})$ if and only if all partial derivatives of u up to order k are uniformly continuous on any bounded $\Omega_0 \subset \Omega$.

[8]For computational aspects see § 1.2.1.

1.1.3 Partial differential equations

Ordinary differential equations involve functions of only one independent variable. On the other hand, partial differential equations contain functions of more independent variables.

A *partial differential equation* on a region $Q \subset \mathbf{R}^n$ $(n \geq 2)$ for a function u determines a functional relation between partial derivatives of u at any point $x \in Q$. The *order* of a partial differential equation is the order of the highest derivative occurring in the relation. For example,

$$\frac{\partial u}{\partial x_1} + a(u)\frac{\partial u}{\partial x_2} = 0, \quad x = (x_1, x_2) \in \mathbf{R}^2, \tag{1.1}$$

where a is a given function, is a partial differential equation (in short: equation) of the first order,

$$\frac{\partial u}{\partial x_3} - \frac{\partial^2 u}{\partial x_1^2} - \frac{\partial^2 u}{\partial x_2^2} = g(x), \quad x = (x_1, x_2, x_3) \in \mathbf{R}^3 \tag{1.2}$$

is an equation of the second order,

$$\frac{\partial^2 u}{\partial x_2^2} + \frac{\partial^4 u}{\partial x_1^4} = 0, \quad x = (x_1, x_2) \in \mathbf{R}^2, \tag{1.3}$$

is an equation of the fourth order. (For economy and lucidity of writing, when no confusion is likely, the arguments of the sought functions are often omitted.)

A general partial differential equation (on a region Q) of order m can be formally written as

$$G\left(x, p_{0\cdots 0}, \ldots, p_{l_1\cdots l_n}, \ldots\right) = 0, \tag{1.4}$$

where G is a certain function of the point x and of variables

$$p_{l_1\cdots l_n} = \frac{\partial^l u}{\partial x_1^{l_1} \cdots \partial x_n^{l_n}},$$

$$l_1, \ldots, l_n \in \mathbf{N} \cup \{0\}, \quad \sum_{j=1}^{n} l_j = l, \quad l = 0, \ldots, m \quad (m \geq 1).$$

For instance, the function G associated with equations (1.1) to (1.3) are given by

$$G = G\left(p_{00}, p_{10}, p_{01}\right) = p_{10} + a\left(p_{00}\right)p_{01},$$
$$G = G\left(x, p_{200}, p_{020}, p_{001}\right) = p_{001} - p_{200} - p_{020} - g(x),$$
$$G = G\left(p_{40}, p_{02}\right) = p_{40} + p_{02},$$

respectively.

By a *classical solution* of a partial differential equation we mean a continuous function u on the region Q such that all its derivatives appearing in the equation are continuous and which satisfies the equation at every point of Q.

If the function G in (1.4) for any fixed $x \in Q$ is a polynomial of degree at most one in variables $p_{l_1\cdots l_n}$, then the partial differential equation (1.4) is said to be *linear* ("the unknown function u and all of its derivatives occur linearly in the equation"). Such is the case in equations (1.2) and (1.3). In the opposite case the equation is said to be *nonlinear* – Eq. (1.1) provided a is not a constant function. In Eq. (1.1) the highest derivatives of the unknown function occur linearly; such nonlinear equations, in general, are called *quasilinear*.

The most general *linear partial differential equation of the second order for a function of two independent variables* assumes the form

$$a(x)\frac{\partial^2 u}{\partial x_1^2} + 2b(x)\frac{\partial^2 u}{\partial x_1 \partial x_2} + c(x)\frac{\partial^2 u}{\partial x_2^2} + d(x)\frac{\partial u}{\partial x_1} + e(x)\frac{\partial u}{\partial x_2} + f(x)u + g(x) = 0,$$
$$x = (x_1, x_2) \in Q, \quad (1.5)$$

where coefficients $a, \dots, g$ are (sufficiently smooth) functions defined on a region $Q \subset \mathbf{R}^2$ and a, b and c do not vanish simultaneously.

In conventional terminology a linear equation is called *homogeneous* if the zero function $u \equiv 0$ is its solution, otherwise it is called *inhomogeneous* in which case it involves a non-vanishing term independent of u and its derivatives – as in Eq. (1.5) the function g provided $g \not\equiv 0$. This term is often placed on the right-hand side of the equation so that it can be written formally as $Lu = g$ – see, for example, Eq. (1.2).

The most important feature of any linear equation is the following property called the *principle of superposition:*

$$\textit{if } g = \sum_{k=1}^{r} \alpha_k g_k, \ \alpha_k \textit{ some constants, and } u_k \textit{ is a solution of } Lu_k = g_k,$$

$$\textit{then } u = \sum_{k=1}^{r} \alpha_k u_k \textit{ is a solution of } Lu = g.$$

In particular, • if u_k $(k = 1, \dots, r)$ are solutions of the homogeneous equation $Lu = 0$ then $\sum_{k=1}^{r} \alpha_k u_k$ is a solution of the same equation for any constants $\alpha_1, \dots, \alpha_r$;
• if u_1 and u_2 are solutions of $Lu = g$ then the difference $u_1 - u_2$ is a solution of the corresponding homogeneous equation $Lu = 0$.
Thus, the general solution of the equation $Lu = g$ can be written in the form of the sum of the general solution of the corresponding homogeneous equation $Lu = 0$ and a particular solution of the inhomogeneous one.

The principle of superposition forms a powerful tool in the theory of linear equations. Starting with some "elementary" solutions we can "superpose" them (as series, integrals depending on parameters, etc.) to build up "new" solutions which satisfy certain subsidiary conditions. We mention the Fourier method as a typical example of this approach. Accordingly, the investigation of linear problems is much easier in comparison with nonlinear problems where the principle of superposition is not available. In this text we shall confine ourselves to linear equations (with a few exceptions in Chapter 2). □

In addition to the classification of partial differential equations according to whether they are stationary or evolutionary, linear or nonlinear, there are a number of classifications by various means (based on formal properties of equations, available methods of solutions and other viewpoints). Without making any attempt to enter into (sometimes intricate) details we restrict ourselves to a rather special, yet considerably important classification according to "types" of linear second order partial differential equations in two independent variables. Another classification concerning a system of m linear first order equations for m functions $u_1, \dots, u_m$ will be discussed at some length in Chapter 2 (§ 2.4.2 and § 2.5.1).

Thus, let us consider Eq. (1.5). With this equation we associate the discriminant

$$D(x) = b^2(x) - a(x)\,c(x), \quad x \in Q \subset \mathbf{R}^2.$$

The partial differential equation (1.5) is called

$$\left.\begin{array}{ll} \bullet & \textit{elliptic} \\ \bullet & \textit{parabolic} \\ \bullet & \textit{hyperbolic} \end{array}\right\} \textit{ at the point } x \text{ if } D(x) \left\{ \begin{array}{l} < 0 \\ = 0 \\ > 0 \end{array}\right.$$

Equation (1.5) is called *elliptic* (*parabolic, hyperbolic,* respectively) *in* Q being such at every $x \in Q$.

This classification has proved helpful and will be vital for our further purposes. Let us note that it depends only on the coefficients of the highest order derivatives appearing in (1.5). The three types of equations show completely different properties and require also specific methods of treatment. Incidentally, this shows that the classification of partial differential equations according to the order is rather coarse-grained.

An example of the elliptic equation (in the whole plane $\mathbf{R}^2$) is the Poisson equation

$$\frac{\partial^2 u}{\partial x_1^2} + \frac{\partial^2 u}{\partial x_2^2} = g(x), \tag{1.6}$$

in the homogeneous case ($g \equiv 0$) called the Laplace equation.

An example of the parabolic equation (in $\mathbf{R}^2$) is the so-called equation for the conduction of heat (simply the heat equation)

$$\frac{\partial u}{\partial x_2} - \frac{\partial^2 u}{\partial x_1^2} = g(x). \tag{1.7}$$

An example of the hyperbolic equation (in $\mathbf{R}^2$) is the so-called wave equation

$$\frac{\partial^2 u}{\partial x_2^2} - \frac{\partial^2 u}{\partial x_1^2} = g(x). \tag{1.8}$$

In spite of their special forms these equations are representatives of the three classes. This means that the properties enjoyed by these equations are shared in an analogous manner by the other equations of the respective class.

The latter two equations are examples of evolutionary ones. In applications the variable x_2 plays a specific role, it is identified with time (in contrast to x_1 which usually means the space variable) and accordingly it is denoted by t (and x_1, in turn, simply by x). It is the goal of Chapters 3 and 4 to investigate these evolutionary equations. The study of (stationary) elliptic equations is beyond the scope of this text.

The analogues of (1.7) and (1.8) in higher dimensions are the equations

$$\frac{\partial u}{\partial t} - \Delta u = g(x,t) \tag{1.9}$$

and

$$\frac{\partial^2 u}{\partial t^2} - \Delta u = g(x,t) \tag{1.10}$$

respectively, where Δ is the so-called *Laplace operator* (well-known from the theory of elliptic equations)

$$\Delta = \sum_{j=1}^{n} \frac{\partial^2}{\partial x_j^2}. \tag{1.11}$$

1.1.4 Initial and boundary conditions

A partial differential equation that is assumed to describe some laws governing the evolution of a certain system possesses in principle many solutions. To obtain a more complete description of the evolution (in particular, a uniquely determined evolution if the process is deterministic) we need some additional information which takes into account the effects from the ***history*** of the system, the influence of the ***surroundings***, etc. We bring out the role of these subsidiary conditions via physically motivated examples.

Let us consider the equation

$$\frac{\partial^2 u}{\partial t^2} - c^2 \frac{\partial^2 u}{\partial x^2} = 0, \quad x \in I,\, t > 0, \tag{1.12}$$

where $I = (0, \ell)$, $\ell > 0$ and $c \in \mathbf{R}$. [The function G in the general relation (1.4) is given here by $G = G(p_{20}, p_{02}) = p_{02} - c^2 p_{20}$ and the region Q is equal to $I \times (0, \infty)$.] This equation describes free (small) longitudinal vibrations of a rod. More precisely, the rod is assumed to be oriented so that the x-axis lies along the axis of the rod, its ends are located at the points $x = 0$ and $x = \ell$ and there are no external forces acting upon the rod. The function $u = u(x,t)$ defines the longitudinal displacement of a point on the cross-section at x at the time t.

The functions $u(x,t) = 1$, $u(x,t) = t$, $u(x,t) = x$ serve as examples of solutions of this equation. The question is how to single out the solution fully describing the vibration of a particular rod.

Our observation starts at a certain instant of time, say, for definiteness, $t = 0$, and we are interested in what happens later at any time $t > 0$. The information on the past history of the system is provided upon stipulating both the displacement and the velocity at the time $t = 0$. In other words, the relations

$$u(x,0) = \varphi(x), \quad \frac{\partial u(x,0)}{\partial t} = \psi(x), \quad x \in I \tag{1.13}$$

define the initial state, φ and ψ are known quantities called the ***initial functions*** (φ initial position, ψ initial velocity).

The effects from outside the system can be set up in numerous ways. If the rod is built-in at both ends then the behaviour of these ends (called fixed or clamped) is defined by the relations

$$u(0,t) = u(\ell,t) = 0, \quad t > 0. \tag{1.14}$$

If the ends of the rod are free the axial force which is proportional to $\partial u/\partial x$ must vanish at these ends and we get the relations

$$\frac{\partial u(0,t)}{\partial x} = \frac{\partial u(\ell,t)}{\partial x} = 0, \quad t > 0. \tag{1.15}$$

If the ends are attached to springs (the so-called springy ends) the pertinent relations are

$$\begin{aligned} -\frac{\partial u(0,t)}{\partial x} + \sigma_0 u(0,t) = 0, & \\ \frac{\partial u(\ell,t)}{\partial x} + \sigma_1 u(\ell,t) = 0, & \qquad t > 0, \end{aligned} \tag{1.16}$$

where σ_0 and σ_1 are positive constants (characterizing the elastic restraints), or more generally positive functions of the variable t. Of course, other possibilities are also physically relevant. For instance,

$$u(0,t) = 0, \quad \frac{\partial u(\ell,t)}{\partial x} = 0, \quad t > 0 \tag{1.17}$$

correspond to a rod one end of which is clamped and the other is free. □

Next, we turn to the equation

$$\frac{\partial u}{\partial t} - a\frac{\partial^2 u}{\partial x^2} = 0, \quad x \in I,\, t > 0,\ (a \in \mathbf{R}^+) \tag{1.18}$$

which governs the conduction of heat in a heat-conducting rod of length ℓ with no heat sources. The rod is assumed to have a perfectly thermally-insulated lateral surface and the heat can flow only in the direction of the x-axis. The sought function $u = u(x,t)$ represents the temperature of a point on the cross-section at x at time t.

The initial state is determined once the temperature distribution is known at an initial instant, that is, if the relation

$$u(x,0) = \varphi(x), \quad x \in I \tag{1.19}$$

is fulfilled, where φ is a given function (initial temperature).

The relations (1.14) correspond to the case when the rod is immersed in a surrounding medium at a constant zero temperature and the ends of the rod, being perfectly conducting, are kept at this temperature. In turn, the relations (1.15) mean that both ends are perfectly insulated so that no heat is allowed to escape through them. The intermediate situation when the ends are neither perfectly conducting nor perfectly insulated, so that the heat on passing through them is absorbed to a certain extent, leads to relations (1.16). □

The constraints of types (1.13) or (1.19) are called the *initial conditions*. The constraints of types (1.14) through (1.16) are called the *boundary conditions*, more specifically, (1.14) the *Dirichlet*, (1.15) the *Neumann* and (1.16) the *Newton boundary conditions*, respectively. All these boundary conditions are termed linear boundary conditions, homogeneous or inhomogeneous according as the right-hand side does, or does not, vanish identically.

A problem given by a partial differential equation and by initial condition(s) is called an *initial-value problem*. A problem given by a partial differential equation, initial condition(s) and boundary condition(s) is called an *initial-boundary value problem*.

The counterparts of (inhomogeneous) boundary conditions (1.14) to (1.16) for equations (1.9) and (1.10) [on the region $Q = \Omega \times (0, \infty)$, where Ω is a bounded regular region in $\mathbf{R}^n$, $n \geq 2$] are

$$u(x,t) = h(x,t), \tag{1.20}$$

$$\frac{\partial u(x,t)}{\partial \nu} = h(x,t) \tag{1.21}$$

and

$$\frac{\partial u(x,t)}{\partial \nu} + \sigma(x,t)\,u(x,t) = h(x,t), \tag{1.22}$$

respectively, where $(x,t) \in \partial\Omega \times (0,\infty)$, h and σ are given functions and $\partial/\partial\nu$ denotes the outward normal derivative on $\partial\Omega$. It might seem that the number of boundary conditions in either case is less than for Eqs. (1.12) and (1.18) with $n = 1$. This is explained by the fact that in the case $n = 1$ the set Ω is an interval whose boundary consists of merely two points and the boundary conditions are commonly written down at each point separately.

In physical terms, roughly speaking, the homogeneous Neumann boundary condition physically means that there is no flux of the quantity under consideration through the boundary. The Newton boundary condition reflects the fact that the quantity is absorbed to a certain degree on passing through the boundary. The Dirichlet condition corresponds to a perfectly absorbing boundary.

It is scarcely necessary to point out that this list of boundary conditions should be complete, yet these are typical ones and cover a great variety of cases that come up in applications. If Ω is not bounded the boundary conditions are usually replaced by the requirement that the solution be bounded or vanish at infinity, or else satisfy a more general decay-rate condition. □

The quantities entering into the problem which are presumed to be known [such as functions φ and ψ in the initial conditions (1.13), (1.19), "material" constants c and a in Eqs. (1.12) and (1.18), the coefficients a to g in Eq. (1.5), the right-hand side g in Eqs. (1.9), (1.10) or functions h and σ from boundary conditions (1.20) to (1.22)] are referred to as the *data* of the problem.

Some care must be taken to make precise the notion of a solution of the initial- (initial-boundary) value problem. In the preceding subsection we spoke of a classical solution of a partial differential equation as a function being defined on a region, hence on an open set. However, the initial and boundary conditions determine the solution and perhaps some of its derivatives on the boundary. Thus, by a *classical solution* (briefly *solution*) *of the initial-* (*initial-boundary*) *value problem* we mean such a classical solution of the given partial differential equation whose partial derivatives involved in initial and boundary conditions can be extended by continuity up to the boundary and the given initial and boundary conditions are satisfied. In the next subsection we elucidate this definition by means of examples.

1.1.5 Existence and uniqueness

We have introduced a definition of the initial-boundary value problem. On rather physical grounds we expect that this definition fits the situation adequately. To bear out this expectation, however, one needs to prove that the problem *has* a solution and that it has *at most one* solution (the latter requirement is plausible if the process in question is deterministic). In other words, the *existence* and the *uniqueness* of a solution have to be established in order to justify the above definition.

The existence can be proved only if certain necessary matching conditions are satisfied. These conditions follow directly from our definition of solution. Let us take, for instance, the problem given by (1.18), (1.19) and (1.14). A function u is a (classical) solution of the problem if it is continuous, together with its partial derivatives u_t and u_{xx} on the semi-infinite strip

$$Q = \{ (x,t) \in \mathbf{R}^2 \mid 0 < x < \ell,\, 0 < t < \infty \},$$

satisfies equation (1.18) and moreover it can be extended by continuity to the closure of Q,

$$\bar{Q} = \{ (x,t) \in \mathbf{R}^2 \mid 0 \le x \le \ell,\, 0 \le t < \infty \},$$

and the relations (1.19) and (1.14) are satisfied. The requirement of the continuity up to the boundary necessitates

$$\varphi(0) = \varphi(\ell) = 0. \tag{1.23}$$

These relations reflect a certain "harmony" between initial and boundary conditions. In this instance (and by analogy in general) they are called the *compatibility conditions*. (What are the compatibility conditions in case of inhomogeneous Dirichlet boundary conditions?)

As we shall see in Chapter 4 for any function φ which is continuous on the interval $[0, \ell]$ and satisfies the compatibility conditions (1.23), the problem has exactly one solution. In physical terms this means: if we investigate the process of the conduction of heat in a rod without sources, insulated against lateral heat conduction, the ends of which are kept at zero temperature, then the knowledge of the temperature distribution at an initial moment enables us to *predict* the temperature distribution within the rod in a unique manner *at any subsequent time.*

In Chapter 3 we shall pursue the initial-boundary value problem given by (1.12), (1.13) and (1.14) and assuming that the functions φ and ψ satisfy some differentiability and compatibility conditions we prove the existence and uniqueness of a (classical) solution [in fact, a slightly "better" solution belonging to $C^2([0, \ell] \times [0, \infty))$]. This means that, when studying the free longitudinal vibration of a rod with clamped ends and no presence of external forces, the *future development* of the rod is *uniquely defined* once the initial displacement and the initial velocity are known. On the one hand, the existence of the solution seems to be self-evident – the rod will surely vibrate in some way. On the other, and this point is important, the result shows that the vibration is uniquely determined upon prescribing *both* the initial functions φ and ψ. □

The existence and uniqueness of solutions to initial-boundary value problems can be proved for a number of partial differential equations. Without any attempt to be precise we put down the result for equations of a particular form, solved for the highest time derivative:

$$\frac{\partial^k u}{\partial t^k} = G\left(x, t, u, \frac{\partial u}{\partial t}, \ldots, \frac{\partial^{k-1} u}{\partial t^{k-1}}, \frac{\partial u}{\partial x_1}, \ldots, \frac{\partial u}{\partial x_n}, \ldots, \frac{\partial^{2m}}{\partial x_1^{l_1} \cdots \partial x_n^{l_n}}\right), \quad \sum_{j=1}^{n} l_j = 2m.$$

The "proper" number of initial conditions (that is, the number of conditions under which a solution exists and is unique) is k (the functions $u, \ldots, \partial^{k-1}u/\partial t^{k-1}$ should be prescribed at time $t = 0$, for example) and the "proper" number of boundary conditions is m. To illustrate: the fourth order equation

$$\frac{\partial^2 u}{\partial t^2} + \frac{\partial^4 u}{\partial x_1^4} + 2\frac{\partial^4 u}{\partial x_1^2\, \partial x_2^2} + \frac{\partial^4 u}{\partial x_2^4} = g(x_1, x_2, t), \quad (x_1, x_2) \in \Omega\ (\subset \mathbf{R}^2),\ t > 0$$

governing free (small) transverse vibrations of a plate (which will be studied in Chapter 5) is to be supplemented by *two* initial conditions

$$u(x_1, x_2, 0) = \varphi(x_1, x_2),$$
$$\frac{\partial u(x_1, x_2, 0)}{\partial t} = \psi(x_1, x_2), \quad (x_1, x_2) \in \Omega$$

and by *two* boundary conditions, say,

$$u(x_1, x_2, t) = \frac{\partial u(x_1, x_2, t)}{\partial \nu} = 0, \quad (x_1, x_2) \in \partial\Omega,\ t > 0$$

(characterizing the clamped edges of the plate), where $\partial/\partial\nu$ denotes the outward normal derivative on $\partial\Omega$. □

Let us add some remarks concerning *linear* initial-boundary value problems in which both the equation and the boundary conditions are linear. The linearity via the principle of superposition is exploited to great advantage in

• existence proofs where it makes it possible to *split up* a given problem into (hopefully) simpler ones which might be treated separately. A typical example: a solution is looked for as the sum $u = u_1 + u_2$, where u_1 is a solution of the problem with homogeneous boundary conditions and u_2 is a solution of the homogeneous equation with zero initial functions and inhomogeneous boundary conditions.

• uniqueness proofs: the uniqueness is ensured as the zero (*trivial*) solution is shown to be *the only* solution of the corresponding problem with homogeneous equation, homogeneous boundary conditions and zero initial functions. □

The study of linear initial- (initial-boundary) value problems with inhomogeneous time dependent data (boundary conditions and/or right-hand sides) may be reduced in a number of cases to the study of similar (much easier) problems with time-independent data. This reduction is possible thanks to a device usually referred to as *Duhamel's principle*. It provides a useful tool in a wide range of both theoretical and practical applications. To illustrate at least briefly the main idea of this formalism let us consider an initial-value problem

$$\begin{aligned} u_t(x,t) + L\,u(x,t) &= g(x,t), \qquad x = (x_1,\dots,x_n) \in \mathbf{R}^n,\ t > 0, \\ u(x,0) &= 0, \qquad x \in \mathbf{R}^n, \end{aligned}$$

where L is a differential expression involving only derivatives with respect to the variables x_j [e.g. $L = -\Delta$ where Δ is introduced in (1.11)]. If the functions $v(x,t,\tau)$ are supposed to be known solutions (for any $\tau \geq 0$) of initial-value problems depending on a parameter τ

$$\begin{aligned} v_t(x,t,\tau) + L\,v(x,t,\tau) &= 0, \qquad x \in \mathbf{R}^n,\ t > \tau, \\ v(x,\tau,\tau) &= g(x,\tau), \qquad x \in \mathbf{R}^n \end{aligned}$$

then

$$u(x,t) = \int_0^t v(x,t,\tau)\,d\tau$$

appears to be a solution of the original problem. This property is not incidental and has many ramifications for more general equations and problems. □

As a final remark it is worth noting that, in a great deal of practical situations, we are not interested in a uniquely determined evolution of systems. We can inquire, for example, if there are initial functions such that the ensuing solution is periodic in time. Such problems have either many solutions or none [for details see, e.g., Vejvoda *et al.* (1982)]. More generally, the development of a solution (or some aspect of its behaviour) can be prescribed and our goal is to determine (modify, control) the data so that a solution behaves as desired. In summary, besides the initial- (initial-boundary) value problems there exist many other problems associated with partial differential equations but we do not go into these matters here.

1.1.6 Continuous dependence on the data. Stability

The data are often determined by experiment and on principle they can never be known with perfect precision. Thus, the mathematical description suitable to supply also quantitative conclusions has meaning only if slight changes in the data due to the inaccuracy in their determination induce slight changes in solution observed during a time interval of interest $[0,T]$. In other words, only if the solution depends continuously on the data. An initial- (initial-boundary) value problem is said to be *well-posed* (*correctly posed*) if

(i) there exists a solution
(ii) the solution is unique
(iii) the solution depends continuously (in some sense) on the data.

Of particular importance is the investigation of changes in the solution caused by changes only in the initial data. In this case, the continuous dependence on data reduces to the continuous dependence on the initial data and this means: the distance of a solution u_0 (considered for $0 \leq t \leq T$, $T < \infty$) from any other solution can be made, for $0 \leq t \leq T$, as small as desired by choosing a sufficiently small distance between their initial data. In general, however, this distance between the initial data can tend to zero as $T \to \infty$ and this brings about the difficulties when studying long-running processes. The purpose of the theory of stability (more precisely Lyapunov stability) is just to cope with the case of infinite intervals or finite but arbitrarily large ones and to classify the tendency of solutions initiating sufficiently near to u_0 to remain near to u_0 *for all times* $t \geq 0$. The solution u_0 (defined for all $t \geq 0$) is said to be ***stable*** if the distance of u_0 from any other solution can be made, for all $t > 0$, as small as desired by choosing a sufficiently small distance between their initial data. The solution u_0 is said to be ***asymptotically stable*** if it is stable and any other solution initiating sufficiently near to the initial data of u_0 converges to u_0 as $t \to \infty$. If this convergence takes place at the rate $\exp(-\alpha t)$ for some $\alpha > 0$, the solution u_0 is said to be ***exponentially stable***.

A lot of questions immediately come to mind. It is natural to ask just what the terms "sufficiently", "near" or "distance" exactly mean. It turns out that the way of measuring the distance plays really a decisive role in this study. The mathematical investigation needs the use of some more advanced tools (such as norms in Banach spaces) and due to the restrained size of this elementary text we are compelled to make no attempt to pursue these topics systematically, though as mentioned earlier they are of considerable practical importance.

In concluding this subsection we remark that all problems to be studied in what follows are well-posed under physically reasonable conditions and under some additional conditions the solutions are stable or asymptotically stable. In many instances the reader familiar with some functional analysis can convince himself of this, especially if explicit formulae for solutions are available.

1.2 SURVEY OF SOME AUXILIARY NOTIONS AND RESULTS

This section collects some material for future reference.

1.2.1 Operators of vector analysis

Let $\Omega \subset \mathbf{R}^n$ ($n \in \mathbf{N}$, $n \geq 2$) be a region and let $\Omega \subset \tilde{\Omega} \subset \bar{\Omega}$.

For any $u \in C^1(\tilde{\Omega})$ the symbol $\operatorname{grad} u$ (or sometimes ∇u) denotes a vector function (called the *gradient* of the function u) defined by

$$\operatorname{grad} u = \left(\frac{\partial u}{\partial x_1}, \ldots, \frac{\partial u}{\partial x_n}\right).$$

If $u \in C^1(\tilde{\Omega})$ and $\alpha = (\alpha_1, \ldots, \alpha_n)$ is an arbitrary unit vector[9] in $\mathbf{R}^n$ then at any point $x \in \tilde{\Omega}$ there the derivative of the function u in the direction α exists and is given by the projection of $\operatorname{grad} u$ on α, that is,

$$\frac{\partial u}{\partial \alpha} = \alpha \cdot \operatorname{grad} u = \alpha_1 \frac{\partial u}{\partial x_1} + \cdots + \alpha_n \frac{\partial u}{\partial x_n}. \tag{1.24}$$

[9] In different terms, for any $j = 1, \ldots, n$, α_j is equal to the cosine of the angle between the direction of α and the positive x_j-axis.

If $u \in C^1(\bar{\Omega})$ then the above result is true at any $x \in \partial\Omega$. In particular, the *outward normal derivative of the function u on $\partial\Omega$ is given by*

$$\frac{\partial u}{\partial \nu} = \nu \cdot \operatorname{grad} u = \nu_1 \frac{\partial u}{\partial x_1} + \cdots + \nu_n \frac{\partial u}{\partial x_n} \tag{1.25}$$

whenever the unit outward normal vector $\nu = (\nu_1, \ldots, \nu_n)$ to $\partial\Omega$ *is defined* (ν varies with $x \in \partial\Omega$). □

It is worth mentioning two other properties of the vector function $\operatorname{grad} u$. Let $x^0 = (x_1^0, \ldots, x_n^0) \in \Omega$ and $\operatorname{grad} u(x^0) \neq 0$.

• In a neighbourhood of the point x^0 the set

$$\Sigma = \{x \mid u(x) = u(x^0)\}$$

determines a surface in $\mathbf{R}^n$ (more precisely, a hypersurface if $n > 3$, a curve if $n = 2$) and the tangent (hyper)plane of Σ (called the level surface) at x^0 is given by the equation

$$(x - x^0) \cdot \operatorname{grad} u(x^0) = 0.$$

In other words, the vector $\operatorname{grad} u(x^0)$ is the normal vector to the level surface Σ at x^0. [Thus, the unit normal vector ν to Σ is given by

$$\nu = \frac{\operatorname{grad} u}{\|\operatorname{grad} u\|}.]$$

• The vector $\operatorname{grad} u(x^0)$ determines the direction of the most rapid growth of the function u at the given point x^0 and the norm $\|\operatorname{grad} u(x^0)\|$ is equal to the greatest directional derivative of u at x^0.[10] □

If $w = (w_1, \ldots, w_n)$ is a vector function all of whose components belong to $C^1(\tilde{\Omega})$ then the symbol $\operatorname{div} w$ denotes a scalar function (called the *divergence* of the function w) given by

$$\operatorname{div} w = \frac{\partial w_1}{\partial x_1} + \cdots + \frac{\partial w_n}{\partial x_n}$$

and if $n = 3$ the symbol $\operatorname{curl} w$ (or sometimes $\operatorname{rot} w$) denotes a vector function (called the *rotation* of w) given by

$$\operatorname{curl} w = \left(\frac{\partial w_3}{\partial x_2} - \frac{\partial w_2}{\partial x_3}, \frac{\partial w_1}{\partial x_3} - \frac{\partial w_3}{\partial x_1}, \frac{\partial w_2}{\partial x_1} - \frac{\partial w_1}{\partial x_2} \right).$$

It may be easily verified that, for any scalar function $u \in C^2(\tilde{\Omega})$, one has

$$\operatorname{div} \operatorname{grad} u = \Delta u = \frac{\partial^2 u}{\partial x_1^2} + \cdots + \frac{\partial^2 u}{\partial x_n^2}, \tag{1.26}$$

$$\operatorname{curl} \operatorname{grad} u = 0 \tag{1.27}$$

and for any vector function $w = (w_1, w_2, w_3)$, $w_j \in C^2(\tilde{\Omega})$, $j = 1, 2, 3$

$$\operatorname{div} \operatorname{curl} w = 0, \tag{1.28}$$

$$\operatorname{curl} \operatorname{curl} w = \operatorname{grad} \operatorname{div} w - \Delta w, \tag{1.29}$$

where $\Delta w = (\Delta w_1, \Delta w_2, \Delta w_3)$.

[10] Any directional derivative at x^0 can be expressed through the outward normal derivative on Σ at the same point, namely $\partial u/\partial \alpha = \cos\psi \, \partial u/\partial \nu$ where ψ is the angle between the directions of α and ν.

1.2.2 Divergence theorem

Integral identities which relate surface (line) integrals over a closed surface to integrals over the enclosed volume constitute an extremely effective tool in the study of equations of mathematical physics. The basic result in this direction is the following identity:

$$\int_\Omega \frac{\partial u}{\partial x_j}\, dx = \int_{\partial\Omega} \nu_j\, u\, dS, \quad j = 1, \ldots, n, \tag{1.30}$$

which is valid for any $u \in C^1(\bar{\Omega})$, $\Omega \subset \mathbf{R}^n$ is a bounded region and $\nu = (\nu_1, \ldots, \nu_n)$ is the unit outward normal vector to $\partial\Omega$.

The relation (1.30) is true for very general bounded regions (we speak of them as being regular). Also, the condition on the function u is sufficient but by no means necessary.

The identity (1.30) is a generalization of the fundamental theorem of Calculus which states that

$$\int_a^b u'(x)\, dx = u(b) - u(a)$$

for any $u \in C^1[a,b], -\infty < a < b < \infty$.

On the basis of formula (1.30) the following result of the utmost importance can be easily derived.

Divergence theorem (Gauss–Ostrogradskiĭ theorem). Let $\Omega \subset \mathbf{R}^n$ be a bounded regular region. Let w be a vector function all of whose components belong to $C^1(\bar{\Omega})$. Then

$$\int_\Omega \operatorname{div} w\, dx = \int_{\partial\Omega} \nu \cdot w\, dS, \tag{1.31}$$

where ν is the unit outward normal vector to the boundary $\partial\Omega$ of the region Ω. □

At this point we make a supplemental remark concerning non-stationary vector fields often encountered in physical applications. Let w be a vector function whose components are functions of both space and time variables x and t, $x \in \bar{\Omega}$, $\Omega \subset \mathbf{R}^n$ a region, $t \in J$, $J \subset \mathbf{R}$ an interval, and have continuous partial derivatives with respect to space variables on $\bar{\Omega} \times J$. For any oriented surface $S \subset \bar{\Omega}$ with a unit normal vector ν the (scalar) quantity

$$\frac{dW}{dt} = \int_S w_\nu\, dS,$$

where w_ν is the projection of w on ν, that is,

$$w_\nu = \nu \cdot w \ \left[= \nu_1(x)\, w_1(x,t) + \cdots + \nu_n(x)\, w_n(x,t)\right],$$

is called the (time) rate of flow of the vector w across the surface S at time t or the *flux of the vector w across S* at time t. In this context, in some instances, we speak of w as the flux-density vector. (Two examples for illustration: i) w magnetic induction or magnetic flux density, dW/dt magnetic flux, ii) w heat flux density, dW/dt heat flux.) In particular, on condition

$$(CV) \qquad \begin{cases} V \subset \Omega \text{ is a bounded regular region ("control volume")} \\ S = \partial V \text{ oriented by the outward normal vector } \nu, \end{cases}$$

we have by virtue of (1.31)[11]

$$\frac{dW}{dt} = \int_V \operatorname{div} w \, dV$$

(the flux of w out of V).

In a number of applications the vector function w, for any t, has both the magnitude and the direction of the most rapid decrease of a scalar (continuously differentiable) function, say u, that is (cf. § 1.2.1)

$$w(x,t) = -\operatorname{grad} u(x,t)^{11}$$

We get upon using (1.25) and (1.26)

$$w_\nu = -\frac{\partial u}{\partial \nu}$$

and on condition (CV)

$$\frac{dW}{dt} = -\int_V \Delta u \, dV.^{11}$$

More generally, if

$$w(x,t) = -\mathcal{A}(x,t) \operatorname{grad} u(x,t)$$

where $\mathcal{A}$ is an $n \times n$ matrix, $\mathcal{A}(x,t) = (a_{jk}(x,t))_{j,k=1}^n$ and $\operatorname{grad} u$ is temporarily regarded as an $n \times 1$ matrix $(\partial u/\partial x_1, \ldots, \partial u/\partial x_n)^T$, then

$$w_\nu = -\frac{\partial u}{\partial \nu_{\mathcal{A}}}$$

where $\partial/\partial \nu_{\mathcal{A}}$ is the so-called *conormal derivative* (with respect to $\mathcal{A}$) on S defined by

$$\frac{\partial u}{\partial \nu_{\mathcal{A}}} = \nu \, \mathcal{A} \operatorname{grad} u \left[= \sum_{j,k=1}^{n} \nu_j \, a_{jk} \, \frac{\partial u}{\partial x_k} \right].$$

On condition (CV) one has

$$\frac{dW}{dt} = -\int_V \operatorname{div}(\mathcal{A} \operatorname{grad} u) \, dV.^{11}$$

1.2.3 Green's theorem

A two-dimensional version of the divergence theorem is Green's theorem in which the line integral of the second kind appears.

Let $\Omega \subset \mathbf{R}^2$ be a bounded regular region. For simplicity assume that it is *simply connected* (without "holes"). Assume further that *the boundary $\partial\Omega$ is positively oriented with respect to Ω* (as the parameter of a given parametrization increases a point of $\partial\Omega$ moves counter-clockwise around $\partial\Omega$). *Let the functions $u, v, \partial u/\partial y$ and $\partial v/\partial x$ belong to $C(\bar{\Omega})$. Then*

$$\int_{\partial\Omega} u \, dx + v \, dy = \int_\Omega \left(\frac{\partial v}{\partial x} - \frac{\partial u}{\partial y} \right) dx dy. \tag{1.32}$$

[The relation (1.32) follows from (1.31) upon setting $w = (w_1, w_2) = (v, -u)$ and taking into account that $dx = -\nu_2 dS$, $dy = \nu_1 dS$.]

---

[11]The operators Δ, div and grad act only on space variables. We write sometimes dV instead of dx in volume integrals.

1.2.4 Two criteria for a continuous function to vanish identically in a region

The following results provide a useful device in many instances, among others, when deriving partial differential equations of mathematical physics (by means of either virtual-work techniques or those of control volumes).

Let $u \in C(\Omega)$, $\Omega \subset \mathbb{R}^n$ a region. Then $u(x) = 0$ everywhere in Ω if either

- $$\int_\Omega u(x)\, v(x)\, dx = 0 \quad \textit{for every } v \in C_0^\infty(\Omega)$$

or

- $$\int_V u(x)\, dx = 0 \quad \textit{for any sphere } V \subset \Omega$$

[instead of spheres other "volumes" (cubes, cylinders) may be used for the same purpose].

1.2.5 Differentiation under the integral sign

Let us consider a function u of variables x and α defined on the set

$$Q = \{\, (x, \alpha) \in \mathbb{R}^2 \mid a < x < b,\ c < \alpha < d \,\}\,.$$

Let a, b, c and d be finite and suppose that u together with its partial derivative $\partial u/\partial \alpha$ belong to $C(\bar{Q})$. *Then the integral*

$$\int_a^b u(x, \alpha)\, dx \tag{1.33}$$

depending on the parameter α is a continuously differentiable function of the variable α on the interval (c, d) and

$$\frac{d}{d\alpha} \int_a^b u(x, \alpha)\, dx = \int_a^b \frac{\partial u(x, \alpha)}{\partial \alpha}\, dx. \tag{1.34}$$

If $a = -\infty$ and/or $b = \infty$, or else the functions u and/or $\partial u/\partial \alpha$ are not continuous up to the boundary, the above conclusion is no longer true (apart from the fact that the integrals may even fail to exist). However, some additional assumptions ensure the validity of the result. One possible set of such hypotheses is:
a) u and $\partial u/\partial \alpha$ belong to $C(Q)$ (a, b not necessarily finite),
b) the (improper) integral (1.33) is convergent at least for one $\alpha \in (c, d)$,
c) there exists a function v on the interval (a, b) such that its (improper) integral is convergent and

$$\left| \frac{\partial u(x, \alpha)}{\partial \alpha} \right| \le v(x), \quad (x, \alpha) \in Q. \ \square$$

Also of importance is the case when the limits of the integral are functions of the parameter α. Let u and $\partial u/\partial\alpha$ belong to $C(\bar{Q})$ (a, b, c, d again finite). Let φ and ψ be two functions of the variable α on the interval (c, d) belonging to $C^1(c, d)$ and moreover $a < \varphi(\alpha) < b,\ a < \psi(\alpha) < b$ for all $c < \alpha < d$. Then the so-called *Leibniz rule* holds:

$$\frac{d}{d\alpha}\int_{\varphi(\alpha)}^{\psi(\alpha)} u(x,\alpha)\,dx = u(\psi(\alpha),\alpha)\,\psi'(\alpha) - u(\varphi(\alpha),\alpha)\,\varphi'(\alpha) + \int_{\varphi(\alpha)}^{\psi(\alpha)} \frac{\partial u(x,\alpha)}{\partial\alpha}\,dx,$$
$$c < \alpha < d. \quad (1.35)$$

Applying this rule twice we obtain, as an illustration,

$$\frac{d}{d\alpha}\int_0^\alpha \int_{\beta-\alpha+\tau}^{\beta+\alpha-\tau} g(\xi,\tau)\,d\xi d\tau = \int_0^\alpha [g(\beta+\alpha-\tau,\tau) + g(\beta-\alpha+\tau,\tau)]\,d\tau,$$
$$0 < \alpha < \infty,\ -\infty < \beta < \infty, \quad (1.36)$$

where g is any function continuous for $-\infty < \xi < \infty$, $0 < \tau < \infty$.

1.2.6 Fourier transform

Let u be a function (in general, complex-valued) defined for $x \in \mathbf{R}$ such that

$$\int_{-\infty}^{\infty} |u(x)|\,dx < \infty.$$

Then the integral

$$\hat{u}(y) = \int_{-\infty}^{\infty} u(x)\exp(-ixy)\,dx \quad (1.37)$$

is convergent for all $y \in \mathbf{R}$ [and $\hat{u} \in C(\mathbf{R})$]. The function $\hat{u}$ is called the *Fourier transform* (image) of the (original) function u.

Under suitable (not too restrictive) hypotheses the functions u and $\hat{u}$ are placed in one-to-one correspondence $\mathcal{F}$ – the Fourier transformation. The function u can be represented in terms of $\hat{u}$ by means of the so-called *Fourier inversion formula*

$$u(x) = \lim_{L\to\infty}\frac{1}{2\pi}\int_{-L}^{L}\hat{u}(y)\exp(iyx)\,dy. \quad (1.38)$$

This formula is true, for example, at any point $x = x_0$ at which the derivative $u'(x_0)$ exists.

Among many important properties only a few needed in the sequel are tabulated below. We leave aside the precise assumptions for their validity referring the reader to a wide range of existing monographs on integral transformations, e.g. Sneddon (1951). The entries in the second column are Fourier transforms of those in the first one:

Original function	*Image function*		
$\alpha u(x) + \beta v(x)$	$\alpha \hat{u}(y) + \beta \hat{v}(y)$	$\alpha, \beta \in \mathbf{R}$	(1.39)
$\frac{1}{c} u(\frac{x}{c})$	$\hat{u}(cy)$	$c > 0$	(1.40)
$\frac{\partial^l u(x)}{\partial x^l}$	$(iy)^l \hat{u}(y)$	$l \in \mathbf{N}$	(1.41)
$\int_{-\infty}^{\infty} u(x-\xi)\, v(\xi)\, d\xi$	$\hat{u}(y)\, \hat{v}(y)$		(1.42_1)
$u(x-\xi)$	$\hat{u}(y) \exp(-i\xi y)$	$\xi \in \mathbf{R}$	(1.42_2)

The first property is the linearity of the Fourier transformation. The second is the scaling property. The third says that the operation of taking derivatives is converted into multiplication by a variable. It is this property that renders the Fourier transformation so useful in solving differential equations. The fourth property states that the Fourier transform of the convolution of two functions is equal to the product of their Fourier transforms. This result is referred to as the *convolution theorem*. The last property shows the relation between the transform of a function and that of its translation a distance ξ in the positive x-direction.

1.2.7 Laplace transform

Let u be a (complex-valued) function defined for $t \in \mathbf{R}$, $u \equiv 0$ for $t < 0$, satisfying

(1) $$\int_0^L |u(t)|\, dt < \infty \qquad \text{for every } L > 0,$$

(for example, u is piecewise continuous on the interval $(0, L)$ for every $L > 0$),

(2) there exist real constants M, t_0 and σ ($M > 0$, $t_0 > 0$) such that
$$|u(t)| \le M \exp(\sigma t), \quad t \ge t_0.$$

Denote
$$\tilde{u}(s) = \int_0^{\infty} u(t) \exp(-st)\, dt. \tag{1.43}$$

Then either a) this integral is (absolutely) convergent for all (complex) values of s and the function $\tilde{u}$ is holomorphic in the whole complex plane, or b) there exists $s_0 \in \mathbf{R}$ such that this integral is (absolutely) convergent for all complex s with the real part greater than s_0 ($\operatorname{Re} s > s_0$) and it is not absolutely convergent for any complex s with $\operatorname{Re} s < s_0$. The function $\tilde{u}$ is holomorphic in the half-plane $\operatorname{Re} s > s_0$ (and in many instances it may be analytically continued in a region larger than this half-plane).

In either case the function $\tilde{u}$ is called the *Laplace transform* (image) of the (original) function u. The correspondence between u and $\tilde{u}$ – the Laplace transformation – is denoted by $\mathcal{L}$; so we write

$$\tilde{u} = \mathcal{L}\{u\} \quad \text{or} \quad \tilde{u}(s) = \mathcal{L}\{u(t)\}(s).$$

We list a few important properties [for details see Doetsch (1961), Churchill (1958)].

Original function	*Image function*		
$\alpha u(t) + \beta v(t)$	$\alpha\tilde{u}(s) + \beta\tilde{v}(s)$	$\alpha, \beta \in \mathbf{R}$	(1.44)
$\frac{1}{c}u(\frac{t}{c})$	$\tilde{u}(cs)$	$c > 0$	(1.45)
$u'(t)$	$s\tilde{u}(s) - \lim_{t\to 0+} u(t)$		(1.46_1)
$u''(t)$	$s^2\tilde{u}(s) - s\lim_{t\to 0+} u(t) - \lim_{t\to 0+} u'(t)$		(1.46_2)
$\int_0^t u(t-\tau)\,v(\tau)\,d\tau$	$\tilde{u}(s)\,\tilde{v}(s)$		(1.47_1)
$u(t-\tau)$	$\tilde{u}(s)\exp(-s\tau)$	$\tau > 0$	(1.47_2)

The function u can be recovered from $\tilde{u}$ by the *Laplace inversion formula*

$$u(t) = \lim_{\omega\to\infty} \frac{1}{2\pi i}\int_{s_1-i\omega}^{s_1+i\omega} \tilde{u}(s)\exp(st)\,ds, \tag{1.48}$$

where the integral is an integral in the complex plane of the variable s along a line $\operatorname{Re} s = s_1$ parallel to the imaginary axis and $s_1 > s_0$ The Laplace inversion formula is valid under suitable (mild) hypotheses on the function u (making $\mathcal{L}$ into a one-to-one correspondence). For example, it is true at $t = t_0$ whenever $u'(t_0)$ exists. (The latter result is an immediate consequence of the sufficient condition mentioned in § 1.2.6 in connection with the Fourier inversion formula.)

Generally speaking, if we look for original functions via the application of the Laplace inversion formula the use of tools of complex analysis (the technique of contour integration, the residue theorem, etc.) seems inevitable. Since we suppose the reader has no experience with this topic we will use instead well-developed collections of tables of the Laplace transformation which are available in the extensive literature on the subject [see, for instance, Erdélyi *et al.* (1954), Roberts and Kaufman (1966)].

Table 1

Original function	*Image function*	
1	$\dfrac{1}{s}$	
$1-\operatorname{erf}\dfrac{x}{2t^{\frac{1}{2}}}$	$\dfrac{\exp(-xs^{\frac{1}{2}})}{s}$	$x>0$
$\dfrac{\partial}{\partial x}\operatorname{erf}\dfrac{x}{2t^{\frac{1}{2}}}=\dfrac{1}{(\pi t)^{\frac{1}{2}}}\exp(-\dfrac{x^2}{4t})$	$\dfrac{\exp(-xs^{\frac{1}{2}})}{s^{\frac{1}{2}}}$	$x>0$
$-\dfrac{\partial^2}{\partial x^2}\operatorname{erf}\dfrac{x}{2t^{\frac{1}{2}}}=\dfrac{x}{2(\pi t^3)^{\frac{1}{2}}}\exp(-\dfrac{x^2}{4t})$	$\exp(-xs^{\frac{1}{2}})$	$x>0$
$\dfrac{1}{\ell}\dfrac{\partial}{\partial x}\Theta(\dfrac{\ell-x}{2\ell},\dfrac{t}{\ell^2})$	$\dfrac{\sinh(xs^{\frac{1}{2}})}{\sinh(\ell s^{\frac{1}{2}})}$	$-\ell<x<\ell$
$-\dfrac{1}{\ell}\dfrac{\partial}{\partial x}\Theta(\dfrac{x}{2\ell},\dfrac{t}{\ell^2})$	$\dfrac{\sinh((\ell-x)s^{\frac{1}{2}})}{\sinh(\ell s^{\frac{1}{2}})}$	$0<x<2\ell$
$\dfrac{1}{2\ell}\Big[\Theta(\dfrac{x-\xi}{2\ell},\dfrac{t}{\ell^2})-$	$\begin{cases}\dfrac{\sinh(\xi s^{\frac{1}{2}})\sinh((\ell-x)s^{\frac{1}{2}})}{s^{\frac{1}{2}}\sinh(\ell s^{\frac{1}{2}})} \\ \end{cases}$	$0\le\xi\le x$
$-\Theta(\dfrac{x+\xi}{2\ell},\dfrac{t}{\ell^2})\Big]$	$\begin{cases}\dfrac{\sinh(xs^{\frac{1}{2}})\sinh((\ell-\xi)s^{\frac{1}{2}})}{s^{\frac{1}{2}}\sinh(\ell s^{\frac{1}{2}})}\end{cases}$	$x\le\xi\le\ell$

The customary notation is used. The function erf is known as the *error function* and is defined by the formula

$$\operatorname{erf} z=\frac{2}{\pi^{\frac{1}{2}}}\int_0^z\exp(-\eta^2)\,d\eta,\quad z\in\mathbf{R}. \tag{1.49}$$

The *theta function* Θ is defined by either of the equivalent representations

$$\begin{aligned}\Theta(v,t)&=\frac{1}{(\pi t)^{\frac{1}{2}}}\sum_{n=-\infty}^{\infty}\exp(-\frac{(v+n)^2}{t})\\ &=1+2\sum_{k=1}^{\infty}\exp(-\pi^2k^2t)\cos(2\pi kv),\\ &\qquad -\frac{1}{2}\le v\le\frac{1}{2}.\end{aligned} \tag{1.50}$$

1.2.8 Bessel's and modified Bessel's functions

It turns out that problems of mathematical physics become more easily tractable if we manage to choose a coordinate system that fits the geometry of the underlying problem adequately. For example, if the region is a right circular cylinder it is suitable to use the cylindrical coordinate system. The functional relationships between the cylindrical coordinates (r, ϑ, z) and the usual rectangular (Cartesian) ones are given by

$$x = r\cos\vartheta, \quad y = r\sin\vartheta, \quad z = z. \tag{1.51}$$

In terms of cylindrical coordinates the Laplace operator (1.11) with $n = 3$ takes the form

$$\Delta u = \frac{1}{r}\frac{\partial}{\partial r}\left(r\frac{\partial u}{\partial r}\right) + \frac{1}{r^2}\frac{\partial^2 u}{\partial \vartheta^2} + \frac{\partial^2 u}{\partial z^2}. \tag{1.52}$$

Similarly, if the intervening region is a circle (a mathematical fiction for plates, membranes etc. in physical applications) then the use of polar coordinates is convenient. (The relations (1.51) and (1.52) remain valid but the coordinate z is not involved.)

In this context we meet with the so-called cylinder and, in particular, Bessel functions. Let $\alpha \in \mathbf{R}$. The second order ordinary differential equation

$$\frac{1}{x}\frac{d}{dx}\left(x\frac{du}{dx}\right) + \left(1 - \frac{\alpha^2}{x^2}\right)u = 0, \quad x \in \mathbf{R} \tag{1.53}$$

is called the Bessel equation of order α. (Generally, α and x may assume complex values.)

Any non-trivial solution of Eq. (1.53) is called a cylinder function. A particular cylinder function given by the formula

$$\mathrm{J}_\alpha(x) = \left(\frac{x}{2}\right)^\alpha \sum_{k=0}^{\infty} \frac{(-1)^k}{k!\,\Gamma(\alpha + k + 1)} \left(\frac{x}{2}\right)^{2k}$$

is called the *Bessel function of the first kind of order* α. [Here Γ is the gamma function

$$\Gamma(x) = \int_0^\infty \exp(-t)\, t^{x-1} dt, \quad x > 0 \ \ (\operatorname{Re} x > 0)$$

and

$$\Gamma(n) = (n-1)! \quad \text{whenever } n \in \mathbf{N}.]$$

Another solution of Eq. (1.53) is the function $\mathrm{J}_{-\alpha}$. The functions J_α and $\mathrm{J}_{-\alpha}$ are linearly independent if and only if α is not an integer. The *Neumann function* (alternatively the *Bessel function of the second kind*) *of order* α is defined by the formula

$$\mathrm{Y}_\alpha(x) = \frac{\cos(\alpha\pi)\mathrm{J}_\alpha(x) - \mathrm{J}_{-\alpha}(x)}{\sin(\alpha\pi)}.$$

[If $\alpha = n$ an integer, then Y_n is defined by taking the limit Y_α as $\alpha \to n$.] The function Y_α and J_α form a fundamental set of solutions of Eq. (1.53) whatever α may be.

The functions

$$\mathrm{H}_\alpha^{(1)}(x) = \mathrm{J}_\alpha(x) + i\mathrm{Y}_\alpha(x), \quad \mathrm{H}_\alpha^{(2)}(x) = \mathrm{J}_\alpha(x) - i\mathrm{Y}_\alpha(x)$$

are called the *Hankel functions of the first and second kind,* respectively, *of order* α. Any two of the functions $\mathrm{J}_\alpha, \mathrm{Y}_\alpha, \mathrm{H}_\alpha^{(1)}$ and $\mathrm{H}_\alpha^{(2)}$ form a fundamental set.

If $\alpha \geq 0$ *then the only (up to an arbitrary multiplicative constant) solution of Eq.* (1.53) *that is bounded in a neighbourhood of the origin is the function* J_α. □

The equation

$$\frac{1}{x}\frac{d}{dx}\left(x\frac{du}{dx}\right) - \left(1 - \frac{\alpha^2}{x^2}\right)u = 0 \tag{1.54}$$

which differs from Eq. (1.53) only by the sign of the term u can be transformed into Eq. (1.53) by the substitution $z = ix$ and this suggests that solutions of (1.54) are cylindrical functions of argument ix.

In the theory of vibrations of plates the following constant multiples of functions $\mathrm{J}_\alpha(ix)$ and $\mathrm{H}_\alpha^{(1)}(ix)$ are commonly used:

$$\mathrm{I}_\alpha(x) = \exp\left(-\frac{i\alpha\pi}{2}\right)\mathrm{J}_\alpha(ix) = \left(\frac{x}{2}\right)^\alpha \sum_{k=0}^{\infty} \frac{1}{k!\,\Gamma(\alpha+k+1)}\left(\frac{x}{2}\right)^{2k},$$

$$\mathrm{K}_\alpha(x) = \frac{i\pi}{2}\exp\left(\frac{i\alpha\pi}{2}\right)\mathrm{H}_\alpha^{(1)}(ix) = \frac{\mathrm{I}_{-\alpha}(x) - \mathrm{I}_\alpha(x)}{\sin(\alpha\pi)}$$

These functions form a fundamental set of solutions of Eq. (1.54) and moreover they possess the important property of being *real* functions for $\alpha \geq 0$ and $x \geq 0$. (At this point let us remark that the function $\mathrm{Y}_\alpha(ix)$ cannot be modified by a simple complex constant multiplication to form together with I_α a fundamental set real for $\alpha \geq 0$ and $x \geq 0$.)

Similarly as the function J_α provides the only (up to a multiplicative constant) solution of (1.53) that is bounded for $x \to 0$, the same property for solutions of Eq. (1.54) is enjoyed by the function I_α.

There is a vast literature on these functions [see, for example, Hochstadt (1964), Watson (1958)].

1.2.9 Generalized solutions and some notation from functional analysis

In § 1.1.4 we defined the notion of a classical solution to an initial- (initial-boundary) value problem for partial differential equations. Many quantities characterizing processes in nature can be described by functions, these being actually classical solutions. However, upon describing some even quite common processes we find that the concept of a classical solution becomes inappropriate, too restrictive (e.g. demanding many derivatives). For that reason we must widen this concept and we turn to the so-called *generalized solutions*. There are a great many ways of introducing them. Consequently many definitions can be encountered in the literature. We shall not pursue these questions systematically since our main concern in this text is to study the classical solutions. Nonetheless, two such typical definitions will be touched upon in

§§ 2.2.8, 2.2.9, 2.5.5, 4.2.8, 4.3.11 (solutions with jump discontinuities) and in §§ 4.4.10, 5.2.4 (solutions introduced by means of Fourier series).

It should be emphasized that the elementary approach we present here to the study of *classical* solutions to *linear* problems needs only elementary means in the sense that no functional-analytical tools are needed to be known to the reader (except for the most elementary ones which will be recalled in the section on the Fourier method). On the other hand, the study of generalized solutions and/or nonlinear problems seems to be troublesome without having recourse to requisite techniques from functional analysis.

In this respect, we shall use occasionally the following customary notation. The symbol $L_2(Q)$, where $Q \subset \mathbf{R}^n$ is a Lebesgue measurable set, denotes the space of square integrable functions, that is, the space of Lebesgue measurable (real) functions u on Q for which the Lebesgue integral $\int_Q u^2 dx$ exists. $L_2(Q)$ is a Hilbert space with the scalar product

$$(u, v)_{L_2(Q)} = \int_Q u(x)\, v(x)\, dx.$$

Let $u \in L_2(Q)$. If there exists a function v defined almost everywhere on Q and integrable on every measurable subset of Q satisfying

$$\int_Q u(x) \frac{\partial^l w}{\partial x_1^{l_1} \cdots \partial x_n^{l_n}}\, dx = (-1)^l \int_Q v(x)\, w(x)\, dx$$

for all $w \in C_0^\infty(Q)$, where $l = l_1 + \cdots + l_n$, then v is called the *generalized derivative* of order l of the function u and is denoted by the same symbol as the classical one,

$$v = \frac{\partial^l u}{\partial x_1^{l_1} \cdots \partial x_n^{l_n}},$$

which is justified by the fact that for continuously differentiable functions the generalized and classical derivatives coincide. □

The symbol $C(J;\, L_2(\Omega))$, where $J \subset \mathbf{R}$ is a closed bounded interval and $\Omega \subset \mathbf{R}^n$ is a Lebesgue measurable set, denotes the space of Lebesgue measurable functions u on $Q = \Omega \times J$ such that for every $t \in J$, u belongs to $L_2(\Omega)$ and

$$u(\cdot, t) \to u(\cdot, t_0) \text{ in the mean on } \Omega \text{ as } t \to t_0, \quad t,\, t_0 \in J,$$

that is

$$\lim_{\substack{t \to t_0 \\ t \in J}} \int_\Omega [u(t,x) - u(t_0, x)]^2\, dx = 0, \quad t_0 \in J.$$

$C(J;\, L_2(\Omega))$ is a Banach space with the norm

$$\max_{t \in J} \left(\int_\Omega u^2(t,x) dx \right)^{\frac{1}{2}}.$$

Any function from $C(J;\, L_2(\Omega))$ belongs also to $L_2(Q)$ and if Ω is bounded then any function $u \in C(\bar{\Omega} \times J)$ belongs to $C(J;\, L_2(\Omega))$.

If Ω is regular and u together with its generalized derivatives $\partial u/\partial x_j$, $j = 1, \ldots, n$ belongs to $C(J; L_2(\Omega))$, then
a) *u is (equivalent to a function) continuous on $\bar{\Omega} \times J$,*
b) *the classical derivatives $\partial u/\partial x_j$ exist almost everywhere on Q [are equal to the generalized ones and belong to $L_2(Q)$].*

If $J_0 \subset \mathbf{R}$ is any interval we put

$$C(J_0; L_2(\Omega)) = \bigcap_{J \subset J_0} C(J; L_2(\Omega)),$$

where J are closed bounded intervals in $\mathbf{R}$. The space $C(J_0; L_2(\Omega))$ is no longer a Banach space (unless J_0 is closed bounded).

1.2.10 The modified implicit functions theorem

For the needs of Chapter 2 we state the following modification of the implicit functions theorem.

Let $Q_t \subset \mathbf{R}^r \times [t_0, \infty)$ be a domain containing the whole plane $t = t_0$. Let $G = (G_1, \ldots, G_s)$, $G_j = G_j(x, t, u) \in C^1(Q_t \times \mathbf{R}^s)$, $\psi = (\psi_1, \ldots, \psi_s)$, $\psi_j = \psi_j(x) \in C^1(\mathbf{R}^r)$, $j = 1, \ldots, s$. Further let $G(x, t_0, \psi(x)) = 0$ and

$$\det \left(\frac{\partial G_j}{\partial u_k}(x, t_0, \psi(x)) \right)_{j,k=1}^{s} = c \neq 0$$

for $x \in \mathbf{R}^r$. Then there exists a domain $Q'_t \subset Q_t$ containing the whole plane $t = t_0$ and a unique vector function $u(x, t)$ the components of which belong to $C^1(Q'_t)$ such that $u(x, t_0) = \psi(x)$, $x \in \mathbf{R}^r$ and

$$G(x, t, u(x, t)) = 0 \quad in \quad Q'_t.$$

CHAPTER 2

First order equations

2.1 SOME EQUATIONS OF PHYSICAL IMPORTANCE

The equations of a compressible fluid are derived. Then the equations governing vibrations of an elastic body and of an electromagnetic field are introduced.

2.1.1 The continuity equation (the law of conservation of mass)

First let us derive the continuity equation for the simplest case of a fluid flowing through a tube of a constant cross-section. For the sake of simplicity we suppose that all quantities are constant in every section and thus they can change in the direction of the tube and in time only. Let the coordinate system be introduced in such a way that the x-axis coincides with the direction of the tube. The amount of mass which flows through the tube at a point x in a time interval (t, t_1) is evidently given by the integral

$$\int_t^{t_1} A\,\rho(x,\tau)\,u(x,\tau)\,d\tau,$$

where A is the area of the cross-section, $\rho(x,t)$ is the mass density and $u(x,t)$ the velocity of the flow at the point x and at time t. (For $u < 0$ this expression is negative which indicates the fact that the fluid moves in the direction of the negative x-half-axis.) Let us consider a portion of the tube between the points x and x_1 $(x < x_1)$ in the time interval (t, t_1). Since the fluid in the tube cannot vanish, the possible difference between the amount of fluid passing through the points x and x_1 at the opposite ends of the portion must be compensated only by the change of the density of the fluid. (Thus, for instance, if the inflow is more powerful than the outflow, the density increases; on the contrary, if the outflow prevails, the density decreases.) The difference of the amount of the fluid due to the inlet and the outlet at the chosen place and time intervals (x, x_1) and (t, t_1) is given by

$$\int_t^{t_1} A\,\rho(x,\tau)\,u(x,\tau)\,d\tau - \int_t^{t_1} A\,\rho(x_1,\tau)\,u(x_1,\tau)\,d\tau. \tag{2.1}$$

The total amount of the mass in the given portion at time t being

$$\int_x^{x_1} A\,\rho(\xi,t)\,d\xi,$$

its change in the time interval (t, t_1) due to the increase or decrease of density is expressed by

$$\int_x^{x_1} A\,\rho(\xi, t_1)\,d\xi - \int_x^{x_1} A\,\rho(\xi, t)\,d\xi. \tag{2.2}$$

Equating (2.1) and (2.2) and dividing by A and $(t_1 - t)$ we get

$$\begin{aligned}\frac{1}{t_1 - t}\int_t^{t_1} [\rho(x,\tau)\,u(x,\tau) - \rho(x_1,\tau)\,u(x_1,\tau)]\,d\tau \\ = \frac{1}{t_1 - t}\left[\int_x^{x_1} \rho(\xi, t_1)\,d\xi - \int_x^{x_1} \rho(\xi, t)\,d\xi\right].\end{aligned}$$

Since this equality is valid for arbitrarily small time increment $t_1 - t$, we can pass to the limit for $t_1 \to t$. By virtue of the relations

$$\begin{aligned}\lim_{t_1 \to t} \frac{1}{t_1 - t}\int_t^{t_1} [\rho(x,\tau)\,u(x,\tau) - \rho(x_1,\tau)\,u(x_1,\tau)]\,d\tau \\ = \rho(x,t)\,u(x,t) - \rho(x_1,t)\,u(x_1,t),\end{aligned}$$

$$\lim_{t \to t_1} \frac{1}{t_1 - t}\left[\int_x^{x_1} \rho(\xi, t_1)\,d\xi - \int_x^{x_1} \rho(\xi, t)\,d\xi\right] = \frac{d}{dt}\int_x^{x_1} \rho(\xi, t)\,d\xi,$$

we obtain the equation

$$\frac{d}{dt}\int_x^{x_1} \rho(\xi, t)\,d\xi + \rho(x_1, t)\,u(x_1, t) - \rho(x, t)\,u(x, t) = 0, \tag{2.3}$$

or, denoting by q the so-called mass-flux ρu, we have

$$\frac{d}{dt}\int_x^{x_1} \rho(\xi, t)\,d\xi + q(x_1, t) - q(x, t) = 0. \tag{2.4}$$

The equation (2.3) or (2.4) expresses the *conservation law of mass*. Under the assumption that ρ and u are continuously differentiable functions we can rewrite it in the differential form

$$\frac{\partial \rho(x,t)}{\partial t} + \frac{\partial}{\partial x}[\rho(x,t)\,u(x,t)] = 0, \tag{2.5}$$

or with the help of the mass-flux in the form

$$\frac{\partial \rho}{\partial t} + \frac{\partial q}{\partial x} = 0. \tag{2.6}$$

In fact, dividing Eq. (2.3) by $x_1 - x$ and making use of the relations

$$\lim_{x_1 \to x} \frac{1}{x_1 - x}\int_x^{x_1} \rho(\xi, t)\,d\xi = \rho(x, t),$$

$$\lim_{x_1 \to x} \frac{1}{x_1 - x}[\rho(x_1,t)\,u(x_1,t) - \rho(x,t)\,u(x,t)] = \frac{\partial}{\partial x}[\rho(x,t)\,u(x,t)],$$

we get Eq. (2.5). It is usually called the *continuity equation.* □

Now, let us deduce the continuity equation in the case of three-dimensional flow. In the flowing fluid, let us consider a fixed volume V with boundary S. The balance equation equates the difference of the mass-flux of the fluid due to the inlet and outlet through the boundary and the difference of the amount of the mass due to the change in density (that is due to the compressibility of the fluid). The increase of the mass in the volume V caused by the inlet and outlet through the boundary per unit time is given by

$$-\int_S \rho(x,t)\,u(x,t)\cdot\nu(x,t)\,dS,$$

where $\nu(x,t)$ is the outward normal to S. (The minus sign in front of the integral says that the increase is positive if the fluid flows inward, that is, in the direction opposite to ν.) On the other hand the increase in mass caused by the change of density is equal to

$$\frac{d}{dt}\int_V \rho(x,t)\,dV.$$

Thus the balance equation has the form

$$-\int_S \rho u\cdot\nu\,dS = \frac{d}{dt}\int_V \rho\,dV.$$

By the Gauss–Ostrogradskiĭ formula (1.31)

$$\int_S \rho u\cdot\nu\,dS = \int_V \operatorname{div}(\rho u)\,dV$$

we have

$$\int_V [\rho_t + \operatorname{div}(\rho u)]\,dV = 0 \tag{2.7}$$

and thus (see § 1.2.4) the arbitrariness of the volume V implies, for smooth functions u, ρ, the equation

$$\rho_t + \operatorname{div}(\rho u) = 0 \tag{2.8}$$

This is the *continuity equation* for the three-dimensional flow.

The continuity equation represents one scalar equation for the unknowns ρ and u. If, for example, the dependence of u on ρ is known from experimental observations, then ρ may be determined from (2.8), see for instance the problem of the traffic flow, § 2.2.9. However, more often, further relations are described by additional equations; for example, in hydrodynamics by the equations of motion, see § 2.1.2, and by the state equation.

If ρ is constant (non-vanishing), the continuity equation reduces to the condition

$$\operatorname{div}\,u = 0. \tag{2.9}$$

We arrive at the same equation when we investigate the motion of an incompressible liquid. In this case the output from an arbitrary volume V bounded by a surface S must be equal to the corresponding input and thus

$$\int_S u(x,t)\cdot\nu(x,t)\,dS = 0$$

which by (1.31) and § 1.2.4 finally yields Eq. (2.9).

Let us mention that the continuity equation is fulfilled also for other physical quantities. So, for instance, if ρ denotes the density of electrical charge, and u its velocity, then

$$\rho_t + \operatorname{div}(\rho u) = 0.$$

When the so-called current density ρu is denoted by J, the continuity equation (the *conservation law of charge*) can be written in the form

$$\rho_t + \operatorname{div} J = 0.$$

2.1.2 Equations of motion (the law of conservation of momentum)

To derive the equations of motion is a task analogous to deriving the equation of continuity. Hence we will proceed faster and start immediately with the three-dimensional flow. Let us consider a fixed control volume V of the fluid with the boundary S and by the second Newton law let us equate the total change of the momentum with the sum of the forces acting on the volume V and on its surface S. The change of the j-th component of the momentum caused by the inlet or outlet of the fluid is given by

$$\int_S \rho u_j u \cdot \nu \, dS,$$

while the time change is given by

$$\frac{d}{dt} \int_V \rho u_j \, dV.$$

Hence the balance equations read

$$\int_S \rho u_j u \cdot \nu \, dS + \frac{d}{dt} \int_V \rho u_j \, dV = - \int_S p \nu_j \, dS + \int_V \rho F_j \, dV, \quad j = 1, 2, 3.$$

Here $F = (F_1, F_2, F_3)$ is a force acting on a unit of mass (often it is the gravity) and p is the pressure. (The minus sign in front of the surface integral is there because we suppose that the pressure is positive if it acts in the direction opposite to ν.) Making use of (1.30) we transform the surface integrals to volume integrals. Due to the arbitrariness of the volume V, for sufficiently smooth functions we obtain the equations

$$\sum_{k=1}^{3} \frac{\partial}{\partial x_k}(\rho u_j u_k) + \frac{\partial}{\partial t}(\rho u_j) = -\frac{\partial p}{\partial x_j} + \rho F_j, \quad j = 1, 2, 3.$$

Calculating the indicated derivatives and applying the continuity equation, we finally get

$$\rho \frac{\partial u_j}{\partial t} + \rho \sum_{k=1}^{3} u_k \frac{\partial u_j}{\partial x_k} = -\frac{\partial p}{\partial x_j} + \rho F_j, \quad j = 1, 2, 3 \tag{2.10}$$

or, in the vector form,

$$\rho\frac{\partial u}{\partial t} + \rho(u \cdot \nabla) \cdot u = -\operatorname{grad} p + \rho F. \tag{2.11}$$

These equations express the *conservation law of momentum* and are called the *equations of motion.* This system together with the continuity equation (2.8) is called the *Euler equations.*

The continuity equation together with the equations of motion form a system of four equations for five unknowns ρ, u_1, u_2, u_3 and p. Thus we still have more unknowns than equations. In some cases, for instance for some gases, the state equation $p = p(\rho)$, determining the relation between ρ and p holds. Another case allowing us to reduce the number of unknowns is the case of a liquid for which ρ may be considered constant. If we can make use of one of these possibilities we have four equations in four unknowns. Of course if the flow takes place in a domain which differs from the whole space, some conditions on the boundary must be added to the equations. Thus, for instance, if the boundary is formed by an impermeable wall, we put $u_\nu = 0$ where u_ν is the normal component of u.

2.1.3 The linearized equations of acoustics

One of the simplest cases of motion is a one-dimensional motion of a liquid on which no volume forces act and whose state equation has the form $p = p(\rho)$, where $p \in C^2(\mathbf{R}^+)$. Such a motion is described by the equations

$$\begin{aligned} u_t + uu_x &= -\frac{p'(\rho)\rho_x}{\rho}, \\ \rho_t + u\rho_x + \rho u_x &= 0. \end{aligned} \tag{2.12}$$

They are often called the *equations of acoustics*, since they describe relatively well the propagation of sound. Nevertheless, even the resolution of this system is rather difficult. Therefore, if the required accuracy admits it, we simplify the system by a so-called linearization. Supposing that the velocity u is small, say $u = \varepsilon\overline{u}$, and that ρ differs slightly from a constant density ρ_0, say $\rho = \rho_0 + \varepsilon\overline{\rho}$, where ε is a small parameter, (2.12) assumes the form

$$\begin{aligned} \varepsilon\overline{u}_t + \varepsilon^2\overline{u}\,\overline{u}_x &= -\frac{p'(\rho_0 + \varepsilon\overline{\rho})}{\rho_0 + \varepsilon\overline{\rho}}\varepsilon\overline{\rho}_x, \\ \varepsilon\overline{\rho}_t + \varepsilon^2\overline{u}\,\overline{\rho}_x + (\rho_0 + \varepsilon\overline{\rho})\varepsilon\overline{u}_x &= 0. \end{aligned}$$

Taking into account that $\dfrac{p'(\rho_0 + \varepsilon\overline{\rho})}{\rho_0 + \varepsilon\overline{\rho}} = \dfrac{p'(\rho_0)}{\rho_0} + O(\varepsilon)$, neglecting the terms of order ε^2 and dividing both equations by ε, we obtain the *linearized equations of acoustics*

$$\begin{aligned} u_t + \frac{c^2}{\rho_0}\rho_x &= 0, \\ \rho_t + \rho_0 u_x &= 0, \end{aligned} \tag{2.13}$$

where we have dropped bars over u and ρ and put $c = [p'(\rho_0)]^{1/2}$. [It may be shown that c has the physical meaning of the *velocity of propagation of sound*, see (2.147).] Of course the linearized system of equations of acoustics approximates well the behaviour of u and ρ provided u and $\rho - \rho_0$ are sufficiently small.

2.1.4 Equations of elasticity

Considering an elastic body let us take a small fixed control volume V with the boundary S. In terms of the stress tensor $\tau = (\tau_{kj})_{k,j=1}^{3}$ the surface force $T = (T_1, T_2, T_3)$ acting on this volume at a point $x \in S$ can be expressed in the form

$$T_j(x) = \sum_{k=1}^{3} \tau_{kj}(x)\nu_k(x), \quad j = 1, 2, 3,$$

where $\nu = (\nu_1, \nu_2, \nu_3)$ is the outward normal to V at the point x. (Let us note that as a consequence of the equilibrium of moments the stress tensor is a symmetric one, that is, $\tau_{12} = \tau_{21}, \tau_{13} = \tau_{31}, \tau_{23} = \tau_{32}$.) The second Newton law for V reads

$$\int_V \rho \frac{\partial^2 u_j}{\partial t^2}\, dV = \int_S \sum_{k=1}^{3} \tau_{kj}\nu_k\, dS + \int_V F_j\, dV, \tag{2.14}$$

where ρ is the density, u is the displacement vector and $F = (F_1, F_2, F_3)$ is the vector of volume forces. Using the divergence theorem we obtain

$$\int_V \rho \frac{\partial^2 u_j}{\partial t^2}\, dV = \int_V \sum_{k=1}^{3} \frac{\partial \tau_{kj}}{\partial x_k}\, dV + \int_V F_j\, dV, \qquad j = 1, 2, 3$$

and consequently

$$\rho \frac{\partial^2 u_j}{\partial t^2} = \sum_{k=1}^{3} \frac{\partial \tau_{kj}}{\partial x_k} + F_j, \qquad j = 1, 2, 3. \tag{2.15}$$

The relation between the stress tensor τ and the displacement vector u is given by the Hooke law

$$\tau_{kj} = \lambda \delta_{kj} \sum_{r=1}^{3} e_{rr} + 2\mu e_{kj}, \quad j, k = 1, 2, 3, \tag{2.16}$$

where $e_{kj} = \dfrac{1}{2}\left(\dfrac{\partial u_k}{\partial x_j} + \dfrac{\partial u_j}{\partial x_k}\right)$ are components of the so-called strain tensor while λ, μ are the Lamé coefficients. Differentiating the last equation with respect to t and putting $v_j = \dfrac{\partial u_j}{\partial t}$, $j = 1, 2, 3$ we get a system of nine equations, called the *equations of elasticity*, for nine unknowns $v_1, v_2, v_3, \tau_{11}, \tau_{12}, \tau_{13}, \tau_{22}, \tau_{23}, \tau_{33}$:

$$\begin{aligned} \rho \frac{\partial v_j}{\partial t} &= \sum_{k=1}^{3} \frac{\partial \tau_{kj}}{\partial x_k} + F_j, \quad j = 1, 2, 3, \\ \frac{\partial \tau_{kj}}{\partial t} &= \lambda \delta_{kj} \sum_{r=1}^{3} \frac{\partial v_r}{\partial x_r} + \mu\left(\frac{\partial v_k}{\partial x_j} + \frac{\partial v_j}{\partial x_k}\right), \quad j, k = 1, 2, 3, \quad k \le j. \end{aligned} \tag{2.17}$$

Inserting (2.16) into (2.15) we obtain three equations of the second order for the unknowns u_1, u_2, u_3:

$$\rho \frac{\partial^2 u}{\partial t^2} = (\lambda + \mu)\, \mathrm{grad}\ \mathrm{div}\, u + \mu \Delta u + F, \tag{2.18}$$

which we shall meet in Chapter 3 (cf. § 3.5.5).

2.1.5 The Maxwell equations of electromagnetic field

Relying primarily on Faraday's and Oersted's experiments, J.C.Maxwell in the seventies of the last century formulated the equations governing an electromagnetic field. In the simplest case they can be written down as a system for seven unknowns $E_1, E_2, E_3, B_1, B_2, B_3, \rho$ in the vector form

$$\varepsilon\frac{\partial E}{\partial t} - \frac{1}{\mu}\operatorname{curl} B + \gamma E = -J, \tag{2.19}$$

$$\frac{\partial B}{\partial t} + \operatorname{curl} E = 0, \tag{2.20}$$

$$\operatorname{div} B = 0, \tag{2.21}$$

$$\varepsilon \operatorname{div} E = \rho, \tag{2.22}$$

where E, B are the vectors of the electric field and the magnetic induction, respectively, J is the vector of the current density of external sources and ε, μ, γ and ρ are the permitivity, the permeability, the conductivity of the medium and the density of free charges, respectively. We suppose that ε, μ and γ are constant scalars although in general they are tensors depending on x, t and E, B. By the way, let us note that the term $\varepsilon\frac{\partial E}{\partial t}$, the so-called displacement current, was added by Maxwell only on the basis of certain theoretical reasoning, which were fully justified by further experiments with high frequency fields in which the displacement current becomes more dominant. To the system (2.19) – (2.22) let us add initial conditions

$$\begin{aligned} E(x_1, x_2, x_3, 0) &= E^0(x_1, x_2, x_3), \\ B(x_1, x_2, x_3, 0) &= B^0(x_1, x_2, x_3). \end{aligned} \tag{2.23}$$

Let us assume that B^0 satisfies the equation (2.21) that is

$$\operatorname{div} B^0 = 0. \tag{2.24}$$

Under this assumption, when investigating the system (2.19) – (2.22) we may restrict ourselves to the first two equations which represent six scalar equations for six unknown components of E and B. Evidently, if E is known, Eq. (2.22) defines ρ. Eq. (2.21) is a consequence of (2.20), (2.24). Indeed, applying the operator div to Eq. (2.20) [and using the identity (1.28)] we get

$$\frac{\partial}{\partial t}\operatorname{div} B = 0$$

and thus $\operatorname{div} B(x,t) = \operatorname{div} B^0(x) = 0$.

An electromagnetic field is usually studied in a domain different from the whole space. Then it is necessary to add some boundary conditions corresponding to the electric and magnetic properties of the boundary. We shall meet with this situation later in Chapter 3.

Let us mention that the equations of an electromagnetic field are often transformed to equations of the second order. Applying the operator curl to Eq. (2.19) and differentiating (2.20) with respect to t, after some simple arrangements we get the relation

$$-\varepsilon B_{tt} - \frac{1}{\mu} \operatorname{curl}\operatorname{curl} B + \gamma \operatorname{curl} E + \operatorname{curl} J = 0.$$

Taking into account the relation (1.29), according to (2.20) and (2.21) we readily conclude that the magnetic induction B fulfills the so-called telegraph equation

$$B_{tt} - \frac{1}{\varepsilon\mu}\Delta B + \frac{\gamma}{\varepsilon} B_t = \frac{1}{\varepsilon} \operatorname{curl} J. \tag{2.25}$$

In vacuum, where $\gamma = 0$, this equation reduces to the wave equation

$$B_{tt} - \frac{1}{\varepsilon_0\mu_0}\Delta B = \frac{1}{\varepsilon_0} \operatorname{curl} J, \tag{2.26}$$

(where ε_0, μ_0 are the values of ε, μ *in vacuum*). If the displacement current $\varepsilon \frac{\partial E}{\partial t}$ in (2.19) may be omitted we get a parabolic equation for B, namely

$$B_t - \frac{1}{\gamma\mu}\Delta B = \frac{1}{\gamma} \operatorname{curl} J. \tag{2.27}$$

Similar equations may be derived for the electric field E.

2.2 ONE EQUATION IN ONE SPACE VARIABLE

The solution to the initial-value problem for a single linear equation in one space variable is derived. For a simple quasilinear homogeneous equation the origin of a shock wave is investigated and the corresponding Rankine–Hugoniot condition is formulated. As an example the traffic flow is studied.

2.2.1 A simple example

Let us deal with the equation

$$u_t + a u_x = f(x,t) \tag{2.28}$$

where f together with f_x belong to $C(\mathbf{R} \times [t_0, \infty))$. It is easy to see that along the straight lines $x = at + c$, which are solutions of the differential equation $\frac{dx}{dt} = a$, the differential operator $\frac{\partial}{\partial t} + a\frac{\partial}{\partial x}$ [defined by the left-hand side of (2.28)], represents a total derivative with respect to t:

$$\frac{d}{dt} = \frac{\partial}{\partial t} + \frac{\partial}{\partial x}\frac{dx}{dt} = \frac{\partial}{\partial t} + a\frac{\partial}{\partial x}.$$

Thus Eq. (2.28) may be rewritten in the form

$$\frac{d}{dt}u(at+c,t) = f(at+c,t) \tag{2.29}$$

[which is an ordinary differential equation for the unknown function $U(t) = u(at+c,t)$]. Integrating Eq. (2.29) from t_0 to t, we get

$$u(at+c,t) - u(at_0+c,t_0) = \int_{t_0}^{t} f(as+c,s)\,ds.$$

Prescribing u on the level $t = t_0$ by

$$u(x,t_0) = \varphi(x), \qquad \varphi \in C^1(\mathbf{R}) \tag{2.30}$$

and bearing in mind that $x = at + c$, we finally get

$$u(x,t) = \varphi(at_0 + x - at) + \int_{t_0}^{t} f(as + x - at, s)\,ds.$$

It can be verified readily that this expression satisfies both the equation (2.28) and the initial condition (2.30).

2.2.2 The general case

Here it is our goal to find a solution $u \in C^1(\mathbf{R} \times [t_0,\infty))$ to the problem

$$u_t + a(x,t)u_x + b(x,t)u = f(x,t), \tag{2.31}$$

$$u(x,t_0) = \varphi(x), \tag{2.32}$$

where a, b, f together with a_x, b_x, f_x belong to $C(\mathbf{R} \times [t_0,\infty))$, $\varphi \in C^1(\mathbf{R})$. Following the idea of the previous subsection, let us introduce curves on which the differential operator $\dfrac{\partial}{\partial t} + a(x,t)\dfrac{\partial}{\partial x}$ may by viewed as a total derivative with respect to t. Reserving the notation (x,t) for the point at which the solution has to be found, let us denote the general point by (ξ,τ). Then the equation of these curves, called *characteristics*, is evidently given by

$$\frac{d\xi}{d\tau} = a(\xi,\tau). \tag{2.33}$$

Its solution passing through the point (x,t) will be denoted by $\chi(\tau;x,t)$. (Thus $x = \chi(t;x,t)$.) By the theory of ordinary differential equations, under our assumption that $a, a_x \in C(\mathbf{R} \times [t_0,\infty))$, there exists a unique solution of (2.33) from the class C^1 in τ, x and t on the domain of its existence, passing through the given point (x,t). From every point (x,t_0), of the straight line $\tau = t_0$ let us lead the characteristic $\chi(\tau;x,t_0)$. Again by the theory of ordinary differential equations it may be continued either for $\tau \to \infty$ or for $x \to \pm\infty$. Let us call the subset of the half-plane $\mathbf{R} \times [t_0,\infty)$, covered by all these characteristics the *domain of influence* of the straight line $\tau = t_0$ and denote it briefly by $\mathcal{DI}(\mathbf{R}; t_0)$. Inserting $\chi(\tau; x, t)$ for x into (2.31) and putting $U(\tau) = u(\chi(\tau; x, t), \tau)$, we see that the equation (2.31) may be rewritten in the form of an ordinary differential equation

$$\frac{dU(\tau)}{d\tau} + b(\chi(\tau; x, t), \tau)\, U(\tau) = f(\chi(\tau; x, t), \tau). \tag{2.34}$$

Integrating it from t_0 to t we get

$$U(t) = U(t_0) \exp(-\int_{t_0}^{t} b(\chi(\sigma; x, t), \sigma)\, d\sigma) + \int_{t_0}^{t} f(\chi(s; x, t), s) \cdot \exp(-\int_{s}^{t} b(\chi(\sigma; x, t), \sigma)\, d\sigma)\, ds$$

and recalling that

$$U(t_0) = u(\chi(t_0; x, t), t_0) = \varphi(\chi(t_0; x, t)),$$
$$U(t) = u(\chi(t; x, t), t) = u(x, t),$$

we readily conclude that the solution has the form

$$\begin{aligned} u(x, t) = \varphi(\chi(t_0; x, t)) \exp(-\int_{t_0}^{t} b(\chi(\sigma; x, t), \sigma)\, d\sigma) \\ + \int_{t_0}^{t} f(\chi(s, x, t), s) \exp(-\int_{s}^{t} b(\chi(\sigma; x, t), \sigma)\, d\sigma)\, ds. \end{aligned} \tag{2.35}$$

It can be again verified readily that this expression satisfies the equation (2.31) with the initial condition (2.32). The method used above is called the *method of characteristics*. We have proved the following theorem.

Let a, b, f together with a_x, b_x, f_x belong to $C(\mathbf{R} \times [t_0, \infty))$, $\varphi \in C^1(\mathbf{R})$. Let $(x, t) \in \mathcal{DI}(\mathbf{R}; t_0)$. Then the problem (2.31), (2.32) *has a unique solution given by the formula* (2.35).

2.2.3 Remarks

2.2.3.1 In particular, in § 2.4.6 we shall make use of the Riemann functions. To be acquainted with them let us prove the above theorem once more using the approach via the Riemann function. Multiplying equality (2.34) by $v(\tau)$ and integrating by parts from t to t_0, we obtain

$$\int_{t}^{t_0} [-\frac{dv(t)}{d\tau} + b(\chi(\tau; x, t), \tau) v(\tau)]\, U(\tau)\, d\tau$$
$$+ [v(\tau)\, U(\tau)]_t^{t_0} = \int_{t}^{t_0} f(\chi(\tau; x, t), \tau)\, v(\tau)\, d\tau.$$

If $v(\tau)$ is chosen in such a way that

$$-\frac{dv(\tau)}{d\tau} + b(\chi(\tau; x, t), \tau)\, v(\tau) = 0, \quad v(t) = 1,$$

it follows readily that

$$v(t_0)U(t_0) - U(t) = \int_t^{t_0} f(\chi(\tau; x, t), \tau) v(\tau)\, d\tau,$$

(the function v is called the *Riemann function*). Since

$$v(\tau) = \exp\left(\int_t^{\tau} b(\chi(\sigma; x, t), \sigma)\, d\sigma\right)$$

and

$$U(t) = u(\chi(t; x, t), t) = u(x, t)$$
$$U(t_0) = u(\chi(t_0; x, t), t_0) = \varphi(\chi(t_0; x, t))$$

we get (2.35) again.

2.2.3.2 The initial condition φ can be given only on an interval J between the points (x_0, t_0) and (x_1, t_0). Then the solution u can be sought in the strip between the characteristics $\chi(\tau; x_0, t_0)$ and $\chi(\tau; x_1, t_0)$, which now bound the set containing the domain of influence of the interval J.

2.2.3.3 Even if a, b, f and φ are defined and sufficiently smooth in the half-plane $\mathbf{R} \times [t_0, \infty)$ and $\mathbf{R}$ respectively, it may happen that $\mathcal{DI}(\mathbf{R}; t_0)$ does not coincide with this whole half-plane. To this goal some supplementary conditions are needed. For example it suffices to suppose that

$$|a(x, t)| \leq (1 + |x|)h(t), \quad \text{where } h \in C[t_0, \infty).$$

2.2.3.4 Let $\mathcal{DI}(\mathbf{R}; t_0)$ be the upper half-plane $\tau \geq t_0$. Let

$$\exp\left(-\int_{t_0}^{t} b(\chi(\sigma; x, t), \sigma)\, d\sigma\right) \leq C \exp(-b_0(t - t_0)), \quad b_0 > 0, C > 0.$$

Then for any two solutions u_0, u of (2.31) the following estimate holds:

$$\sup_{x \in \mathbf{R}} |u(x, t) - u_0(x, t)| \leq C \sup_{x \in \mathbf{R}} |u(x, t_0) - u_0(x, t_0)| \exp(-b_0(t - t_0)),$$

which means that every solution u_0 is globally exponentially stable in the norm $\|u(\cdot, t)\| = \sup_{x \in \mathbf{R}} |u(x, t)|$.

Our assertion is an immediate consequence of the relation

$$\begin{aligned} |u(x, t) - u_0(x, t)| &= |u(\chi(t_0; x, t), t_0) - u_0(\chi(t_0; x, t), t_0)| \\ &\cdot \exp\left(-\int_{t_0}^{t} b(\chi(\sigma; x, t), \sigma)\, d\sigma\right) \leq C |u(\chi(t_0; x, t), t_0) - u_0(\chi(t_0; x, t), t_0)| \\ &\cdot \exp(-b_0(t - t_0)). \end{aligned}$$

2.2.3.5 If $b \equiv 0$ and $f \equiv 0$, then by (2.35) $u(x, t) = \varphi(\chi(t_0; x, t))$, which means that the solution is constant on every characteristic. This is not surprising since the equation $u_t + a u_x = 0$ says that the derivative along each characteristic vanishes.

2.2.4 Examples

2.2.4.1 Let us solve the problem

$$u_t + x\cos t\ u_x = 0, \qquad u(x,0) = \frac{1}{1+x^2}, \qquad -\infty < x < \infty,\ t > 0.$$

The equation of the characteristics reads

$$\frac{d\xi}{d\tau} = \xi\cos\tau.$$

Its solution passing through the point (x,t) has the form $\chi(\tau;x,t) = x\exp(\sin\tau - \sin t)$. Thus the domain of influence of the straight line $\tau = 0$ is the half-plane $\tau \geq 0$ and the solution is given by

$$u(x,t) = \frac{1}{1 + x^2\exp(-2\sin t)}.$$

2.2.4.2 Let us solve the problem

$$u_t + x^2 u_x = 0, \qquad u(x,0) = \varphi(x), \qquad -\infty < x < \infty,\ t > 0 \tag{2.36}$$

where $\varphi \in C^1(\mathbf{R})$.

The solution of the equation of the characteristic passing through the point (x,t) being

$$\chi(\tau;x,t) = \frac{x}{1 + x(t-\tau)},$$

we easily find that the domain of influence of the straight line $\tau = 0$ consists of all points of the first quadrant and of the points of the second quadrant lying below the hyperbola

$$t = -1/x.$$

Exactly one characteristic passes through every point of this set. It intersects the straight line $\tau = 0$ at the point $(x/(1+xt), 0)$. Thus the solution in $\mathcal{DI}(\mathbf{R};0)$ is given by

$$u(x,t) = \varphi\left(\frac{x}{1+xt}\right).$$

2.2.4.3 Let us solve the problem

$$u_t + au_x + (1+\cos t)u = f(x,t), \qquad u(x,0) = \varphi(x), \qquad -\infty < x < \infty,\ t > 0$$

where $f, f_x \in C(\mathbf{R}\times\mathbf{R}^+)$, $\varphi \in C^1(\mathbf{R})$.

The solution of the equation of the characteristic passing through the point (x,t) being

$$\chi(\tau;x,t) = x + a(\tau - t),$$

the domain of influence $\mathcal{DI}(\mathbf{R};0)$ is the upper half-plane $\mathbf{R}\times\mathbf{R}^+$. Then the solution reads

$$u(x,t) = \varphi(x - at)\exp(-t - \sin t) + \int_0^t f(x + a(s-t), s) \cdot \exp(-t - \sin t + s + \sin s)\, ds.$$

Note that every solution is globally exponentially stable in the norm $\|u(\cdot,t)\| = \sup_{x\in\mathbf{R}}|u(x,t)|$, since $\exp(-t - \sin t) \leq e\ \exp(-t)$.

2.2.5 Exercises

Solve the problems (with $\varphi \in C^1(\mathbf{R})$).

2.2.5.1

$$u_t + xtu_x = x^2t, \qquad u(x,0) = \varphi(x), \qquad -\infty < x < \infty,\ t > 0.$$

Answer.

$$u(x,t) = \varphi(x\exp(-t^2/2)) + \frac{1}{2}x^2\exp(-t^2)(\exp t^2 - 1).$$

2.2.5.2

$$u_t + \frac{1}{1+|x|}u_x = 0, \qquad u(x,0) = \varphi(x), \qquad -\infty < x < \infty,\ t > 0.$$

Answer.

$$\begin{aligned} u(x,t) &= \varphi(1-(1+2t-2x-x^2)^{1/2}) && \text{for } 0 \le x < -1+(1+2t)^{1/2},\\ u(x,t) &= \varphi(-1+(1-2t+2x-x^2)^{1/2}) && \text{for } x \ge -1+(1+2t)^{1/2},\\ u(x,t) &= \varphi(1-(1+2t-2x+x^2)^{1/2}) && \text{for } x < 0. \end{aligned}$$

2.2.5.3

$$u_t + (x+t)u_x + t(x+1)u = 0, \qquad u(x,0) = \varphi(x), \qquad -\infty < x < \infty,\ t > 0.$$

Answer.

$$u(x,t) = \varphi\left((x+t+1)\exp(-t) - 1\right)\exp\left(\frac{t^3}{3} + (1-t-\exp(-t))(x+t+1)\right).$$

2.2.6 The equation $u_t + a(u)u_x = 0$

Although this book is mainly devoted to linear problems, we will make an exception and study some quasilinear equations of the first order, since it is possible to master them by methods similar to those of the linear theory. This deviation from the framework of linear problems may be justified also by the extraordinary importance of these equations in applications. We will deal with the particular case of quasilinear equations of the first order of the form

$$u_t + a(u)u_x = 0, \qquad -\infty < x < \infty,\ t > 0. \tag{2.37}$$

Let us look for a solution of (2.37) fulfilling the initial condition

$$u(x,0) = \varphi(x), \qquad -\infty < x < \infty. \tag{2.38}$$

Let us suppose that a, $\varphi \in C^1(\mathbf{R})$. Further, let us suppose that we already know a solution $u(x,t)$ to (2.37). If (x,t) is a chosen fixed point, the equation

$$\frac{d\xi}{d\tau} = a(u(\xi,\tau)) \tag{2.39}$$

has a solution, at least in some vicinity of this point, which for $\tau = t$ attains the value x. Let us denote it by the symbol $\chi(\tau; x, t)$. In this subsection let us call this solution a *characteristic*, though the definition of a characteristic for the general quasilinear equation in several space variables introduced in § 2.3.3 is different. (We do not follow here the general scheme of resolution from § 2.3.3, since the homogeneity of the equation and the independence of the coefficient a on x, t enables us to find the solution in a simpler way.)

Clearly we have

$$\frac{d}{d\tau}u(\chi(\tau;x,t),\tau) = u_t(\chi(\tau;x,t),\tau) + a(u(\chi(\tau;x,t),\tau))\,u_x(\chi(\tau;x,t),\tau) = 0,$$

which means that u is constant along the curve $\xi = \chi(\tau; x, t)$. Let us denote by c this constant value (depending evidently on the choice of the point (x,t)). Therefore the solutions of (2.39) are straight lines

$$\chi(\tau;x,t) = x + a(c(x,t))(\tau - t),$$

which may be continued up to the axis $\tau = 0$. Thus the x-coordinate of the point of intersection is $x - a(c(x,t))t$, and at this point the solution of the initial problem must attain the value $\varphi(x - a(c(x,t))t)$. Since u is constant along the whole curve $\xi = \chi(\tau; x, t)$, the value of $c(x,t)$ has to satisfy the equation

$$G(x,t,c) \equiv c - \varphi(x - a(c)t) = 0. \tag{2.40}$$

Let us suppose that in some domain $Q \subset \mathbf{R} \times \mathbf{R}^+$ part of whose boundary is formed by the whole x-axis this equation has a unique continuously differentiable solution $c = c^*(x,t)$. By virtue of the Modified Implicit Functions Theorem with $r = 1$, $s = 1$, $\psi = \varphi$ and $Q_t = \mathbf{R} \times \mathbf{R}^+$ such a domain surely exists. Indeed, the equation (2.40) has a solution $c^*(x,0) = \varphi(x)$ for $t = 0$ and the derivative of the left-hand side of (2.40) with respect to c equals unity for $t = 0$.

Differentiating the identity

$$c^*(x,t) - \varphi(x - a(c^*(x,t))t) = 0 \tag{2.41}$$

successively by x and t we get

$$\begin{aligned} c_x^* &= \frac{\varphi'(x - a(c^*)t)}{1 + \varphi'(x - a(c^*)t)\,a'(c^*)t}, \\ c_t^* &= \frac{-a(c^*)\varphi'(x - a(c^*)t)}{1 + \varphi'(x - a(c^*)t)\,a'(c^*)t}. \end{aligned} \tag{2.42}$$

Herefrom we see that the function $u(x,t) = c^*(x,t)$ is in fact the solution of (2.37) and according to (2.41) it satisfies the initial condition (2.38), too. From the formulae (2.42) it is clear that a C^1-solution exists provided

$$1 + \varphi'(x - a(c^*)t)\, a'(c^*)t > 0.$$

A simple consideration shows that a sufficient condition guaranteeing the existence of a unique global solution of (2.37), (2.38) is that the inequality

$$\varphi' a' \geq 0 \tag{2.43}$$

holds. In this case the curves $y = c$ and $y = \varphi(x - a(c)t)$ intersect in the plane (c, y) at exactly one point. If the condition (2.43) is not satisfied then for some t the denominator of the fractions (2.42) may vanish and the derivatives of the solution cease to be finite. Let us note that the equation (2.40), being nonlinear, may have more than one solution. In the last case the value of $u(x,t)$ at this point ceases to be uniquely determined and the theory of shock waves, explained later on, must be applied.

2.2.7 Example

Let us solve the initial-value problem

$$u_t + u^2 u_x = 0, \qquad u(x,0) = x, \qquad -\infty < x < \infty,\ t > 0.$$

In this case the equation (2.40) reads

$$c - x + c^2 t = 0.$$

Therefore

$$c(x,t) = \frac{-1 \pm (1 + 4xt)^{1/2}}{2t} \qquad \text{for } t \neq 0,$$
$$c(x,0) = x.$$

Evidently the minus sign is not acceptable since it does not give a continuous solution. On the other hand, the function with plus defines a C^1-solution

$$u(x,t) = \frac{2x}{1 + (1 + 4xt)^{1/2}}$$

on the set

$$Q = \{(x,t) \in \mathbf{R} \times \mathbf{R}^+ \mid 1 + 4xt > 0\}.$$

2.2.8 Rankine–Hugoniot condition

On a domain Q with a sufficiently smooth boundary let us investigate the equation (the so-called *conservation law*)

$$v_t + (f(v))_x = 0. \tag{2.44}$$

In the sequel we shall meet solutions to (2.44) which will not be everywhere continuous or will not have continuous derivatives. Therefore let us introduce the following definition of a generalized solution. We say that (2.44) has a *generalized solution* v, if for an arbitrary function $\Phi \in C^1(Q)$ with a compact support (that is, Φ may be non-zero on some bounded set and vanishes near the boundary ∂Q) the following equality holds:

$$\int_Q (v\Phi_t + f(v)\Phi_x)\, dxdt = 0.$$

(This definition is reasonable since, for $v \in C^1(Q)$, by integrating by parts we get (2.44) again.) Let the domain Q be separated into two parts Q_1, Q_2 by a curve C defined by the equation $\sigma(x,t) = 0$, $\sigma \in C^1$. Let us suppose that v is a generalized solution to (2.44) on Q such that its restrictions on Q_j are classical solutions there and such that the vector $(f(v), v) = V$ belongs to $C^1(Q_j)$, $j = 1, 2$, so that V has both one-sided finite limits. Making use of the Gauss–Ostrogradskiĭ theorem, we obtain

$$\begin{aligned}
0 &= \int_Q (v\Phi_t + f(v)\Phi_x)\, dxdt = \int_Q V\cdot \operatorname{grad}\Phi\, dxdt \\
&= \int_{Q_1} V\cdot \operatorname{grad}\Phi\, dxdt + \int_{Q_2} V\cdot \operatorname{grad}\Phi\, dxdt \\
&= \int_{Q_1} \operatorname{div}(\Phi V)\, dxdt + \int_{Q_2} \operatorname{div}(\Phi V)\, dxdt - \int_{Q_1} \Phi \operatorname{div} V\, dxdt - \int_{Q_2} \Phi \operatorname{div} V\, dxdt \\
&= \int_{\partial Q_1} \Phi V\cdot \nu\, ds + \int_{\partial Q_2} \Phi V\cdot \nu\, ds - \int_{Q_1} \Phi \operatorname{div} V\, dxdt - \int_{Q_2} \Phi \operatorname{div} V\, dxdt \\
&= \int_C \Phi(V_1 - V_2)\cdot \nu\, dC,
\end{aligned}$$

where V_j represents the limit of V from Q_j, $j = 1, 2$ (note that the minus sign between V_1 and V_2 is due to the fact that when integrating over ∂Q_1 and ∂Q_2 the curve C has opposite orientation). Due to the arbitrariness of Φ this implies

$$(V_1 - V_2)\cdot \nu = 0 \qquad \text{on } C.$$

Since

$$\nu = \left(\frac{\sigma_x}{\sqrt{\sigma_x^2 + \sigma_t^2}}, \frac{\sigma_t}{\sqrt{\sigma_x^2 + \sigma_t^2}} \right)$$

and $\sigma_x dx + \sigma_t dt = 0$, we finally get

$$-\frac{dx}{dt}\,[v] + [f(v)] = 0 \qquad \text{on } C, \tag{2.45}$$

where the brackets denote the difference of the limits

$$[v(x_0,t_0)] = v_1(x_0,t_0) - v_2(x_0,t_0), \quad v_j(x_0,t_0) = \lim_{\substack{(x,t)\to(x_0,t_0)\\ (x,t)\in Q_j}} v(x,t) \quad \text{for } (x_0,t_0)\in C.$$

The relation (2.45) is called the *Rankine–Hugoniot condition*. It connects the velocity $\frac{dx}{dt}$ of propagation of the discontinuity with the jumps of v and $f(v)$. If the equation (2.44) represents the law of conservation of mass, then the equation (2.45) takes the form

$$-\frac{dx}{dt}\,[\rho] + [\rho u] = 0, \tag{2.46}$$

or equivalently

$$\rho_1(u_1 - \frac{dx}{dt}) = \rho_2(u_2 - \frac{dx}{dt}).$$

Denoting this value by m, we usually distinguish two cases: if $m = 0$ we speak about a *contact discontinuity*, while if $m \neq 0$, that is, if the mass crosses the curve of the discontinuity, we speak about a *shock wave*.

2.2.9 Example: traffic flow

In a number of publications, see for instance Whitham (1974), the so-called problem of *traffic flow* is studied. We consider a stretch of a highway without entries and exists, which we identify with the x-axis, on which cars move in the positive direction. We define the density $\rho(x,t)$ of traffic as the number of cars per unit length, and with some degree of simplification hold it for a function defined at every point x. Under these conditions it may be supposed that the flow of traffic is governed by the equation of continuity

$$\rho_t + (\rho u)_x = 0, \tag{2.47}$$

where u is the car velocity. Evidently the velocity u depends on the density ρ. Let us suppose that

$$\begin{aligned} u(0) &= c > 0, \\ u(\rho) &= 0 \quad \text{for } \rho \geq d > 0 \end{aligned}$$

and u is strictly decreasing for $\rho \in (0,d)$, so that $u' < 0$. Moreover, let us suppose that $u'' \leq 0$. Finally, let us suppose that the initial state of the density of cars is such that

$$\begin{aligned} \rho_0(x) &= \rho_1 \quad \text{for } x \in (-\infty, 0] \\ \rho_0(x) &= \rho_2 \quad \text{for } x \in (0, \infty) \end{aligned}$$

and $0 < \rho_1 < \rho_2 < d$. Writing Eq. (2.47) in the form

$$\rho_t + (u(\rho) + \rho u'(\rho))\rho_x = 0,$$

we see that the equation for the characteristics reads

$$\frac{d\xi}{d\tau} = u(\rho) + \rho u'(\rho) \equiv a(\rho).$$

Obviously $a(\rho_1) > a(\rho_2)$ since $a'(\rho) = (\rho u)'' = 2u' + \rho u'' < 0$. Characteristics starting for $t = 0$ from the negative x-half-axis which are straight lines with the slope $a(\rho_1)$

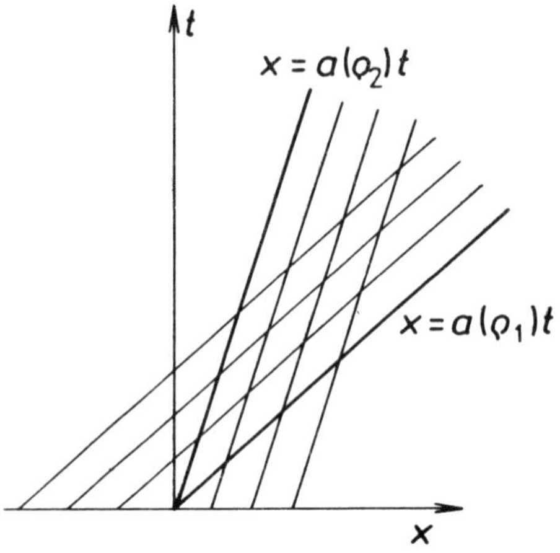

Fig. 1

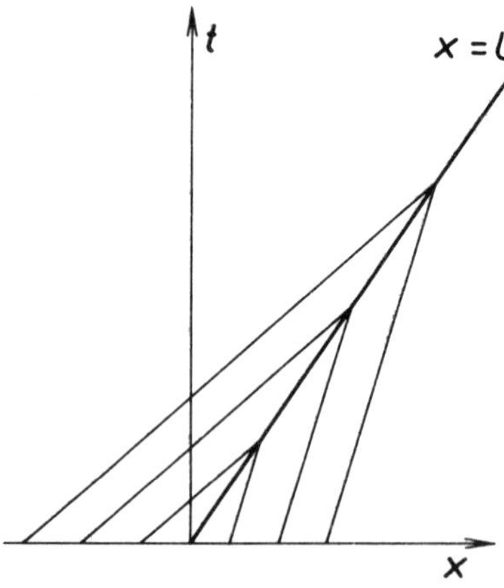

Fig. 2

intersect the characteristics starting for $t = 0$ from the positive x-half-axis which are straight lines with the slope $a(\rho_2)$. Therefore in the angle between the characteristics $x = a(\rho_1)t$ and $x = a(\rho_2)t$ the solution is not determined uniquely, see Fig. 1.

But this is acceptable neither from the mathematical nor from the physical point of view. Thus let us look for a more convenient solution among the discontinuous ones. Denoting

$$U = \frac{[\rho u]}{[\rho]} = \frac{\rho_2 u(\rho_2) - \rho_1 u(\rho_1)}{\rho_2 - \rho_1}$$

we get by (2.46)

$$\frac{dx}{dt} = U,$$

so that for the shock wave starting at the origin we have

$$x(t) = Ut = \frac{\rho_2 u(\rho_2) - \rho_1 u(\rho_1)}{\rho_2 - \rho_1} t. \tag{2.48}$$

It may be easily verified that

$$a(\rho_2) < U < a(\rho_1).$$

Therefore the characteristics starting on the axis $t = 0$ terminate on the shock wave, see Fig. 2. Hence the discontinuous solution is given by

$$\rho(x,t) = \rho_1 \quad \text{for } x \le Ut,$$
$$\rho(x,t) = \rho_2 \quad \text{for } x > Ut.$$

If $U < 0$, that is, if the shock wave propagates backward, the cars must slow down to the velocity $u(\rho_2)$ already at $x = x(t) < 0$. In the opposite case, if $U > 0$, that is,

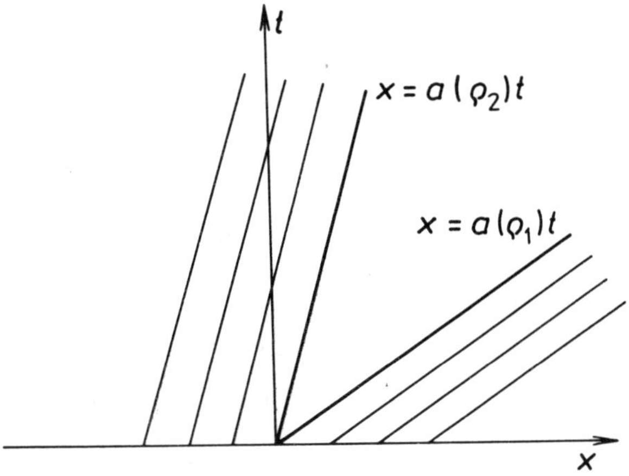

Fig. 3

the shock wave propagates forward, the cars going behind the slower ones may continue with their original velocity $u(\rho_1)$ after passing the origin for some positive time.

If $\rho_2 < \rho_1$ then $a(\rho_1) < a(\rho_2)$ and in the angle between the characteristics $x = a(\rho_1)t$ and $x = a(\rho_2)t$ no characteristic lies and no solution is defined, see Fig. 3.

In this case we complete the solution by the so-called *centred rarefaction* or *expansion wave* which is defined in the following way: on the straight line $x = kt, a(\rho_1) < k < a(\rho_2)$ the solution is determined by the equation

$$a(\rho) = k.$$

Evidently the origin is a singular point of this solution. Except for this point, the solution is continuous, but both its derivatives of the first order have jumps on the straight lines $x = a(\rho_j)t,\ j = 1, 2$. This solution is physically fully conceivable since it says that the slower cars successively accelerate till they reach the velocity $u(\rho_2)$. Of course in a real case, the places with different car density alternate, which means that the solution may include several shock waves and expansion waves that again may intersect one another which leads to a rather complicated pattern which cannot be dealt with here.

2.2.10 Exercises

2.2.10.1 Investigate in detail the foregoing example with the function

(a) $$u(\rho) = c - \frac{c}{d}\rho,$$

(b) $$u(\rho) = (c^2 - \frac{c^2}{d^2}\rho^2)^{1/2},$$

$c, d > 0$.

2.2.10.2 Investigate the traffic flow under the assumption that $u(\rho) = c - c\rho/d$, $c,\ d > 0$ and

$$\begin{aligned} \rho(x,0) &= \rho_0 > 0 && \text{for} \quad x \in (-\infty, 0),\\ \rho(x,0) &= \rho_0 + x && \text{for} \quad x \in [0,1],\\ \rho(x,0) &= \rho_0 + 1 && \text{for} \quad x \in (1,\infty). \end{aligned}$$

Answer.
Until the time $t^* = \dfrac{d}{2c}$ there exists a continuous uniquely determined solution defined by

$$\begin{aligned} \rho(x,t) &= \rho_0 && \text{for} \quad x \le a(\rho_0)t, \quad t < t^*,\\ \rho(x,t) &= \frac{\rho_0 + x - c}{1 - 2ct/d} && \text{for} \quad x \in (a(\rho_0)t, 1 + a(\rho_1)t), \quad t < t^*,\\ \rho(x,t) &= \rho_0 + 1 && \text{for} \quad x \ge 1 + a(\rho_1)t, \quad t < t^*, \end{aligned}$$

where $a(\rho) = c - 2c\rho/d$. For $t = t^*$ the solution reads

$$\begin{aligned} \rho(x,t^*) &= \rho_0 && \text{for} \quad x < x^* = \frac{d}{2} - \rho_0,\\ \rho(x,t^*) &= \rho_0 + 1 && \text{for} \quad x > x^* \end{aligned}$$

and at the point (x^*, t^*) a shock wave arises similar to that in Example 2.2.9.

2.2.11 The initial-boundary value problem

On $Q = \mathbf{R}^+ \times \mathbf{R}^+$ let us investigate the problem

$$u_t + a(x,t)u_x + b(x,t)u = f(x,t), \tag{2.49}$$
$$u(x,0) = \varphi(x), \quad x \in \mathbf{R}^+, \tag{2.50}$$
$$u(0,t) = h(t), \quad t \in \mathbf{R}^+. \tag{2.51}$$

Let $\varphi,\ h \in C^1(\mathbf{R}^+)$, $a, b, g, a_x, b_x, g_x \in C(Q)$, $a > 0$ in Q. Let us suppose that a characteristic passing through every point $(x,t) \in Q$ can be continued up to one of the positive half-axes. If the characteristic intersects the positive x-half-axis, the solution is given by

$$\begin{aligned} u(x,t) &= \varphi(\chi(0;x,t)) \exp(-\int_0^t b(\chi(\sigma;x,t),\sigma)\,d\sigma)\\ &+ \int_0^t f(\chi(s;x,t),s) \exp(-\int_s^t b(\chi(\sigma;x,t),\sigma)\,d\sigma)\,ds. \end{aligned}$$

If the characteristic intersects the positive t-half-axis, let us denote by $\vartheta = \vartheta(x,t)$ the solution of the equation

$$0 = \chi(\vartheta; x, t).$$

Then the solution takes the form

$$u(x,t) = h(\vartheta(x,t)) \exp(-\int_{\vartheta(x,t)}^{t} b(\chi(\sigma;x,t),\sigma)\,d\sigma)$$

$$+ \int_{\vartheta(x,t)}^{t} f(\chi(s;x,t),s) \exp(-\int_{s}^{t} b(\chi(\sigma;x,t),\sigma)\,d\sigma)\,ds.$$

We easily find that the solution u is continuous in Q if $\varphi(0) = h(0)$ and is classical, (that is it has continuous derivatives of the first order) provided

$$h'(0) + a(0,0)\,\varphi'(0) + b(0,0)\,\varphi(0) = f(0,0).$$

Nonetheless, let us note that in some practical problems, the requirements on the smoothness are superfluous, since the discontinuous solutions may describe the real situation better.

In the previous considerations we have supposed that $a > 0$. If negative values of a are admitted, we must require that characteristics starting at positive x-half-axis do not intersect the positive t-half-axis. If this happens, the value of u at the point of intersection is determined by the initial condition, and it is no longer possible to prescribe it by the boundary condition.

2.2.12 Exercise

The propagation of a one-sided undamped wave in an undisturbed medium, excited at the origin with intensity h and spreading with a velocity a in the direction of the positive x-half-axis is described by the initial-boundary value problem

$$u_t + au_x = 0,\ (x,t) \in \mathbf{R}^+ \times \mathbf{R}^+, \quad u(x,0) = 0,\ x \geq 0, \quad u(0,t) = h(t),\ t \geq 0,$$

$a > 0$ being a constant. Solve this problem under the assumption $h(0) = 0$.

Answer.

$$u(x,t) = 0 \text{ for } x \geq at, \quad u(x,t) = h(t - x/a) \text{ for } x \leq at.$$

2.3 ONE EQUATION IN SEVERAL SPACE VARIABLES

First the linear and quasilinear cases of a single equation with several space variables are examined. Further, the general nonlinear equation is studied in some detail and the results are applied for determination of characteristic surfaces.

2.3.1 The linear case

Let us solve the problem

$$u_t + \sum_{j=1}^{n} a_j(x,t) u_{x_j} + b(x,t)u = f(x,t), \tag{2.52}$$

$$u(x,t_0) = \varphi(x), \tag{2.53}$$

where $x = (x_1, \ldots, x_n)$, a_j, b, f together with their first derivatives with respect to x_j, $j = 1, \ldots, n$ are of the class $C(\mathbf{R}^n \times [t_0, \infty))$, $\varphi \in C^1(\mathbf{R}^n)$. Since the approach to the resolution of this problem is the same as in the case of a single space variable, we proceed faster.

Let any solution to the system

$$\frac{d\xi_j}{d\tau} = a_j(\xi, \tau), \quad j = 1, \ldots, n,$$

be called a *characteristic* of (2.52). The characteristic passing through the point (x,t) will be denoted by $\chi(\tau; x, t) = (\chi_1(\tau; x, t), \ldots, \chi_n(\tau; x, t))$.

Let us construct all characteristics which start at the points (x, t_0), $x \in \mathbf{R}^n$. The subset of the half-space $\mathbf{R}^n \times [t_0, \infty)$ covered by all these characteristics will be called the *domain of influence* of the plane $\tau = t_0$ and will be denoted by $\mathcal{DI}(\mathbf{R}^n; t_0)$.

The following theorem can be proved in the same way as Theorem 2.2.2.

Let a_j, b, f together with their first derivatives with respect to x_j, $j = 1, \ldots, n$ be of the class $C(\mathbf{R}^n \times [t_0, \infty))$, $\varphi \in C^1(\mathbf{R}^n)$. Let $(x,t) \in \mathcal{DI}(\mathbf{R}^n; t_0)$. Then the problem (2.52), (2.53) *has a unique solution u given by the formula*

$$\begin{aligned} u(x,t) &= \varphi(\chi(t_0;, x, t)) \exp(-\int_{t_0}^{t} b(\chi(\sigma; x, t), \sigma) d\sigma) \\ &+ \int_{t_0}^{t} f(\chi(s; x, t), s) \exp(-\int_{s}^{t} b(\chi(\sigma; x, t), \sigma) d\sigma) ds. \end{aligned} \tag{2.54}$$

Let us derive the foregoing result once more with help of the *Riemann function* v. To this end write Eq. (2.52) on the characteristic $\chi(\tau; x, t)$:

$$\frac{d}{d\tau} u(\chi(\tau; x, t), \tau) + b(\chi(\tau; x, t), \tau)\, u(\chi(\tau; x, t), \tau) = f(\chi(\tau; x, t), \tau).$$

Multiplying this equation by $v(\tau)$ and integrating by parts from t to t_0 we obtain

$$\begin{aligned} &\int_{t}^{t_0} [-\frac{dv(\tau)}{d\tau} + b(\chi(\tau; x, t), \tau)\, v(\tau)] u(\chi(\tau; x, t), \tau)\, d\tau + \big[v(\tau)\, u(\chi(\tau; x, t), \tau)\big]_t^{t_0} \\ &= \int_{t}^{t_0} f(\chi(\tau; x, t), \tau)\, v(\tau)\, d\tau. \end{aligned} \tag{2.55}$$

Choosing $v(\tau)$ in such a way that

$$-\frac{dv(\tau)}{d\tau} + b(\chi(\tau; x, t), \tau)\, v(\tau) = 0, \ v(t) = 1,$$

it follows readily that

$$v(\tau) = \exp(\int_{t}^{\tau} b(\chi(\sigma; x, t), \sigma) d\sigma),$$

and inserting it into (2.55) we again get formula (2.54). □

Remarks similar to Remarks 2.2.3.2 – 2.2.3.5 could be formulated, too.

2.3.2 Example

Many physical problems are described by the so-called *Boltzmann equation*

$$\frac{\partial f}{\partial t} + \sum_{j=1}^{3} v_j \frac{\partial f}{\partial x_j} + \sum_{j=1}^{3} F_j \frac{\partial f}{\partial v_j} = 0, \tag{2.56}$$

where $f(x, v, t)$ denotes the number of particles investigated which at time t have position x and velocity v; $F = F(x, v, t, f)$ represents the forces acting in the system.

In particular, if we study the motion of charged particles bearing a charge e in an electromagnetic field (the terms representing the interactions of particles among themselves being omitted and the field being considered as uninfluenced by the particles), then the force F is given by $F = e(E + v \times B)$ and f satisfies the equation

$$f_t + \sum_{j=1}^{3} v_j f_{x_j} + \sum_{j=1}^{3} e(E + v \times B)_j f_{v_j} = 0, \tag{2.57}$$

with E, B constant.

The system for the characteristics corresponding to this equation reads

$$\frac{d\xi_j}{d\tau} = \eta_j, \quad j = 1, 2, 3,$$

$$\frac{d\eta_j}{d\tau} = e(E + \eta \times B)_j, \quad j = 1, 2, 3,$$

where we write η_j instead of ξ_{j+3}. Writing the equation for η in the form

$$\frac{d\eta}{d\tau} = e(E + \mathcal{B}\eta),$$

where

$$\mathcal{B} = \begin{pmatrix} 0 & B_3 & -B_2 \\ -B_3 & 0 & B_1 \\ B_2 & -B_1 & 0 \end{pmatrix},$$

taking into account that the solution of the corresponding homogeneous system is given by

$$\eta(\tau; x, v, t) = v \exp(\mathcal{B}(\tau - t)) = v \sum_{n=0}^{\infty} \frac{1}{n!} \mathcal{B}^n (\tau - t)^n$$

and making use of the circumstance that in virtue of the Cayley–Hamilton theorem the matrix $\mathcal{B}$ satisfies the relation $\mathcal{B}^3 + \alpha^2 \mathcal{B} = 0$, with $\alpha = B_1^2 + B_2^2 + B_3^2$ we find that

$$\begin{aligned} \eta(\tau; x, v, t) &= \left(\frac{v\mathcal{B}}{\sqrt{\alpha}} - \frac{E\mathcal{B}^2}{\alpha^{3/2}}\right) \sin(e\sqrt{\alpha}(\tau - t)) - \left(\frac{E\mathcal{B}}{\alpha} + \frac{v\mathcal{B}^2}{\alpha}\right) \cos(e\sqrt{\alpha}(\tau - t)) \\ &\quad + eE\left(\mathcal{I} + \frac{\mathcal{B}^2}{\alpha}\right)(\tau - t) + v\left(\mathcal{I} + \frac{\mathcal{B}^2}{\alpha}\right) + \frac{E\mathcal{B}}{\alpha}, \\ \xi(\tau; x, v, t) &= x + \int_t^\tau \eta(s; x, v, t) ds. \end{aligned}$$

If the initial state is given by

$$f(x, v, t_0) = \varphi(x, v), \tag{2.58}$$

the solution to (2.57), (2.58) is

$$f(x, t, v) = \varphi(\xi(t_0; x, v, t), \eta(t_0; x, v, t)).$$

2.3.3 The quasilinear case

Let us deal with the problem

$$u_t + \sum_{j=1}^{n} a_j(x, t, u) u_{x_j} = f(x, t, u), \tag{2.59}$$

$$u(x, t_0) = \varphi(x), \tag{2.60}$$

where a_j, f together with their first derivatives with respect to x_j, $j = 1, \ldots, n$ and u are of the class $C(\mathbf{R}^n \times [t_0, \infty) \times \mathbf{R})$, $j = 1, \ldots, n$, $\varphi \in C^1(\mathbf{R}^n)$.

In this case a *characteristic* must be defined in a way somewhat different from that used in the linear case, that is, as a solution $(\xi(\tau), v(\tau)) = (\xi_1(\tau), \ldots, \xi_n(\tau), v(\tau))$ to the system

$$\frac{d\xi_j}{d\tau} = a_j(\xi, \tau, v), \quad j = 1, \ldots, n, \tag{2.61}$$

$$\frac{dv}{d\tau} = f(\xi, \tau, v). \tag{2.62}$$

Here we have to add the last equation to complete the system to $n+1$ equations for $n+1$ unknowns. This means that the search for the characteristics is inseparably joined with the resolution of the whole problem.

Another consequence of the fact that the coefficients depend on u is that it is impossible to define the domain of dependence of the plane $\tau = t_0$ independently of the initial function φ.

The existence of a solution of (2.59), (2.60) in the neighbourhood of the plane $\tau = t_0$ may be proved by the method of successive approximations.

Since the existence result is included in the existence theorem for a more general problem in § 2.3.5 we omit the proof here.

Of course, it is also possible to proceed analogously as we did in § 2.2.6. Let $(x, t) \in \mathbf{R}^n \times [t_0, \infty)$ be an arbitrary point and let u_0 be the supposed value (obviously depending on (x, t), that is $u_0 = u_0(x, t)$) of the solution at this point (this value, provided it exists, will be determined later). Let us suppose that there is a unique characteristic $(\chi(\tau; x, t, u_0), w(\tau; x, t, u_0))$ passing through the point (x, t, u_0), which means that

$$\begin{aligned} \chi_j(t; x, t, u_0) &= x_j, \quad j = 1, \ldots, n, \\ w(t; x, t, u_0) &= u_0. \end{aligned} \tag{2.63}$$

Supposing that this characteristic may be continued till $\tau = t_0$, that is to the point $(\chi(t_0; x, t, u_0), w(t_0; x, t, u_0))$, it is obvious that the constant u_0 must satisfy the relation

$$u_0 = u(x,t) = \varphi(\chi(t_0; x, t, u_0)) + \int_{t_0}^{t} f(\chi(\tau; x, t, u_0), \tau, w(\tau; x, t, u_0))\, d\tau$$

or, in another form

$$G(x,t,u_0) \equiv u_0 - \varphi(\chi(t_0; x, t, u_0)) - \int_{t_0}^{t} f(\chi(\tau; x, t, u_0), \tau, w(\tau; x, t, u_0))d\tau = 0 \quad (2.64)$$

and the problem reduces to the investigation for which points (x,t) this equation has a solution $u_0 = u_0(x,t)$.

Since $G(x, t_0, \varphi(x)) = 0$ and $\frac{\partial G}{\partial u_0}(x, t_0, \varphi(x)) = 1$, it follows readily by the Modified Implicit Functions Theorem (see § 1.2.10) with $r = n$, $s = 1$, $\psi = \varphi$ and $Q_t = \mathbf{R}^n \times [t_0, \infty)$ that there exists a unique solution $u_0 = u_0(x,t)$ for every point (x,t), t lying sufficiently close to the plane $t = t_0$.

In particular, the problem is easy if the equation is homogeneous. Then it suffices to solve the shortened system of equations

$$\frac{d\xi_j}{d\tau} = a_j(\xi, \tau, u_0)$$

and to investigate the equation

$$\varphi(\chi(t_0; x, t, u_0)) = u_0. \quad (2.65)$$

Let us show how the inhomogeneous equation (2.59) can be transformed to a homogeneous one. To this end let us look for a solution in an implicit form

$$V(x,t,u) = 0.$$

Then evidently $\frac{\partial V}{\partial x_j} + \frac{\partial V}{\partial u}\frac{\partial u}{\partial x_j} = 0$, $j = 1, \ldots, n$, $\frac{\partial V}{\partial t} + \frac{\partial V}{\partial u}\frac{\partial u}{\partial t} = 0$ and thus $\frac{\partial u}{\partial x_j} = -\frac{\partial V}{\partial x_j}/\frac{\partial V}{\partial u}$, $\frac{\partial u}{\partial t} = -\frac{\partial V}{\partial t}/\frac{\partial V}{\partial u}$. Inserting this into (2.59) we get

$$\frac{\partial V}{\partial t} + \sum_{j=1}^{n} a_j \frac{\partial V}{\partial x_j} + f\frac{\partial V}{\partial u} = 0,$$

which is a linear homogeneous equation for V. By the definition from § 2.3.1 its characteristics are given by (2.61), (2.62), which once more justifies our definition of the characteristics in the quasilinear case.

2.3.4 Example

Let us solve the initial-value problem

$$u_t + (x+u)u_x + (y+u)u_y = 0,$$

$$u(x,y,0) = x + y.$$

The equations of the characteristics being

$$\frac{d\xi}{d\tau} = \xi + v,$$
$$\frac{d\eta}{d\tau} = \eta + v,$$
$$\frac{dv}{d\tau} = 0$$

we readily obtain that

$$\chi_1(\tau; x, y, t, u_0) = (x + u_0)\exp(\tau - t) - u_0,$$
$$\chi_2(\tau; x, y, t, u_0) = (y + u_0)\exp(\tau - t) - u_0,$$
$$w(\tau; x, y, t, u_0) = u_0.$$

The equation (2.65) defining u_0 has the form

$$(x + u_0)\exp(-t) - u_0 + (y + u_0)\exp(-t) - u_0 = u_0,$$

hence

$$u(x,y,t) = u_0(x,y,t) = \frac{(x+y)\exp(-t)}{3 - 2\exp(-t)},$$

which is defined for every $(x,y,t) \in \mathbf{R}^2 \times \mathbf{R}^+$.

2.3.5 General equation of the first order: the initial-value problem

Let us consider the general equation of the first order in the form

$$p_0 + f(x_1, \ldots, x_n, t, u, p_1, \ldots, p_n) = 0, \tag{2.66}$$

where $p_0 = \partial u/\partial t, p_j = \partial u/\partial x_j,\ j = 1, \ldots, n$, together with the initial condition

$$u(x_1, \ldots, x_n, t_0) = \varphi(x_1, \ldots, x_n). \tag{2.67}$$

Let us differentiate the equation (2.66) with respect to $x_1, \ldots, x_n$ and t. We get respectively

$$\begin{aligned}
&\frac{\partial p_0}{\partial x_j} + \frac{\partial f}{\partial x_j} + \frac{\partial f}{\partial u}p_j + \sum_{k=1}^{n} \frac{\partial f}{\partial p_k}\frac{\partial p_k}{\partial x_j} = 0, \quad j = 1, \ldots, n, \\
&\frac{\partial p_0}{\partial t} + \frac{\partial f}{\partial t} + \frac{\partial f}{\partial u}p_0 + \sum_{k=1}^{n} \frac{\partial f}{\partial p_k}\frac{\partial p_k}{\partial t} = 0.
\end{aligned} \tag{2.68}$$

Supposing that u has continuous second order derivatives, we conclude that

$$\frac{\partial p_k}{\partial x_j} = \frac{\partial^2 u}{\partial x_j \partial x_k} = \frac{\partial^2 u}{\partial x_k \partial x_j} = \frac{\partial p_j}{\partial x_k}, \quad j,k = 1,\dots,n,$$
$$\frac{\partial p_0}{\partial x_j} = \frac{\partial^2 u}{\partial x_j \partial t} = \frac{\partial^2 u}{\partial t \partial x_j} = \frac{\partial p_j}{\partial t}, \quad j = 1,\dots,n,$$
$$\frac{\partial p_k}{\partial t} = \frac{\partial^2 u}{\partial t \partial x_k} = \frac{\partial^2 u}{\partial x_k \partial t} = \frac{\partial p_0}{\partial x_k}, \quad k = 1,\dots,n.$$

Hence (2.68) may be rewritten in the form

$$\begin{aligned} &\frac{\partial p_j}{\partial t} + \frac{\partial f}{\partial x_j} + \frac{\partial f}{\partial u}p_j + \sum_{k=1}^{n} \frac{\partial f}{\partial p_k}\frac{\partial p_j}{\partial x_k} = 0, \quad j = 1,\dots,n, \\ &\frac{\partial p_0}{\partial t} + \frac{\partial f}{\partial t} + \frac{\partial f}{\partial u}p_0 + \sum_{k=1}^{n} \frac{\partial f}{\partial p_k}\frac{\partial p_0}{\partial x_k} = 0. \end{aligned} \tag{2.69}$$

If we regard the j-th equation as the equation for p_j and the last equation as the equation for p_0, we see that the equations of characteristics for all of them coincide, giving

$$\frac{dx_k}{dt} = \frac{\partial f}{\partial p_k}, \quad k = 1,\dots,n. \tag{2.70}$$

Inserting this into (2.69) we get

$$\begin{aligned} &\frac{dp_j}{dt} + \frac{\partial f}{\partial u}p_j = -\frac{\partial f}{\partial x_j}, \quad j = 1,\dots,n, \\ &\frac{dp_0}{dt} + \frac{\partial f}{\partial u}p_0 = -\frac{\partial f}{\partial t}. \end{aligned} \tag{2.71}$$

Adding to the system (2.70), (2.71) of $2n+1$ equations the equation

$$\frac{du}{dt} = p_0 + \sum_{k=1}^{n} \frac{\partial f}{\partial p_k}p_k \tag{2.72}$$

(which is a consequence of $\frac{du}{dt} = \sum_{k=1}^{n} \frac{\partial u}{\partial x_k}\frac{dx_k}{dt} + \frac{\partial u}{\partial t} = \sum_{k=1}^{n} \frac{\partial f}{\partial p_k}p_k + p_0$), we obtain a system of $2n+2$ equations for $2n+2$ unknowns $x_1,\dots,x_n$, u, p_0, $p_1,\dots,p_n$ which is called the *characteristic system.*

Let $x_1(t),\dots,x_n(t)$, $u(t)$, $p_0(t)$, $p_1(t),\dots,p_n(t)$ solve the characteristic system. Then evidently

$$\frac{d}{dt}[p_0(t) + f(x_1(t),\dots,x_n(t),t,u(t),p_1(t),\dots,p_n(t))] = 0,$$

so that $p_0(t) + f(x_1(t),\dots,x_n(t),t,u(t),p_1(t),\dots,p_n(t)) = \text{const.}$ Denoting $y_j = x_j(t_0)$, putting

$$u(t_0) = \varphi(x_1(t_0),\dots,x_n(t_0)) = \varphi(y_1,\dots,y_n),$$
$$p_j(t_0) = \varphi_{x_j}(x_1(t_0),\dots,x_n(t_0)) = \varphi_{x_j}(y_1,\dots,y_n)$$

and choosing

$$p_0(t_0) = -f(y_1,\ldots,y_n,t_0,\varphi(y_1,\ldots,y_n),\varphi_{x_1}(y_1,\ldots,y_n),\ldots,\varphi_{x_n}(y_1,\ldots,y_n))$$

we conclude that

$$p_0(t) + f(x_1(t),\ldots,x_n(t),t,u(t),p_1(t),\ldots,p_n(t)) = 0 \tag{2.73}$$

for all $t \geq t_0$. The solution $x_j = X_j(t,y),\ u = U(t,y), p_0 = P_0(t,y), p_j = P_j(t,y),\ j = 1,\ldots,n$ fulfilling (2.73) is called a *characteristic strip*. Its existence in the class C^2 for $f \in C^2$ is guaranteed at least in a neighbourghood of a plane $t = t_0$, by the existence theorem for systems of ordinary differential equations. Since for $t = t_0$ the Jacobian $\dfrac{\partial(X_1,\ldots,X_n)}{\partial(y_1,\ldots,y_n)}$ equals unity, there exist (according to the Implicit Functions Theorem) functions Y_j such that $y_j = Y_j(x_1,\ldots,x_n,t),\ j = 1,\ldots,n$. Inserting these functions into U and P_j we see that the functions

$$u(x,t) = U(t,Y(x,t)),\ p_j(x,t) = P_j(t,Y(x,t)) \tag{2.74}$$

satisfy the equation (2.66) and the initial condition (2.67). To prove that $u(x,t)$ is a solution to (2.66), (2.67) it remains to show that the functions p_j and p_0 are really the derivatives of u with respect to $x_j,\ j = 1,\ldots,n$ and t, respectively.

To avoid some difficulties in the proof let us introduce an auxiliary parameter s instead of t by the relation

$$s = t - t_0$$

and rewrite the characteristic system in the form

$$\begin{aligned}
\frac{dx_k}{ds} &= \frac{\partial f}{\partial p_k}, \quad k = 1,\ldots,n,\\
\frac{dp_j}{ds} + \frac{\partial f}{\partial u}p_j &= -\frac{\partial f}{\partial x_j}, \quad j = 1,\ldots,n,\\
\frac{dp_0}{ds} + \frac{\partial f}{\partial u}p_0 &= -\frac{\partial f}{\partial t},\\
\frac{du}{ds} &= p_0 + \sum_{k=1}^{n} \frac{\partial f}{\partial p_k}p_k.
\end{aligned}$$

The initial conditions for $x_j(s), p_j(s), u(s)$ are

$$\begin{aligned}
x_j(0) &= y_j,\\
p_j(0) &= \varphi_{x_j}(y), \quad j = 1,\ldots,n,\\
p_0(0) &= -f(y,t_0,\varphi(y),\ \varphi_{x_1}(y),\ldots,\varphi_{x_n}(y)),\\
u(0) &= \varphi(y).
\end{aligned}$$

Let the solution to this system be denoted by

$$X_j(s,y),\ P_j(s,y),\ P_0(s,y),\ U(s,y), \quad j = 1,\ldots,n.$$

Its local existence is again guaranteed by the existence theorems for ordinary differential equations. Let us introduce expressions G_j and H by

$$G_j(s,y) = \frac{\partial U}{\partial y_j} - \sum_{k=1}^{n} P_k \frac{\partial X_k}{\partial y_j}, \quad j = 1,\ldots,n,$$
$$H(s,y) = \frac{\partial U}{\partial s} - \sum_{k=1}^{n} P_k \frac{\partial X_k}{\partial s} - P_0 \frac{dt}{ds}.$$

Evidently $H = 0$ by virtue of the characteristic system. To prove that $G_j = 0$, let us evaluate:

$$\begin{aligned}\frac{\partial G_j}{\partial s} &= \frac{\partial G_j}{\partial s} - \frac{\partial H}{\partial y_j} = -(\sum_{k=1}^{n} \frac{\partial P_k}{\partial s}\frac{\partial X_k}{\partial y_j} - \sum_{k=1}^{n} \frac{\partial P_k}{\partial y_j}\frac{\partial X_k}{\partial s} - \frac{\partial P_0}{\partial y_j}) \\ &= \sum_{k=1}^{n} (\frac{\partial f}{\partial x_k} + \frac{\partial f}{\partial u}P_k)\frac{\partial X_k}{\partial y_j} + \sum_{k=1}^{n} \frac{\partial f}{\partial p_k}\frac{\partial P_k}{\partial y_j} + \frac{\partial P_0}{\partial y_j}. \end{aligned} \tag{2.75}$$

Differentiating the identity

$$0 = P_0(s,y) + f(X_1(s,y),\ldots,X_n(s,y), s+t_0, U(s,y), P_1(s,y),\ldots,P_n(s,y))$$

with respect to y_j and subtracting the result from (2.75), we get

$$\frac{\partial G_j}{\partial s} = -\frac{\partial f}{\partial u}[\frac{\partial U}{\partial y_j} - \sum_{k=1}^{n} P_k \frac{\partial X_k}{\partial y_j}] = -\frac{\partial f}{\partial u} G_j.$$

Since

$$G_j(0,y) = \frac{\partial U(0,y)}{\partial y_j} - \sum_{k=1}^{n} P_k(0,y)\frac{\partial X_k(0,y)}{\partial y_j} = \varphi_{x_j}(y) - \sum_{k=1}^{n} \varphi_{x_k}(y)\delta_{jk} = 0$$

and

$$G_j(s,y) = G_j(0,y)\exp(-\int_0^s f_u(X(\sigma,y),\sigma+t_0,U(\sigma,y),P(\sigma,y))d\sigma$$

we have

$$G_j(s,y) \equiv 0, \quad j = 1,\ldots,n.$$

Hence

$$\begin{aligned} 0 &= \frac{\partial U}{\partial y_j} - \sum_{k=1}^{n} P_k \frac{\partial X_k}{\partial y_j}, \quad j = 1,\ldots,n, \\ 0 &= \frac{\partial U}{\partial s} - \sum_{k=1}^{n} P_k \frac{\partial X_k}{\partial s} - P_0 \frac{dt}{ds}. \end{aligned} \tag{2.76}$$

On the other hand, differentiating the equality

$$U(s,y) = u(X(s,y),s)$$

(which is equivalent to the first relation in (2.74)) with respect to $y_1, \ldots, y_n$ and s, we get respectively

$$0 = \frac{\partial U}{\partial y_j} - \sum_{k=1}^{n} \frac{\partial u}{\partial x_k} \frac{\partial X_k}{\partial y_j},$$

$$0 = \frac{\partial U}{\partial s} - \sum_{k=1}^{n} \frac{\partial u}{\partial x_k} \frac{\partial X_k}{\partial s} - \frac{\partial u}{\partial t} \frac{dt}{ds}.$$

The system (2.76) can be viewed as a system of linear algebraic equations in unknowns $P_0, P_1, \ldots, P_n$ with the determinant of the system equal to $\frac{dt}{ds} \frac{\partial(X_1, \ldots, X_n)}{\partial(y_1, \ldots, y_n)}$ which is near to unity for s near to zero. Thus this system possesses locally a unique solution. We have shown above that both $P_0, P_1, \ldots, P_n$ and $\frac{\partial u}{\partial t}, \frac{\partial u}{\partial x_1}, \ldots, \frac{\partial u}{\partial x_n}$ solve the system (2.76). Hence due to the uniqueness, necessarily,

$$\frac{\partial u}{\partial t}(X(s,y),s) = P_0(s,y) = p_0(X(s,y),s), \quad \frac{\partial u}{\partial x_k}(X(s,y),s) = P_k(s,y) = p_k(X(s,y),s),$$
$$k = 1, \ldots, n$$

which was to be proved. We have thus obtained the following theorem.

Let $f \in C^2(\mathbf{R}^{2n+2})$, $\varphi \in C^2(\mathbf{R}^n)$. Then near to the plane $t = t_0$ there exists a solution $u(x,t) \in C^2(\mathbf{R}^{n+1})$ to the problem (2.66), (2.67) *given by* (2.74).

2.3.6 Remarks

2.3.6.1 Sometimes the general equation of the first order is given in the form

$$g(x_1, \ldots, x_n, t, u, p_0, p_1, \ldots, p_n) = 0. \tag{2.77}$$

If the equation is hyperbolic then there exist r real roots

$$p_0 = -f_j(x_1, \ldots, x_n, t, u, p_1, \ldots, p_n), \quad j = 1, \ldots, r$$

of this equation. However, it is often difficult or even impossible to find these roots explicitly. Then we introduce the auxiliary parameter s and write the characteristic system corresponding to (2.77) as follows:

$$\frac{dx_j}{ds} = \frac{\partial g}{\partial p_j}, \quad j = 1, \ldots, n, \tag{2.78}$$

$$\frac{dt}{ds} = \frac{\partial g}{\partial p_0}, \tag{2.79}$$

$$\frac{dp_0}{ds} + \frac{\partial g}{\partial u} p_0 = -\frac{\partial g}{\partial s}, \tag{2.80}$$

$$\frac{dp_j}{ds} + \frac{\partial g}{\partial u} p_j = -\frac{\partial g}{\partial x_j}, \quad j = 1, \ldots, n, \tag{2.81}$$

$$\frac{du}{ds} = \sum_{j=0}^{n} \frac{\partial g}{\partial p_j} p_j. \tag{2.82}$$

Let us note that as the equation (2.66) as (2.77) are often called the *Hamilton–Jacobi equation.*

2.3.6.2 Note that if g does not depend on u then the terms $\frac{\partial g}{\partial u}p_j$ disappear and the last equation can be solved separately. The remaining part of $2n+2$ equations forms a Hamiltonian system of ordinary differential equations.

2.3.6.3 If the function g is a *homogeneous function* in $p_0, \ldots, p_n$ of the r-th degree, $r \in \mathbb{N}$ that is if

$$g(x_1, \ldots, x_n, t, u, \lambda p_0, \lambda p_1, \ldots, \lambda p_n) = \lambda^r g(x_1, \ldots, x_n, t, u, p_0, p_1, \ldots, p_n)$$

for an arbitrary real number λ then $\frac{du}{ds} = \sum_{j=0}^{n} \frac{\partial g}{\partial p_j} p_j = rg = 0$, which says that u is constant along the characteristic strip.

2.3.7 Example

Let us solve the problem

$$u_t - \sqrt{u_{x_1}^2 + u_{x_2}^2} = 0, \; u(x_1, x_2, t_0) = \psi(x_1^2 + x_2^2), \tag{2.83}$$

where $\psi' > 0$. The corresponding characteristic system reads

$$\frac{dx_1}{dt} = \frac{-p_1}{\sqrt{p_1^2 + p_2^2}},$$
$$\frac{dx_2}{dt} = \frac{-p_2}{\sqrt{p_1^2 + p_2^2}},$$
$$\frac{dp_1}{dt} = 0, \frac{dp_2}{dt} = 0, \frac{dp_0}{dt} = 0, \frac{du}{dt} = p_0 - \frac{p_1^2 + p_2^2}{\sqrt{p_1^2 + p_2^2}} = 0,$$

and the initial conditions are

$$x_1(t_0) = y_1, \; x_2(t_0) = y_2, \; u(t_0) = \psi(y_1^2 + y_2^2),$$
$$p_1(t_0) = 2y_1\psi'(y_1^2 + y_2^2), \; p_2(t_0) = 2y_2\psi'(y_1^2 + y_2^2),$$
$$p_0(t_0) = 2\sqrt{y_1^2 + y_2^2}\,\psi'(y_1^2 + y_2^2).$$

Hence $x_1(t) = y_1 - \frac{y_1}{\sqrt{y_1^2 + y_2^2}}(t - t_0)$, $x_2(t) = y_2 - \frac{y_2}{\sqrt{y_1^2 + y_2^2}}(t - t_0)$. Looking for y_1, y_2 in the form $y_1 = r\cos\varphi$, $y_2 = r\sin\varphi$ we get

$$\mathrm{tg}\varphi = \frac{x_2}{x_1}, \; r = \frac{x_1}{\cos\varphi} + t - t_0 = \frac{x_2}{\sin\varphi} + t - t_0,$$
$$y_1 = r\cos\varphi = x_1 + (t - t_0)\cos\varphi,$$
$$y_2 = r\sin\varphi = x_2 + (t - t_0)\sin\varphi.$$

Because

$$\cos\varphi = \frac{\operatorname{sgn} x_1}{\sqrt{1+\operatorname{tg}^2\varphi}} = \frac{\operatorname{sgn} x_1}{\sqrt{1+x_2^2/x_1^2}} = \frac{x_1}{\sqrt{x_1^2+x_2^2}},$$
$$\sin\varphi = \frac{\operatorname{sgn} x_2}{\sqrt{1+\operatorname{cotg}^2\varphi}} = \frac{\operatorname{sgn} x_2}{\sqrt{1+x_1^2/x_2^2}} = \frac{x_2}{\sqrt{x_1^2+x_2^2}},$$

we get

$$y_1 = x_1\big(1+\frac{t-t_0}{\sqrt{x_1^2+x_2^2}}\big),$$
$$y_2 = x_2\big(1+\frac{t-t_0}{\sqrt{x_1^2+x_2^2}}\big)$$

and finally

$$u(x_1,x_2,t) = \psi(y_1^2+y_2^2) = \psi\big(\big(\sqrt{x_1^2+x_2^2}+t-t_0\big)^2\big).$$

We can directly verify that this function is really a solution to our problem except for the straight line $x_1 = 0$, $x_2 = 0$, where the first and second derivatives are not defined. This is evidently caused by the fact that our function does not satisfy the assumption of Theorem 2.3.5 at the point $p_1 = p_2 = 0$.

2.3.8 General equation of the first order: conoidal solutions

In the sequel we will see that the problem of finding a characteristic surface to the corresponding system of equations of the first order or to a hyperbolic equation of the second order leads to the nonlinear equation of the first order

$$g(x_1,\dots,x_n,t,p_0,p_1,\dots,p_n) = 0, \tag{2.84}$$

where g is a homogeneous function in $p_0,\dots,p_n$ of degree r. In the case that g may be solved with respect to p_0, (2.84) reduces to

$$p_0 + f(x_1,\dots,x_n,t,p_1,\dots,p_n) = 0, \tag{2.85}$$

where f is a homogeneous function of the degree one in $p_1,\dots,p_n$.

Note that the equations (2.84) and (2.85) never depend explicitly on u, so that in the corresponding equivalent system of ordinary differential equations the equation (2.72) or (2.82), respectively, may be omitted.

Since the equations (2.84) and (2.85) are already equations for characteristic surfaces, the solutions to (2.78) and (2.70), earlier called characteristics, are in this case called *bicharacteristics*. To find a *characteristic surface* means to find a solution $u(x,t)$ to (2.84) or (2.85) and put $u(x,t) = 0$. Since no initial conditions are given, there are always many such solutions but as we will see, only those among them are of importance which have the form of a conoid or, in simpler cases when g or f do not depend on x and t, of a cone with its vertex at a given point $(a_1,\dots,a_n,t_0)$. It is so because the solutions of this form determine the front of a wave or the domains of dependence and influence. To construct such a conoid or cone means to find a surface enveloping all bicharacteristics originating at the point $(a_1,\ldots,a_n,t_0)$. This is achieved by finding the solution to (2.78) or to (2.70) not depending on the initial conditions $p_{j,0}$ for $p_j, j=0,\ldots,n$. If g or f do not depend on x and t, one way to obtain such a surface is to construct all planes passing through the vertex whose normals are defined by (2.84) or (2.85) and to find their envelope, since then these planes are the tangent planes to the cone.

2.3.9 Examples

2.3.9.1 According to the § 2.5.2.4 the system of equations

$$v_{1t}+\alpha_{11}v_{1x_1}+\alpha_{12}v_{2x_1}+\beta_{11}v_{1x_2}+\beta_{12}v_{2x_2}+\gamma_{11}v_1+\gamma_{12}v_2=f_1(x_1,x_2,t),$$
$$v_{2t}+\alpha_{21}v_{1x_1}+\alpha_{22}v_{2x_1}+\beta_{21}v_{1x_2}+\beta_{22}v_{2x_2}+\gamma_{21}v_1+\gamma_{22}v_2=f_2(x_1,x_2,t), \tag{2.86}$$

is strictly hyperbolic if

$$\det\mathcal{D}>0,\ d_{11}>0,$$

where $\mathcal{D}=(d_{jk})^2_{j,k=1}$, $d_{11}=(\alpha_{11}-\alpha_{22})^2+4\alpha_{12}\alpha_{21}$, $d_{22}=(\beta_{11}-\beta_{22})^2+4\beta_{12}\beta_{21}$, $d_{12}=d_{21}=\alpha_{11}\beta_{11}-\alpha_{22}\beta_{11}-\alpha_{11}\beta_{22}+\alpha_{22}\beta_{22}+2\alpha_{21}\beta_{12}+2\alpha_{12}\beta_{21}$. Then the corresponding equation for characteristic surfaces reads (see § 2.5.5)

$$p_0^2+p_0((\alpha_{11}+\alpha_{22})p_1+(\beta_{11}+\beta_{22})p_2)+(\alpha_{11}\alpha_{22}-\alpha_{12}\alpha_{21})p_1^2$$
$$+(\beta_{11}\beta_{22}-\beta_{12}\beta_{21})p_2^2+(\alpha_{11}\beta_{22}+\alpha_{22}\beta_{11}-\alpha_{21}\beta_{12}-\alpha_{12}\beta_{21})p_1p_2=0.$$

This equation is equivalent to the two equations

$$p_0-c_1p_1-c_2p_2+\sqrt{d_{11}p_1^2+(d_{12}+d_{21})p_1p_2+d_{22}p_2^2}=0, \tag{2.87}$$

$$p_0-c_1p_1-c_2p_2-\sqrt{d_{11}p_1^2+(d_{12}+d_{21})p_1p_2+d_{22}p_2^2}=0, \tag{2.88}$$

where $c_1=-\alpha_{11}-\alpha_{22}$, $c_2=-\beta_{11}-\beta_{22}$. The characteristic system for the former equation has the form

$$\frac{dx_1}{dt}=-c_1+\frac{d_{11}p_1+d_{12}p_2}{\sqrt{d_{11}p_1^2+(d_{12}+d_{21})p_1p_2+d_{22}p_2^2}},$$
$$\frac{dx_2}{dt}=-c_2+\frac{d_{21}p_1+d_{22}p_2}{\sqrt{d_{11}p_1^2+(d_{12}+d_{21})p_1p_2+d_{22}p_2^2}},$$
$$\frac{dp_1}{dt}=0,$$
$$\frac{dp_2}{dt}=0.$$

Hence $p_1=p_{1,0}$, $p_2=p_{2,0}$ and the bicharacteristics originating at the point (a_1,a_2,t_0) are

$$x_1-a_1+c_1(t-t_0)=\frac{d_{11}p_{1,0}+d_{12}p_{2,0}}{\sqrt{d_{11}p_{1,0}^2+(d_{12}+d_{21})p_{1,0}p_{2,0}+d_{22}p_{2,0}^2}}(t-t_0),$$

$$x_2 - a_2 + c_2(t - t_0) = \frac{d_{21}p_{1,0} + d_{22}p_{2,0}}{\sqrt{d_{11}p_{1,0}^2 + (d_{12} + d_{21})p_{1,0}p_{2,0} + d_{22}p_{2,0}^2}}(t - t_0).$$

Denoting by r_1, r_2 the right-hand sides of these two equalities, we may verify that

$$D_{11}r_1^2 + (D_{12} + D_{21})r_1 r_2 + D_{22}r_2^2 = (t - t_0)^2,$$

where D_{jk} are the elements of the matrix $\mathcal{D}^{-1}$:

$$\mathcal{D}^{-1} = \frac{1}{\det \mathcal{D}} \begin{pmatrix} d_{22} & -d_{12} \\ -d_{21} & d_{11} \end{pmatrix}.$$

Thus

$$d_{22}(x_1 - a_1 + c_1(t - t_0))^2 - (d_{12} + d_{21})(x_1 - a_1 + c_1(t - t_0))(x_2 - a_2 + c_2(t - t_0)) + d_{11}(x_2 - a_2 + c_2(t - t_0))^2 = \det \mathcal{D}(t - t_0)^2. \quad (2.89)$$

This leads us to putting

$$u(x_1, x_2, t) = \sqrt{\det \mathcal{D}}(t - t_0) \pm \{d_{22}(x_1 - a_1 + c_1(t - t_0))^2 - (d_{12} + d_{21})(x_1 - a_1 + c_1(t - t_0))(x_2 - a_2 + c_2(t - t_0)) + d_{11}(x_2 - a_2 + c_2(t - t_0))^2\}^{1/2}, \quad (2.90)$$

which in fact is the solution to (2.87). Now $u(x_1, x_2, t) = 0$ yields the desired characteristic surface in the form of a cone. Its cross-section with the plane $t = t_1$ is an ellipse, since due to the condition $\det \mathcal{D} > 0$ the left hand side of (2.89) is a positive definite form. Of course the investigation of (2.88) leads to the same result.

Now let us look for a characteristic surface in the form of a cone with the aid of the theory of envelopes. Evidently the tangent planes to the sought cone are given by the equation

$$p_1(x_1 - a_1) + p_2(x_2 - a_2) + \{c_1p_1 + c_2p_2 - \sqrt{d_{11}p_1^2 + (d_{12} + d_{21})p_1p_2 + d_{22}p_2^2}\}(t - t_0) = 0. \quad (2.91)$$

Differentiating it with respect to p_1 and p_2 we get

$$x_1 - a_1 + \{c_1 - \frac{2d_{11}p_1 + (d_{12} + d_{21})p_2}{2\sqrt{d_{11}p_1^2 + (d_{12} + d_{21})p_1p_2 + d_{22}p_2^2}}\}(t - t_0) = 0,$$

$$x_2 - a_2 + \{c_2 - \frac{(d_{12} + d_{21})p_1 + 2d_{22}p_2}{2\sqrt{d_{11}p_1^2 + (d_{12} + d_{21})p_1p_2 + d_{22}p_2^2}}\}(t - t_0) = 0.$$

Our task is to eliminate p_1, p_2 from the two equations. We could use the same procedure as above but let us apply here another approach. Denoting

$$r = \sqrt{d_{11}p_1^2 + (d_{12} + d_{21})p_1p_2 + d_{22}p_2^2}$$

and viewing it for a while as a known expression we can evaluate p_1, p_2 from the last two equations. We get

$$p_1 = 2\frac{2d_{22}(x_1 - a_1 + c_1(t-t_0)) - (x_2 - a_2 + c_2(t-t_0))(d_{12}+d_{21})}{4d_{11}d_{22}(t-t_0) - (d_{12}+d_{21})^2(t-t_0)}r,$$

$$p_2 = 2\frac{2d_{11}(x_2 - a_2 + c_2(t-t_0)) - (x_1 - a_1 + c_1(t-t_0))(d_{12}+d_{21})}{4d_{11}d_{22}(t-t_0) - (d_{12}+d_{21})^2(t-t_0)}r.$$

Inserting these expressions into the equation (2.91) we obtain after dividing by $2r$:

$$\frac{2d_{22}(x_1 - a_1 + c_1(t-t_0)) - (x_2 - a_2 + c_2(t-t_0))(d_{12}+d_{21})}{4d_{11}d_{22}(t-t_0) - (d_{12}+d_{21})^2(t-t_0)}(x_1 - a_1 + c_1(t-t_0))$$
$$+\frac{2d_{11}(x_2 - a_2 + c_2(t-t_0)) - (x_1 - a_1 + c_1(t-t_0))(d_{12}+d_{21})}{4d_{11}d_{22}(t-t_0) - (d_{12}+d_{21})^2(t-t_0)}(x_2 - a_2 + c_2(t-t_0))$$
$$-\frac{1}{2}(t-t_0) = 0.$$

Simple arrangements give

$$4\{d_{22}(x_1 - a_1 + c_1(t-t_0))^2 - (d_{12}+d_{21})(x_1 - a_1 + c_1(t-t_0))(x_2 - a_2 + c_2(t-t_0)) + d_{11}(x_2 - a_2 + c_2(t-t_0))^2\} = (t-t_0)^2\left[4d_{11}d_{22} - (d_{12}+d_{21})^2\right].$$

Taking into account the relation $d_{12} = d_{21}$, we have

$$4d_{11}d_{22} - (d_{12}+d_{21})^2 = 4d_{11}d_{22} - 4d_{12}^2 = 4(d_{11}d_{22} - d_{12}^2) = 4(d_{11}d_{22} - d_{12}d_{21}) = 4\det \mathcal{D}$$

and thus this procedure yields the equation (2.89) again.

2.3.9.2 Let us look for a conoidal characteristic surface given by the equation

$$p_0 - \alpha(t)\sqrt{p_1^2 + p_2^2 + p_3^2} = 0, \tag{2.92}$$

where $\alpha > 0$ has a continuous first derivative.

The corresponding characteristic system (2.70), (2.71) reads

$$\frac{dx_j}{dt} = -\alpha(t)\frac{p_j}{\sqrt{p_1^2 + p_2^2 + p_3^2}}, \quad j = 1,2,3,$$
$$\frac{dp_0}{dt} = \alpha'(t)\sqrt{p_1^2 + p_2^2 + p_3^2},$$
$$\frac{dp_j}{dt} = 0, \quad j = 1,2,3.$$

The characteristic strip passing through the point (a_1, a_2, a_3, t_0) is given by

$$p_j(t) = p_{j,0}, \ (p_{j,0} \text{ constants }), \quad j = 1,2,3,$$
$$p_0(t) = p_{0,0} + (\alpha(t) - \alpha(t_0))\sqrt{p_{1,0}^2 + p_{2,0}^2 + p_{3,0}^2} = \alpha(t)\sqrt{p_{1,0}^2 + p_{2,0}^2 + p_{3,0}^2},$$
$$x_j(t) - a_j = (A(t_0) - A(t))\frac{p_{j,0}}{\sqrt{p_{1,0}^2 + p_{2,0}^2 + p_{3,0}^2}}, \quad j = 1,2,3,$$

(since by (2.92) $p_{0,0} = \alpha(t_0)\sqrt{p_{1,0}^2 + p_{2,0}^2 + p_{3,0}^2}$), where $A(t)$ is a primitive function to $\alpha(t)$. Squaring the last equalities and summing them together we get

$$\sum_{j=1}^{3}(x_j(t) - a_j)^2 = (A(t) - A(t_0))^2.$$

It may easily be verified that

$$u(x,t) = A(t) - A(t_0) \pm \sqrt{\sum_{j=1}^{3}(x_j - a_j)^2}$$

is a solution to (2.92) and

$$u(x,t) = 0$$

is the desired characteristic surface in the form of a conoid the cross-section of which with the plane $t = t_1$ is a sphere with radius $|A(t_1) - A(t_0)|$.

2.3.9.3 Let us look for a cone-like characteristic surface given by the equation

$$p_0^2 - c^2(p_1^2 + p_2^2 + p_3^2) = 0$$

with $c > 0$ a constant. Since the components of a normal along the whole surface to a cone-like characteristic surface satisfy the relation

$$p_{0,0} = \pm c\sqrt{p_{1,0}^2 + p_{2,0}^2 + p_{3,0}^2},$$

every tangent plane to it passing through the point (a_1, a_2, a_3, t_0) has the equation

$$\pm c\sqrt{p_{1,0}^2 + p_{2,0}^2 + p_{3,0}^2}(t - t_0) + p_{1,0}(x_1 - a_1) + p_{2,0}(x_2 - a_2) + p_{3,0}(x_3 - a_3) = 0.$$

The envelope of these tangent planes must satisfy the following equations

$$\frac{\pm c p_{j,0}}{\sqrt{p_{1,0}^2 + p_{2,0}^2 + p_{3,0}^2}}(t - t_0) + x_j - a_j = 0, \quad j = 1, 2, 3.$$

This gives

$$c^2(t - t_0)^2 = \sum_{j=1}^{3}(x_j - a_j)^2,$$

which is the equation of the desired cone.

2.3.10 Exercise

Find the characteristic surface in the form of a conoid given by the equation

$$p_0 + \beta(x_1^2 + x_2^2)\sqrt{p_1^2 + p_2^2} = 0,$$

where $\beta \in C^1(\mathbf{R}^+)$.

2.4 A SYSTEM IN ONE SPACE VARIABLE

Hyperbolic systems are defined and the existence theorem for the initial-value problem is introduced via the method of characteristics. Another approach to the solution of initial-value problems via the Riemann functions is illustrated in three representative examples. Using the concept of Riemann invariants the motion of a gas in a tube is investigated. Finally, the initial-boundary value problem is considered briefly.

2.4.1 Preliminaries

Let us say at the very beginning of the next two sections that the problem of the resolution of a system of partial differential equations of the first order is substantially more difficult than the problems of the preceding two sections since it is no longer reducible to the resolution of one or several ordinary differential equations. Let us illustrate it on a very simple initial-value problem for the system with constant coefficients

$$\begin{gathered} u_{1t} + a_1 u_{1x} + b_{11} u_1 + b_{12} u_2 = 0, \\ u_{2t} + a_2 u_{2x} + b_{21} u_1 + b_{22} u_2 = 0, \\ u_1(x, t_0) = \varphi_1(x), \qquad u_2(x, t_0) = \varphi_2(x). \end{gathered}$$

Proceeding similarly as in Sections 2.2 and 2.3 we define the *characteristics* to this system as the solutions of the equations

$$\frac{d\xi_1}{d\tau} = a_1, \qquad \frac{d\xi_2}{d\tau} = a_2.$$

Then the solutions passing through the point (x, t) are $\chi_1(\tau; x, t) = x + a_1(\tau - t)$, $\chi_2(\tau; x, t) = x + a_2(\tau - t)$ and we may write

$$\frac{d}{d\tau} u_1(x + a_1(\tau - t), \tau) + b_{11} u_1(x + a_1(\tau - t), \tau) + b_{12} u_2(x + a_1(\tau - t), \tau) = 0,$$

$$\frac{d}{d\tau} u_2(x + a_2(\tau - t), \tau) + b_{21} u_1(x + a_2(\tau - t), \tau) + b_{22} u_2(x + a_2(\tau - t), \tau) = 0.$$

Now it is already easy to see that, in general, that is if $a_1 \neq a_2$, this system is not a system of ordinary differential equations because arguments in the first and second equations are different. (Of course, in the very special case when $a_1 = a_2 = a$ the system can be solved in the framework of the theory of ordinary differential equations with the solution

$$\begin{pmatrix} u_1(x,t) \\ u_2(x,t) \end{pmatrix} = \exp\left(- \begin{pmatrix} b_{11} & b_{12} \\ b_{21} & b_{22} \end{pmatrix} (t - t_0) \right) \begin{pmatrix} \varphi_1(x + a(t_0 - t)) \\ \varphi_2(x + a(t_0 - t)) \end{pmatrix}.)$$

We will not follow further the solution of the above system since it will be included in a more general existence theorem.

2.4.2 Hyperbolic systems

We will investigate systems of differential equations of the form

$$u_t + \mathcal{A}(x,t)u_x + \mathcal{B}(x,t)u = f(x,t), \tag{2.93}$$

where

$$u = \begin{pmatrix} u_1 \\ \vdots \\ u_m \end{pmatrix}, \quad u_t = \begin{pmatrix} u_{1t} \\ \vdots \\ u_{mt} \end{pmatrix}, \quad u_x = \begin{pmatrix} u_{1x} \\ \vdots \\ u_{mx} \end{pmatrix}, \quad f = \begin{pmatrix} f_1 \\ \vdots \\ f_m \end{pmatrix}$$

and $\mathcal{A}(x,t)$, $\mathcal{B}(x,t)$ are $m \times m$ square matrices. For a while let us suppose that $\mathcal{A}$, $\mathcal{B}$ and f are as smooth as necessary. Our next task is to find new dependent variables $v_1, \dots, v_m$ and some significant curves, which we will again call characteristics, such that the differential part in the new variables in the j-th equation be expressed as the derivative of v_j along the j-th characteristic. This is not always possible in the field of real numbers and real functions. Nevertheless it is possible for the so-called hyperbolic and strictly hyperbolic systems. We say that a system (2.93) is *hyperbolic* in a domain $Q \subset \mathbf{R}^2$ if all eigenvalues $\lambda_1(x,t), \dots, \lambda_m(x,t)$ (each of them being written as many times as its multiplicity indicates) of the matrix $\mathcal{A}(x,t)$ are real for all $(x,t) \in Q$ and there exist m linearly independent left eigenvectors $\ell_1(x,t), \dots, \ell_m(x,t)$. We say that the system is *strictly hyperbolic* if the eigenvalues $\lambda_1(x,t), \dots, \lambda_m(x,t)$ are real and distinct. (The last property already guarantees the existence of m linearly independent eigenvectors.) Finally, we say that a system is *weakly hyperbolic* if the matrix $\mathcal{A}$ has real eigenvalues only. Although the theory for weakly hyperbolic systems is not much more difficult than that for the former ones, we leave it aside. (Let us note that the classification in the theory of hyperbolic systems is not fully unified.)

For a hyperbolic or a strictly hyperbolic system (2.93) let us define the *j-th characteristic* as the j-th component of a solution to the system

$$\frac{d\xi_j}{d\tau} = \lambda_j(\xi,\tau), \quad j = 1, \dots, m.$$

The corresponding solution passing through the point (x,t) will be denoted by $\chi_j(\tau; x, t)$ and the vector $(\chi_1, \dots, \chi_m)$ briefly by χ.

The quasilinear system

$$u_t + \mathcal{A}(x,t,u)u_x = f(x,t,u)$$

is said to be *hyperbolic* or *strictly hyperbolic* if the matrix $\mathcal{A}$ has the same property for all real u as we required in the linear case.

Let us show how the hyperbolic or strictly hyperbolic system (2.93) may be simplified to the required form. Let us denote by $\mathcal{L}(x,t)$ the matrix whose k-th row is formed by the left eigenvector $\ell_k(x,t)$. Because of the linear independence of the vectors $\ell_1, \dots, \ell_m$ the matrix $\mathcal{L}$ is regular and therefore the inverse matrix $\mathcal{L}^{-1}$ exists. Multiplying the equation (2.93) by $\mathcal{L}$, denoting $v = \mathcal{L}u$, $\mathcal{D} = \mathcal{D}(\lambda_1, \dots, \lambda_m) = \mathcal{L}\mathcal{A}\mathcal{L}^{-1}$, we obtain after some arrangements

$$v_t + \mathcal{D}(x,t)v_x + \mathcal{B}_0(x,t)v = f_0(x,t), \tag{2.94}$$

where $\mathcal{B}_0 = (\mathcal{L}\mathcal{B} - \mathcal{D}\mathcal{L}_x - \mathcal{L}_t)\mathcal{L}^{-1}$, $f_0 = \mathcal{L}f$ and $\mathcal{D} = \mathcal{D}(\lambda_1, \ldots, \lambda_m)$ is the diagonal matrix with entries $\lambda_1, \ldots, \lambda_m$. (This notation for a diagonal matrix will be used also in what follows.) We say that the system (2.94) is the system *in diagonal form*. In the sequel we will suppose that our system is already transformed in the upper half-plane $t \geq t_0$ to diagonal form and we will write simply $\mathcal{B}$ and f instead of $\mathcal{B}_0$ and f_0, respectively. It is easy to see that, in some special cases, for instance if $\mathcal{B}$ is a diagonal or a triangular matrix, the system in diagonal form may be integrated without difficulties.

2.4.3 Exercises

2.4.3.1 Show that the system

$$\begin{aligned} u_{1t} + u_{1x} \qquad &= 0, \\ u_{2t} - u_{1x} + u_{2x} &= 0, \end{aligned}$$

is weakly hyperbolic, but not hyperbolic, in an arbitrary domain $Q \subset \mathbf{R}^2$. Find its solution fulfilling $u_1(x, t_0) = \varphi_1(x)$, $u_2(x, t_0) = \varphi_2(x)$.

Answer.

$\lambda_1 = \lambda_2 = 1$; every left eigenvector is up to a constant factor equal to $(1, 0)$; $u_1(x, t) = \varphi_1(x - t + t_0)$, $u_2(x, t) = \varphi_2(x - t + t_0) + (t - t_0)\varphi_1'(x - t + t_0)$.

2.4.3.2 Show that the system

$$\begin{aligned} u_{1t} - u_{1x} \qquad &= 0, \\ u_{2t} - u_{2x} + u_1 &= 0, \end{aligned}$$

is hyperbolic but not strictly hyperbolic in an arbitrary domain $Q \subset \mathbf{R}^2$.

Answer.

$\lambda_1 = \lambda_2 = -1$; two linearly independent eigenvectors are, for instance, $(1, 0)$, $(0, 1)$.

2.4.3.3 Show that the system

$$\begin{aligned} u_{1t} - u_{2x} &= 0, \\ u_{2t} - u_{1x} &= 0, \end{aligned}$$

is strictly hyperbolic in every domain $Q \subset \mathbf{R}^2$.

Answer.

$\lambda_1 = 1, \lambda_2 = -1$.

2.4.3.4 Show that the system (the so-called Cauchy–Riemann equations)

$$\begin{aligned} u_{1t} - u_{2x} &= 0, \\ u_{2t} + u_{1x} &= 0, \end{aligned}$$

is not even weakly hyperbolic in any domain $Q \subset \mathbf{R}^2$.

Answer.

$\lambda_{1,2} = \pm i$.

2.4.3.5 Show that the system (the so-called *equations of long-distance transmission line*)

$$\begin{aligned} LI_t &+ u_x &+ RI &= 0, \\ Cu_t &+ I_x &+ Gu &= 0, \end{aligned}$$

governing the current I and the voltage u in long-distance transmission line is strictly hyperbolic and transform it to the diagonal form (2.94).
Answer.

$$\lambda_{1,2} = \pm(LC)^{-1/2}; \quad \begin{pmatrix} v_1 \\ v_2 \end{pmatrix}_t + \begin{pmatrix} -(LC)^{-1/2} & 0 \\ 0 & (LC)^{-1/2} \end{pmatrix} \begin{pmatrix} v_1 \\ v_2 \end{pmatrix}_x$$
$$+ \frac{1}{2} \begin{pmatrix} RL^{-1} + GC^{-1} & RL^{-1} - GC^{-1} \\ RL^{-1} - GC^{-1} & RL^{-1} + GC^{-1} \end{pmatrix} \begin{pmatrix} v_1 \\ v_2 \end{pmatrix} = \begin{pmatrix} 0 \\ 0 \end{pmatrix},$$

where

$$\begin{pmatrix} v_1 \\ v_2 \end{pmatrix} = \begin{pmatrix} \sqrt{L} & -\sqrt{C} \\ \sqrt{L} & \sqrt{C} \end{pmatrix} \begin{pmatrix} I \\ u \end{pmatrix}.$$

2.4.3.6 Show that the quasilinear system

$$\begin{aligned} \rho_t &+ u\rho_x &+ \rho u_x &= 0, \\ u_t &+ \frac{c^2}{\rho}\rho_x &+ uu_x &= 0, \end{aligned}$$

$c \neq 0$, is strictly hyperbolic.
Answer.
The two different eigenvalues are $\lambda_1 = u + c, \ \ \lambda_2 = u - c$.

2.4.4 The initial-value problem: existence theorem

Let us investigate the existence of a solution to the initial-value problem

$$u_t + \mathcal{D}(\lambda_1(x,t), \ldots, \lambda_m(x,t))\, u_x + \mathcal{B}(x,t)u = f(x,t), \qquad x \in \mathbf{R}, \quad t \geq t_0, \tag{2.95}$$

$$u\,(x, t_0) = \varphi(x), \qquad x \in \mathbf{R}. \tag{2.96}$$

For $j = 1, \ldots, m$ let us construct all characteristics $\chi_j(\tau; x, t_0)$ for $x \in \mathbf{R}$ and $\tau \geq t_0$. Denote the sets covered by these characteristics by Q_j and define the *domain of influence* of the straight line $t = t_0$ as the set

$$\mathcal{DI}\,(\mathbf{R}; t_0) = \bigcap_{j=1}^{m} Q_j.$$

Let $(x,t) \in \mathcal{DI}\,(\mathbf{R}; t_0)$. Rewriting (2.95) in the form

$$\frac{d}{d\tau} u_j\,(\chi_j(\tau; x,t), \tau) + \sum_{k=1}^{m} b_{jk}\,(\chi_j(\tau; x,t), \tau)\, u_k\,(\chi_j(\tau; x,t), \tau)$$
$$- f_j\,(\chi_j(\tau; x,t), \tau) = 0, \qquad j = 1, \ldots, m$$

and integrating it from t_0 to t, we get

$$G_j(x,t,u) \equiv u_j(x,t) - \varphi_j(\chi_j((t_0;x,t)) + \int_{t_0}^{t} \sum_{k=1}^{m} b_{jk}(\chi_j(\tau;x,t),\tau)\, u_k(\chi_j(\tau;x,t),\tau)\, d\tau - \int_{t_0}^{t} f_j(\chi_j(\tau;x,t),\tau)\, d\tau = 0,$$

$$j = 1,\ldots,m.$$

Putting $r = 1$, $s = m$, $Q_t = \mathcal{DI}(\mathbf{R};t_0)$, $\psi = \varphi$ in the Modified Implicit Functions Theorem (see § 1.2.10) and bearing in mind that

$$G_j(x,t_0,\varphi(x)) = 0$$

and

$$\det\left(\frac{\partial G_j(x,t_0,\varphi(x))}{\partial u_k}\right)_{j,k=1}^{m} = 1,$$

we readily get the following theorem.

Let λ_j, b_{jk}, f_j *and* φ_j, $j,k = 1,\ldots,m$ *together with their first derivatives with respect to* x *be continuous in* $\mathbf{R}\times[t_0,\infty)$ *and* $\mathbf{R}$ *respectively. Then there exist a domain* $Q'_t \subset \mathcal{DI}(\mathbf{R};t_0)$ *containing the straight line* $t = t_0$ *and functions* $u_j(x,t)$, $j = 1,\ldots,m$ *defined and continuous together with their first derivatives with respect to* x *and* t *in* Q'_t *which form a unique solution to the problem* (2.95), (2.96) *in* Q'_t.

2.4.5 Example

Let us solve the system

$$u_{1t} - au_{2x} = f_1(x,t),$$
$$u_{2t} - bu_{1x} = f_2(x,t),$$

with the initial conditions

$$u_1(x,0) = \varphi_1(x), \quad u_2(x,0) = \varphi_2(x),$$

where a, b are positive constants, $f_1, f_2 \in C^1(\mathbf{R}\times\mathbf{R}^+)$, $\varphi_1,\varphi_2 \in C^1(\mathbf{R})$. Putting

$$\mathcal{A} = \begin{pmatrix} 0 & -a \\ -b & 0 \end{pmatrix}$$

we find easily that its two eigenvalues are $\lambda_1 = -\sqrt{ab}$, $\lambda_2 = \sqrt{ab}$ and its two linearly independent left eigenvectors are for instance

$$(\sqrt{b},\sqrt{a}),\ (\sqrt{b},-\sqrt{a}).$$

Thus the system may be transformed to the following system in diagonal form

$$v_{1t} - \sqrt{ab}\, v_{1x} = f_{1,0}(x,t),$$
$$v_{2t} + \sqrt{ab}\, v_{2x} = f_{2,0}(x,t),$$

with the initial conditions

$$v_1(x,0) = \psi_1(x), \ v_2(x,0) = \psi_2(x),$$

where $v_1 = \sqrt{b}\,u_1 + \sqrt{a}\,u_2$, $v_2 = \sqrt{b}\,u_1 - \sqrt{a}\,u_2$, $f_{1,0} = \sqrt{b}\,f_1 + \sqrt{a}\,f_2$, $f_{2,0} = \sqrt{b}\,f_1 - \sqrt{a}\,f_2$, $\psi_1 = \sqrt{b}\,\varphi_1 + \sqrt{a}\,\varphi_2$, $\psi_2 = \sqrt{b}\,\varphi_1 - \sqrt{a}\,\varphi_2$. The characteristics of this system being

$$\chi_1(\tau; x, t) = x - \sqrt{ab}(\tau - t), \quad \chi_2(\tau; x, t) = x + \sqrt{ab}(\tau - t),$$

its solution is

$$v_1(x,t) = \psi_1(x + \sqrt{ab}\,t) + \int_0^t f_{1,0}(x - \sqrt{ab}(\tau - t), \tau)\,d\tau,$$

$$v_2(x,t) = \psi_2(x - \sqrt{ab}\,t) + \int_0^t f_{2,0}(x + \sqrt{ab}(\tau - t), \tau)\,d\tau$$

and

$$u_1(x,t) = \frac{1}{2\sqrt{b}}(v_1(x,t) + v_2(x,t)),$$

$$u_2(x,t) = \frac{1}{2\sqrt{a}}(v_1(x,t) - v_2(x,t)).$$

This result enables us to find the solution of the so-called *wave equation*

$$u_{tt} - c^2 u_{xx} = f(x,t)$$

with initial conditions $u(x,0) = \varphi(x)$, $u_t(x,0) = \psi(x)$. Putting $u_1 = u_t$, $u_2 = u_x$ the wave equation may be rewritten in the form

$$u_{1t} - c^2 u_{2x} = f(x,t).$$

Adding the equation $u_{2t} - u_{1x} = 0$ which is equivalent to the condition $u_{xt} = u_{tx}$ we get the original system with $a = c^2$, $b = 1$, $f_1 = f$, $f_2 = 0$, $\varphi_1 = \psi$, $\varphi_2 = \varphi'$. Using the identity

$$u(x,t) = \varphi(x) + \int_0^t u_t(x,\tau)\,d\tau,$$

we get immediately

$$u(x,t) = \frac{\varphi(x+ct) + \varphi(x-ct)}{2} + \frac{1}{2c}\int_{x-ct}^{x+ct} \psi(\xi)\,d\xi + \frac{1}{2c}\int_0^t \int_{x-ct+c\tau}^{x+ct-c\tau} f(\xi,\tau)\,d\xi d\tau.$$

2.4.6 Riemann functions: examples

The above existence theorem does not suggest any idea about the global structure of the solution to the problem (2.95), (2.96). This can be achieved for example by introducing the Riemann functions. The theory of construction of Riemann functions to linear hyperbolic systems even with any number of space variables is thoroughly elaborated [see Courant (1962)]. However it is rather complicated and makes use of the theory of distributions which we wish to avoid. Recently, for systems with one space variable a somewhat simpler method was suggested in Romanovskiĭ (1985). But in its general form even this method is too involved to be introduced here. Hence we restrict ourselves to explain its substance on three typical examples. Let us note that in Romanovskiĭ (1985), the reader can also find existence proofs for the Gourşat problem the solution of which is needed in the construction of the Riemann functions.

2.4.6.1 *Two dimensional system with multiple characteristics*

Let us investigate the initial-value problem

$$\begin{aligned} u_{1t} + au_{1x} + b_{11}(x,t)u_1 + b_{12}(x,t)u_2 &= f_1(x,t), \\ u_{2t} + au_{2x} + b_{21}(x,t)u_1 + b_{22}(x,t)u_2 &= f_2(x,t), \\ u_1(x,t_0) = \varphi_1(x), \quad u_2(x,t_0) &= \varphi_2(x), \end{aligned} \tag{2.97}$$

with b_{jk}, $f_j \in C^1(\mathbf{R} \times [t_0, \infty))$ and $\varphi_j \in C^1(\mathbf{R})$.

Multiplying the first equation by $v_1(t)$, the second by $v_2(t)$ and integrating their sum along the characteristic $\chi(\tau; x, t) = x + a(\tau - t)$ passing through the point (x, t), we get

$$\begin{aligned} &\int_{t_0}^{t} [v_1(\tau)\, f_1(x + a(\tau - t), \tau) + v_2(\tau)\, f_2(x + a(\tau - t), \tau)]\, d\tau \\ &\quad = \int_{t_0}^{t} [v_1(\tau) \frac{d}{d\tau} u_1(x + a(\tau - t), \tau) + v_2(\tau) \frac{d}{d\tau} u_2(x + a(\tau - t), \tau) \\ &\quad + u_1(x + a(\tau - t), \tau)\,(b_{11}(x + a(\tau - t), \tau)\, v_1(\tau) + b_{21}(x + a(\tau - t), \tau)\, v_2(\tau)) \\ &\quad + u_2(x + a(\tau - t), \tau)\,(b_{12}(x + a(\tau - t), \tau)\, v_1(\tau) + b_{22}(x + a(\tau - t), \tau)\, v_2(\tau))]\, d\tau. \end{aligned}$$

Assuming that v_1, v_2 satisfy the system

$$\begin{aligned} \frac{dv_1}{d\tau} - b_{11}(x + a(\tau - t), \tau)\, v_1(\tau) - b_{21}(x + a(\tau - t), \tau)\, v_2(\tau) &= 0, \\ \frac{dv_2}{d\tau} - b_{12}(x + a(\tau - t), \tau)\, v_1(\tau) - b_{22}(x + a(\tau - t), \tau)\, v_2(\tau) &= 0, \end{aligned} \tag{2.98}$$

we have

$$\begin{aligned} &\int_{t_0}^{t} [v_1(\tau)\, f_1(x + a(\tau - t), \tau) + v_2(\tau)\, f_2(x + a(\tau - t), \tau)]\, d\tau \\ &\quad = \int_{t_0}^{t} \frac{d}{d\tau} [v_1(\tau)\, u_1(x + a(\tau - t), \tau) + v_2(\tau)\, u_2(x + a(\tau - t), \tau)]\, d\tau \\ &= v_1(t)\, u_1(x,t) + v_2(t)\, u_2(x,t) - v_1(t_0)\, \varphi_1\, (x + a(t_0 - t)) - v_2(t_0)\, \varphi_2\, (x + a(t_0 - t))\,. \end{aligned}$$

Denoting by v_{11}, v_{12} the solution of (2.98) fulfilling the initial condition

$$v_{11}(t) = 1 \qquad v_{12}(t) = 0 \tag{2.99}$$

and by v_{21}, v_{22} the solution of (2.98) fulfilling the initial condition

$$v_{21}(t) = 0, \qquad v_{22}(t) = 1, \tag{2.100}$$

we obtain for the solution to (2.97) the formulae

$$\begin{aligned} u_1(x,t) &= v_{11}(t_0)\,\varphi_1\left(x + a(t_0 - t)\right) + v_{12}(t_0)\,\varphi_2\left(x + a(t_0 - t)\right) \\ &\quad + \int_{t_0}^{t} \left(v_{11}(\tau)\, f_1\left(x + a(\tau - t), \tau\right) + v_{12}(\tau)\, f_2\left(x + a(\tau - t), \tau\right)\right) d\tau, \end{aligned}$$

$$\begin{aligned} u_2(x,t) &= v_{21}(t_0)\,\varphi_1\left(x + a(t_0 - t)\right) + v_{22}(t_0)\,\varphi_2\left(x + a(t_0 - t)\right) \\ &\quad + \int_{t_0}^{t} \left(v_{21}(\tau)\, f_1\left(x + a(\tau - t), \tau\right) + v_{22}(\tau)\, f_2\left(x + a(\tau - t), \tau\right)\right) d\tau. \end{aligned}$$

2.4.6.2 *Two dimensional system with simple characteristics*

Let us investigate the initial-value problem

$$\begin{aligned} u_{1t} + a_1 u_{1x} + b_{11}(x,t)u_1 + b_{12}(x,t)u_2 &= f_1(x,t), \\ u_{2t} + a_2 u_{2x} + b_{21}(x,t)u_1 + b_{22}(x,t)u_2 &= f_2(x,t), \\ u_1(x,t_0) = \varphi_1(x), \qquad u_2(x,t_0) &= \varphi_2(x), \end{aligned} \tag{2.101}$$

with $a_1 > a_2$ and $b_{jk}, f_j \in C^1(\mathbf{R} \times [t_0, \infty))$ and $\varphi_j \in C^1(\mathbf{R})$. Multiplying the first equation by $v_1(x,t)$, the second by $v_2(x,t)$ with v_1 and v_2 satisfying the system of equations

$$\begin{aligned} v_{1t} + a_1 v_{1x} - b_{11}v_1 - b_{21}v_2 &= 0, \\ v_{2t} + a_2 v_{2x} - b_{12}v_1 - b_{22}v_2 &= 0, \end{aligned} \tag{2.102}$$

integrating the sum over the triangle T with vertices at the points $P_1 = (x + a_1(t_0 - t), t_0), P_2 = (x + a_2(t_0 - t), t_0)$, $P = (x,t)$ and using the Green formula, we get

$$\begin{aligned} &\int_T (v_1 f_1 + v_2 f_2)(\xi,\tau)\, d\xi d\tau \\ &\quad = \int_T \left[\frac{\partial}{\partial t}(v_1u_1 + v_2u_2) + \frac{\partial}{\partial x}(a_1v_1u_1 + a_2v_2u_2)\right](\xi,\tau)\, d\xi d\tau \\ &\quad = \int_C \left[(-v_1u_1 - v_2u_2)(\xi,\tau)\, d\xi + (a_1v_1u_1 + a_2v_2u_2)(\xi,\tau)\, d\tau\right] \\ &\quad = -\int_{x+a_1(t_0-t)}^{x+a_2(t_0-t)} \left[v_1(\xi,t_0)\,\varphi_1(\xi) + v_2(\xi,t_0)\,\varphi_2(\xi)\right] d\xi \\ &\qquad + (a_1 - a_2)\int_{t_0}^{t} (v_1u_1)(x + a_2(\tau - t), \tau)\, d\tau \\ &\qquad + (a_1 - a_2)\int_{t_0}^{t} (v_2u_2)(x + a_1(\tau - t), \tau)\, d\tau, \end{aligned} \tag{2.103}$$

where C denotes the positively oriented boundary of T.

In order to be able to determine the component u_1, let us modify the term $(v_2u_2)\,(x+a_1(\tau-t),\tau)$ which cannot be integrated directly, on the first characteristic, with the aim of rewriting it in the form of an expression the principal part of which is a total derivative of a suitable function. Putting

$$(a_1-a_2)\,v_2\,(x+a_1(\tau-t),\tau)=\frac{dw_{12}(\tau)}{d\tau} \tag{2.104}$$

and making use of an auxiliary function $w_{11}(\tau)$, we have the equalities

$$\begin{aligned}
(a_1-a_2)\,v_2u_2\,(x+a_1(\tau-t),\tau) &= \frac{dw_{12}(\tau)}{d\tau}u_2\,(x+a_1(\tau-t),\tau)\\
&+[\frac{d}{d\tau}\,(w_{11}(\tau)\,u_1\,(x+a_1(\tau-t),\tau))-\frac{dw_{11}(\tau)}{d\tau}u_1\,(x+a_1(\tau-t),\tau)\\
&-w_{11}(\tau)\frac{d}{d\tau}u_1\,(x+a_1(\tau-t),\tau)] = \frac{dw_{12}(\tau)}{d\tau}u_2\,(x+a_1(\tau-t),\tau)\\
&+\frac{d}{d\tau}\,(w_{11}(\tau)\,u_1\,(x+a_1(\tau-t),\tau))-\frac{dw_{11}(\tau)}{d\tau}u_1\,(x+a_1(\tau-t),\tau)\\
&\quad -w_{11}(\tau)\,[f_1-b_{11}u_1-b_{12}u_2]\,(x+a_1(\tau-t),\tau)\\
&=\frac{d}{d\tau}\,(w_{11}(\tau)\,u_1\,(x+a_1(\tau-t),\tau))-w_{11}(\tau)\,f_1\,(x+a_1(\tau-t),\tau)\\
&+u_1\,(x+a_1(\tau-t),\tau)\left[-\frac{dw_{11}(\tau)}{d\tau}+b_{11}\,(x+a_1(\tau-t),\tau)\,w_{11}(\tau)\right]\\
&\quad +u_2\,(x+a_1(\tau-t),\tau)\left[\frac{dw_{12}(\tau)}{d\tau}+b_{12}\,(x+a_1(\tau-t),\tau)\,w_{11}(\tau)\right].
\end{aligned}$$

Now requiring that the functions w_{11}, w_{12} satisfy the initial-value problem

$$\begin{aligned}
&\frac{dw_{11}(\tau)}{d\tau}-b_{11}\,(x+a_1(\tau-t),\tau)\,w_{11}(\tau)=0,\\
&\frac{dw_{12}(\tau)}{d\tau}+b_{12}\,(x+a_1(\tau-t),\tau)\,w_{11}(\tau)=0,\\
&\qquad w_{11}(t)=1, \qquad w_{12}(t)=0,
\end{aligned}$$

we find

$$\begin{aligned}
w_{11}(\tau)&=\exp\left(\int_t^\tau b_{11}\,(x+a_1(\sigma-t),\sigma)\,d\sigma\right)\\
w_{12}(\tau)&=-\int_t^\tau b_{12}\,(x+a_1(s-t),s)\,\exp\left(\int_t^s b_{11}\,(x+a_1(\sigma-t),\sigma)\,d\sigma\right)ds
\end{aligned} \tag{2.105}$$

and thus

$$\begin{aligned}
(a_1-a_2)\int_{t_0}^t v_2u_2\,(x+a_1(\tau-t),\tau)\,d\tau &= u_1(x,t)-w_{11}(t_0)\,\varphi_1\,(x+a_1(t_0-t))\\
&\quad -\int_{t_0}^t w_{11}(\tau)\,f_1\,(x+a_1(\tau-t),\tau)\,d\tau.
\end{aligned}$$

Moreover, putting

$$v_1\left(x+a_2(\tau-t),\tau\right)=0, \tag{2.106}$$

from the identity (2.103) with $v_1=v_{11}, v_2=v_{12}$ we finally get

$$\begin{aligned}u_1(x,t)=w_{11}(t_0)\,\varphi_1\left(x+a_1(t_0-t)\right)+\int_{x+a_1(t_0-t)}^{x+a_2(t_0-t)}&(v_{11}(\xi,t_0)\,\varphi_1(\xi)\\+v_{12}(\xi,t_0)\,\varphi_2(\xi))\,d\xi+\int_{t_0}^{t}w_{11}(\tau)\,f_1\left(x+a_1(\tau-t),\tau\right)d\tau&\\+\int_T\left(v_{11}(\xi,\tau)\,f_1(\xi,\tau)+v_{12}(\xi,\tau)\,f_2(\xi,\tau)\right)d\xi d\tau,&\end{aligned}$$

where (v_{11},v_{12}) is a solution to the system (2.102) satisfying the supplementary conditions (2.104) and (2.106) with w_{12} given by (2.105), that is

$$\begin{aligned}&v_{11}\left(x+a_2(\tau-t),\tau\right)=0,\\&v_{12}\left(x+a_1(\tau-t),\tau\right)\\&\quad=-\left(a_1-a_2\right)^{-1}b_{12}\left(x+a_1(\tau-t),\tau\right)\exp\left(\int_t^{\tau}b_{11}\left(x+a_1(\sigma-t),\sigma\right)d\sigma\right),\\&\qquad\qquad\tau\in[0,t].\end{aligned}$$

It means that (v_{11},v_{12}) is a solution of the so-called Goursat problem, the existence of which can be proved similarly as the existence of a solution to the initial-value problem, see Romanovskiĭ (1985).

Proceeding similarly as above, for the second component u_2 we get the formula

$$\begin{aligned}u_2(x,t)=w_{22}(t_0)\,\varphi_2\left(x+a_2(t_0-t)\right)+\int_{x+a_1(t_0-t)}^{x+a_2(t_0-t)}&(v_{21}(\xi,t_0)\,\varphi_1(\xi)\\+v_{22}(\xi,t_0)\,\varphi_2(\xi))\,d\xi+\int_{t_0}^{t}w_{22}(\tau)\,f_2\left(x+a_2(\tau-t),\tau\right)d\tau&\\+\int_T\left(v_{21}(\xi,\tau)\,f_1(\xi,\tau)+v_{22}(\xi,\tau)\,f_2(\xi,\tau)\right)d\xi d\tau,&\end{aligned}$$

where $w_{22}(\tau)=\exp\left(\int_t^{\tau}b_{22}\left(x+a_2(\sigma-t),\sigma\right)d\sigma\right)$ and (v_{21},v_{22}) solves the system (2.102) with the following supplementary conditions on the characteristics:

$$\begin{aligned}&v_{21}\left(x+a_2(\tau-t),\tau\right)\\&\quad=-\left(a_1-a_2\right)^{-1}b_{21}\left(x+a_2(\tau-t),\tau\right)\exp\left(\int_t^{\tau}b_{22}\left(x+a_2(\sigma-t),\sigma\right)d\sigma\right),\\&v_{22}\left(x+a_1(\tau-t),\tau\right)=0,\\&\qquad\qquad\tau\in[0,t].\end{aligned}$$

2.4.6.3 *Three dimensional system with simple characteristics*

Finally, let us investigate the problem

$$\begin{aligned}
&u_{1t} + a_1 u_{1x} + b_{11}(x,t)u_1 + b_{12}(x,t)u_2 + b_{13}(x,t)u_3 = f_1(x,t),\\
&u_{2t} + a_2 u_{2x} + b_{21}(x,t)u_1 + b_{22}(x,t)u_2 + b_{23}(x,t)u_3 = f_2(x,t),\\
&u_{3t} + a_3 u_{3x} + b_{31}(x,t)u_1 + b_{32}(x,t)u_2 + b_{33}(x,t)u_3 = f_3(x,t),\\
&u_1(x,t_0) = \varphi_1(x), \quad u_2(x,t_0) = \varphi_2(x), \quad u_3(x,t_0) = \varphi_3(x),
\end{aligned} \tag{2.107}$$

with $a_1 > a_2 > a_3$ and b_{jk}, $f_j \in C^1(\mathsf{R} \times [t_0, \infty))$ and $\varphi_j \in C^1(\mathsf{R})$. Let us denote by T_1 and T_2 the triangles with vertices $P_1 = (x + a_1(t_0 - t), t_0)$, $P_2 = (x + a_2(t_0 - t), t_0)$, $P = (x,t)$ and $P_2 = (x + a_2(t_0 - t), t_0)$, $P_3 = (x + a_3(t_0 - t), t_0)$, $P = (x,t)$, respectively.

Multiplying the j-th equation by v_j^k in T_k where v_j^k satisfy the system of equations

$$\begin{aligned}
v_{1t} + a_1 v_{1x} - b_{11}v_1 - b_{21}v_2 - b_{31}v_3 = 0,\\
v_{2t} + a_2 v_{2x} - b_{12}v_1 - b_{22}v_2 - b_{32}v_3 = 0,\\
v_{3t} + a_3 v_{3x} - b_{13}v_1 - b_{23}v_2 - b_{33}v_3 = 0,
\end{aligned} \tag{2.108}$$

integrating the sum over T_k, applying the Green formula for $k = 1,2$ and adding the two results we get

$$\begin{aligned}
&\sum_{k=1}^{2} \int_{T_k} \left(v_1^k f_1 + v_2^k f_2 + v_3^k f_3\right)(\xi,\tau)\, d\xi d\tau\\
&= -\sum_{k=1}^{2} \int_{x+a_k(t_0-t)}^{x+a_{k+1}(t_0-t)} \left(v_1^k(\xi,t_0)\varphi_1(\xi) + v_2^k(\xi,t_0)\varphi_2(\xi) + v_3^k(\xi,t_0)\varphi_3(\xi)\right) d\xi\\
&\quad + \int_{t_0}^{t} \left[(a_1 - a_2)\, v_2^1 u_2 + (a_1 - a_3)\, v_3^1 u_3\right]_{|\chi_1(\tau)}\, d\tau\\
&\quad + \int_{t_0}^{t} \left[(a_1 - a_2)\left(v_1^1 - v_1^2\right) u_1 + (a_3 - a_2)\left(v_3^1 - v_3^2\right) u_3\right]_{|\chi_2(\tau)}\, d\tau\\
&\quad + \int_{t_0}^{t} \left[(a_1 - a_3)\, v_1^2 u_1 + (a_2 - a_3)\, v_2^2 u_2\right]_{|\chi_3(\tau)}\, d\tau.
\end{aligned} \tag{2.109}$$

Analogously, as in the preceding example, to determine the component u_1 of the solution we rewrite the integrand on the first characteristic in (2.109) in such a way that its principal part is formed by a total derivative of a suitable function. Introducing functions $w_{12}(\tau), w_{13}(\tau)$ by

$$\begin{aligned}
(a_1 - a_2)\, v^1_{2|\chi_1(\tau)} &= \frac{dw_{12}}{d\tau},\\
(a_1 - a_3)\, v^1_{3|\chi_1(\tau)} &= \frac{dw_{13}}{d\tau},
\end{aligned} \tag{2.110}$$

and making use of an auxiliary function $w_{11}(\tau)$ we get after some rearrangements

$$\begin{aligned}
&\left[(a_1 - a_2)\, v_2^1 u_2 + (a_1 - a_3)\, v_3^1 u_3\right]_{|\chi_1(\tau)}\\
&\quad = \frac{d}{d\tau}(w_{11}u_1) - w_{11}f_1 + u_1\left[-\frac{dw_{11}}{d\tau} + b_{11}w_{11}\right]\\
&\qquad + u_2\left[\frac{dw_{12}}{d\tau} + b_{12}w_{11}\right] + u_3\left[\frac{dw_{13}}{d\tau} + b_{13}w_{11}\right].
\end{aligned}$$

Requiring that the coefficients at u_1, u_2 and u_3 vanish and adding the initial conditions

$$w_{11}(t) = 1, \qquad w_{12}(t) = w_{13}(t) = 0,$$

we obtain

$$\begin{aligned}
w_{11}(\tau) &= \exp\left(\int_t^\tau b_{11}\left(x + a_1(\sigma - t), \sigma\right) d\sigma\right),\\
w_{12}(\tau) &= -\int_t^\tau b_{12}\left(x + a_1(s - t), s\right) \exp\left(\int_t^s b_{11}\left(x + a_1(\sigma - t), \sigma\right) d\sigma\right) ds, \qquad (2.111)\\
w_{13}(\tau) &= -\int_t^\tau b_{13}\left(x + a_1(s - t), s\right) \exp\left(\int_t^s b_{11}\left(x + a_1(\sigma - t), \sigma\right) d\sigma\right) ds
\end{aligned}$$

and thus

$$\begin{aligned}
\int_{t_0}^t &\left[(a_1 - a_2)\, v_2^1 u_2 + (a_1 - a_3)\, v_3^1 u_3\right]_{|\chi_1(\tau)}\, d\tau\\
&= u_1(x, t) - w_{11}(t_0)\, \varphi_1\left(x + a_1(t_0 - t)\right)\\
&\qquad - \int_{t_0}^t w_{11}(\tau) f_1\left(x + a_1(\tau - t), \tau\right) d\tau.
\end{aligned}$$

Moreover, putting

$$\begin{aligned}
\left(v_1^1 - v_1^2\right)_{|\chi_2(\tau)} = \left(v_3^1 - v_3^2\right)_{|\chi_2(\tau)} &= 0,\\
v_1^2{}_{|\chi_3(\tau)} = v_2^2{}_{|\chi_3(\tau)} &= 0,
\end{aligned} \qquad (2.112)$$

from the identity (2.109) with $v_1^k = v_{11}^k$, $v_2^k = v_{12}^k$, $v_3^k = v_{13}^k$ we finally get

$$\begin{aligned}
u_1(x, t) &= w_{11}(t_0)\, \varphi_1\left(x + a_1(t_0 - t)\right)\\
&+ \sum_{k=1}^2 \int_{x + a_k(t_0 - t)}^{x + a_{k+1}(t_0 - t)} \left[v_{11}^k(\xi, t_0)\, \varphi_1(\xi) + v_{12}^k(\xi, t_0)\, \varphi_2(\xi) +.v_{13}^k(\xi, t_0)\, \varphi_3(\xi)\right] d\xi\\
&\qquad + \int_{t_0}^t w_{11}(\tau)\, f_1\left(x + a_1(\tau - t), \tau\right) d\tau\\
&\qquad\qquad + \sum_{k=1}^2 \int_{T_k} \left(v_{11}^k f_1 + v_{12}^k f_2 + v_{13}^k f_3\right)(\xi, \tau)\, d\xi d\tau,
\end{aligned}$$

where $v_{11}^k, v_{12}^k, v_{13}^k$ is a solution to the system (2.108) satisfying the supplementary conditions (2.110) and (2.112) on characteristics with w_{12}, w_{13} given by (2.111). Since the determination of u_3 goes in the same lines, let us state only the result

$$\begin{aligned}
u_3(x, t) &= w_{33}(t_0)\, \varphi_3\left(x + a_3(t_0 - t)\right)\\
&+ \sum_{k=1}^2 \int_{x + a_k(t_0 - t)}^{x + a_{k+1}(t_0 - t)} \left[v_{31}^k(\xi, t_0)\, \varphi_1(\xi) + v_{32}^k(\xi, t_0)\, \varphi_2(\xi) + v_{33}^k(\xi, t_0)\, \varphi_3(\xi)\right] d\xi\\
&\qquad + \int_{t_0}^t w_{33}(\tau) f_3\left(x + a_3(\tau - t), \tau\right) d\tau\\
&\qquad\qquad + \sum_{k=1}^2 \int_{T_k} \left(v_{31}^k f_1 + v_{32}^k f_2 + v_{33}^k f_3\right)(\xi, \tau)\, d\xi d\tau,
\end{aligned}$$

where v_{31}^k, v_{32}^k and v_{33}^k solve the system (2.108) and satisfy the supplementary conditions

$$\begin{aligned} v^1_{32|\chi_1(\tau)} &= v^1_{33|\chi_1(\tau)} = 0, \\ \left(v^1_{31} - v^2_{31}\right)_{|\chi_2(\tau)} &= \left(v^1_{33} - v^2_{33}\right)_{|\chi_2(\tau)} = 0, \\ (a_1 - a_3)\, v^2_{31|\chi_3(\tau)} &= \frac{dw_{31}}{d\tau}, \\ (a_2 - a_3)\, v^2_{32|\chi_3(\tau)} &= \frac{dw_{32}}{d\tau}, \end{aligned}$$

with w_{31}, w_{32}, w_{33} given by

$$\begin{aligned} w_{31}(\tau) &= -\int_t^\tau b_{31}\left(x + a_3(s-t), s\right) \exp\left(\int_t^s b_{33}\left(x + a_3(\sigma - t), \sigma\right) d\sigma\right) ds, \\ w_{32}(\tau) &= -\int_t^\tau b_{32}\left(x + a_3(s-t), s\right) \exp\left(\int_t^s b_{33}\left(x + a_3(\sigma - t), \sigma\right) d\sigma\right) ds, \\ w_{33}(\tau) &= \exp\left(\int_t^\tau b_{33}\left(x + a_3(\sigma - t), \sigma\right) d\sigma\right). \end{aligned}$$

On the other hand since the evaluation of u_2 is somewhat different, let us indicate it. Choosing appropriately an auxiliary function $w_{22}(\tau)$ and introducing $w_{21}(\tau)$ and $w_{23}(\tau)$ by

$$\begin{aligned} (a_1 - a_2)\left(v^1_1 - v^2_1\right)_{|\chi_2(\tau)} &= \frac{dw_{21}}{d\tau}, \\ (a_3 - a_2)\left(v^1_3 - v^2_3\right)_{|\chi_2(\tau)} &= \frac{dw_{23}}{d\tau}, \end{aligned} \tag{2.113}$$

we obtain

$$\begin{aligned} &\left[(a_1 - a_2)\left(v^1_1 - v^2_1\right) u_1 + (a_3 - a_2)\left(v^1_3 - v^2_3\right) u_3\right]_{|\chi_2(\tau)} \\ &\quad = \frac{d}{d\tau}\left(w_{22} u_2\right) - w_{22} f_2 + u_1 \left[\frac{dw_{21}}{d\tau} + b_{21} w_{22}\right] \\ &\qquad + u_2 \left[-\frac{dw_{22}}{d\tau} + b_{22} w_{22}\right] + u_3 \left[\frac{dw_{23}}{d\tau} + b_{23} w_{22}\right]. \end{aligned}$$

If w_{21}, w_{22}, w_{23} satisfy the system of ordinary differential equations

$$\begin{aligned} \frac{dw_{21}}{d\tau} + b_{21}\left(x + a_2(\tau - t), \tau\right) w_{22} &= 0, \\ -\frac{dw_{22}}{d\tau} + b_{22}\left(x + a_2(\tau - t), \tau\right) w_{22} &= 0, \\ \frac{dw_{23}}{d\tau} + b_{23}\left(x + a_2(\tau - t), \tau\right) w_{22} &= 0, \end{aligned} \tag{2.114}$$

with initial conditions

$$w_{21}(t) = w_{23}(t) = 0, \qquad w_{22}(t) = 1, \tag{2.115}$$

the above expression may be integrated easily. We get

$$\int_{t_0}^{t} \left[(a_1 - a_2)\left(v_1^1 - v_1^2\right) u_1 + (a_3 - a_2)\left(v_3^1 - v_3^2\right) u_3\right]_{|\chi_2(\tau)}$$

$$= u_2(x,t) - w_{22}(t_0)\,\varphi_2\left(x + a_2(t_0 - t)\right) - \int_{t_0}^{t} w_{22}(\tau)\, f_2\left(x + a_2(\tau - t), \tau\right) d\tau.$$

Adding the remaining conditions

$$\begin{aligned} v^1_{2\,|\chi_1(\tau)} &= v^1_{3\,|\chi_1(\tau)} = 0, \\ v^2_{1\,|\chi_3(\tau)} &= v^2_{2\,|\chi_3(\tau)} = 0, \end{aligned} \tag{2.116}$$

we get readily from (2.109) with $v_1^k = v_{21}^k$, $v_2^k = v_{22}^k$, $v_3^k = v_{23}^k$ the formula

$$\begin{aligned} u_2(x,t) &= w_{22}(t_0)\,\varphi_2\left(x + a_2(t_0 - t)\right) \\ &+ \sum_{k=1}^{2} \int_{x+a_k(t_0-t)}^{x+a_{k+1}(t_0-t)} \left[v_{21}^k(\xi,t_0)\,\varphi_1(\xi) + v_{22}^k(\xi,t_0)\,\varphi_2(\xi) + v_{23}^k(\xi,t_0)\,\varphi_3(\xi)\right] d\xi \\ &+ \int_{t_0}^{t} w_{22}(\tau) f_2\left(x + a_2(\tau - t), \tau\right) d\tau \\ &+ \sum_{k=1}^{2} \int_{T_k} \left(v_{21}^k f_1 + v_{22}^k f_2 + v_{23}^k f_3\right)(\xi,\tau)\, d\xi d\tau, \end{aligned}$$

where $v_{21}^k, v_{22}^k, v_{23}^k$ satisfy the Goursat problem for the system (2.108) with the conditions on the characteristics (2.113), (2.116) and with w_{21}, w_{22}, w_{23} solving (2.114), (2.115). Let us close this subsection by the note that the functions w_{ij} and the functions v_{ij}^k, respectively are called the *Riemann functions of the first and of the second kind*, respectively.

2.4.7 Riemann invariants

The theory of quasilinear systems in one space variable has been studied at length. Whole books are devoted to this problem, e.g. Rozhdestvenskiĭ and Yanenko (1983), Chorin and Marsden (1984), Jeffrey (1976), Smoller (1983). We have to restrict ourselves only to showing in an example, namely on the movement of a gas in a tube, how this problem can be mastered with the use of the so-called Riemann invariants. We have already seen in the case of one equation that it is quite useful to know that a certain quantity is constant along a characteristic. Therefore, also in the case of a system, we will try to find quantities which are constant along a characteristic. Such quantities are called *Riemann invariants*. We look for the Riemann invariants to the system

$$\begin{aligned} \rho_t + u\rho_x + \rho u_x &= 0, \\ u_t + \frac{c^2}{\rho}\rho_x + u u_x &= 0, \end{aligned} \tag{2.117}$$

with $c = c(\rho)$.

Similarly as in the linear case we define a j-th characteristic as any of the solutions to the equation

$$\frac{d\xi_j}{d\tau} = \lambda_j, \qquad j = 1, 2,$$

where $\lambda_j = \lambda_j(\rho, u)$ are the eigenvalues of the matrix

$$\mathcal{A} = \begin{pmatrix} u & \rho \\ c^2/\rho & u \end{pmatrix}.$$

Since $\lambda_{1,2} = u \pm c$, the characteristics are defined by the equations

$$\frac{d\xi_1}{d\tau} = u + c, \qquad \frac{d\xi_2}{d\tau} = u - c.$$

Let us denote by $\chi_+(\tau; x, t)$ and $\chi_-(\tau; x, t)$ the characteristics passing through the point (x, t). Requiring that the Riemann invariants $I(\rho, u)$ be constant along a characteristic and differentiating the relations $I(\rho, u)_{|\chi\pm} =$ const. with respect to τ we get

$$\begin{aligned} \frac{d}{d\tau} I(\rho, u)_{|\chi\pm} &= I_\rho \left[\rho_t + \rho_x \frac{d\chi\pm}{d\tau}\right] + I_u \left[u_t + u_x \frac{d\chi\pm}{d\tau}\right] \\ &= I_\rho \left[-u\rho_x - \rho u_x + \rho_x(u \pm c)\right] + I_u \left[-\frac{c^2}{\rho}\rho_x - uu_x + u_x(u \pm c)\right] \\ &= I_\rho \left(-\rho u_x \pm c\rho_x\right) + I_u \left(-\frac{c^2}{\rho}\rho_x \pm cu_x\right) = 0. \end{aligned}$$

Thus putting $I_u = 1$ for instance we get

$$I_\rho = \frac{c}{\rho} \quad \text{on } \chi_+ \qquad \text{and} \qquad I_\rho = -\frac{c}{\rho} \quad \text{on } \chi_-.$$

We see readily that the function

$$I_+(\rho, u) = u + \int_a^\rho \frac{c(\sigma)}{\sigma} d\sigma$$

is constant on χ_+ while the function

$$I_-(\rho, u) = u - \int_a^\rho \frac{c(\sigma)}{\sigma} d\sigma$$

is constant on χ_- where $a > 0$ is arbitrary. These are the two desired Riemann invariants for (2.117).

2.4.8 The motion of a gas in a tube

Let us investigate the motion of a gas in an infinite straight tube. Let the x-axis coincide with the axis of the tube. For $x > 0$ let the tube be filled with a gas while at the point $x = 0$ it is closed by a piston. We will investigate the expansion of the gas when the piston starts to move at time $t = 0$ to the left with a constant velocity $-v$ ($v > 0$). The motion of the barotropic gas with $p = p(\rho)$ is described by the Euler equations, that is by the continuity equation and by the equation of motion

$$\begin{aligned} \rho_t + u\rho_x + \rho u_x &= 0, \\ u_t + \frac{c^2}{\rho}\rho_x + uu_x &= 0, \end{aligned} \tag{2.118}$$

where $c = \sqrt{p'(\rho)}$. It is natural to require that $p'(\rho) > 0$. Further, let us suppose that before $t = 0$ the gas was at rest and had a constant density ρ_0, that is,

$$u(x,0) = 0, \qquad \rho(x,0) = \rho_0.$$

Besides, we suppose that the particles of the gas stick to the piston, that is, they move with its velocity

$$u(-vt,t) = -v.$$

In the foregoing subsection we have found the Riemann invariants of the system (2.118):

$$I_+(\rho,u) = u + \int_a^\rho \frac{c(\sigma)}{\sigma}d\sigma,$$
$$I_-(\rho,u) = u - \int_a^\rho \frac{c(\sigma)}{\sigma}d\sigma.$$

Since on the positive x-half-axis both invariants are known, they are known also at all points which can be connected with the positive x-half-axis by both characteristics. Let us denote this set by Q_1. At all its points we have

$$u + \int_a^\rho \frac{c(\sigma)}{\sigma}d\sigma = \int_a^{\rho_0} \frac{c(\sigma)}{\sigma}d\sigma,$$
$$u - \int_a^\rho \frac{c(\sigma)}{\sigma}d\sigma = -\int_a^{\rho_0} \frac{c(\sigma)}{\sigma}d\sigma.$$

Adding and subtracting the two relations we get $u = 0$ and $\int_a^\rho \frac{c(\sigma)}{\sigma}d\sigma = \int_a^{\rho_0} \frac{c(\sigma)}{\sigma}d\sigma$. The positivity of the integrand readily implies that $\rho = \rho_0$ in Q_1. Consequently, the equations of the characteristics in Q_1 have the form

$$\frac{d\xi}{d\tau} = \pm c(\rho_0) = \pm\sqrt{p'(\rho_0)}.$$

Hence Q_1 is the angle between the positive x-half-axis and the characteristic $\xi = \tau\sqrt{p'(\rho_0)}$.

Now let us deal with the situation in the vicinity of the piston. First, let us show that on the trajectory of the piston the density is constant. Indeed, let us choose an arbitrary point P on the trajectory. By the hypothesis it moves with constant velocity $u(P) = -v$. Supposing that the velocity of the piston is not too high the domain between the positive x-half-axis and the path of the piston can be covered by the characteristics χ_-, and in this domain the invariant I_- is constant and equal to

$$-\int_a^{\rho_0} \frac{c(\sigma)}{\sigma} d\sigma.$$

Thus in particular

$$I_-(\rho(P), u(P)) = -v - \int_a^{\rho(P)} \frac{c(\sigma)}{\sigma} d\sigma = -\int_a^{\rho_0} \frac{c(\sigma)}{\sigma} d\sigma.$$

Hence the value $\rho(P)$ does not depend on P and its common value $\bar{\rho}$ on the whole path of the piston may be evaluated from the equation

$$\int_{\rho_0}^{\bar{\rho}} \frac{c(\sigma)}{\sigma} d\sigma = -v.$$

Evidently $\bar{\rho} < \rho_0$.

Further, let us prove that every characteristic χ_+ is a straight line. Indeed, since I_- is constant in the whole domain and I_+ is constant on χ_+ we can conclude that ρ, u and c are constant on every χ_+. Our assertion follows immediately from the equation of the χ_+ characteristics $\dfrac{d\xi_+}{d\tau} = u + c =$ const .

Let us denote by Q_3 the domain between the trajectory of the piston and the straight line $\xi = \left(-v + \sqrt{p'(\bar{\rho})}\right)\tau$, which is covered by χ_+ issuing from the path of the piston. In the domain Q_3 the quantities u and ρ are constant and $u = -v, \rho = \bar{\rho}$.

Finally, let us investigate the motion in the domain Q_2 separating the domains Q_1, Q_3. As there are two characteristics χ_+ issuing from the origin, namely

$$\xi = \tau\sqrt{p'(\rho_0)}, \qquad \xi = \left(-v + \sqrt{p'(\bar{\rho})}\right)\tau,$$

all the other χ_+ characteristics in Q_2 issue naturally from the origin. Then their equations read $\chi_+(\tau; x, t) = \dfrac{x}{t}\tau$ and thus $\dfrac{d\chi_+}{d\tau} = \dfrac{x}{t}$ and finally

$$\frac{x}{t} = u(x,t) + c(\rho(x,t)).$$

Adding the equation for I_-, that is

$$u(x,t) - \int_a^{\rho(x,t)} \frac{c(\sigma)}{\sigma} d\sigma = -\int_a^{\rho_0} \frac{c(\sigma)}{\sigma} d\sigma,$$

we get a system determining u and ρ.

Evidently, the above solution is not a classical one. The origin is obviously a singular point and it may be shown that on the straight lines $\xi = \tau\sqrt{p'(\rho_0)}$ and $\xi = \left(-v + \sqrt{p'(\bar{\rho})}\right)\tau$ the solution is continuous while its first derivatives have jumps. Nevertheless, it describes quite well the process studied.

The solution found may be interpreted as follows: at the time $t = t_1$ in the interval $(t_1\sqrt{p'(\rho_0)}, \infty)$ the gas has its original density ρ_0, in the interval $((-v + \sqrt{p'(\bar{\rho})})t_1, t_1\sqrt{p'(\rho_0)})$ the actual rarefaction takes place and in the interval $(-vt_1, (-v + \sqrt{p'(\bar{\rho})})t_1)$ the tube is filled with the gas which assumed its final density $\bar{\rho}$.

The opposite case when the piston moves against the gas is more difficult due to the appearance of shock waves.

2.4.9 Initial-boundary value problem

We will not formulate any existence theorem for initial-boundary value problems since for a sufficiently general case such an existence theorem would be too cumbersome [see, for instance, Godunov (1979)]. In this subsection we will restrict ourselves only to some elementary remarks while in the next one we will find the solution of a particular problem.

An initial-boundary value problem in $Q \subset \mathbb{R} \times [t_0, \infty)$ is given by a system of differential equations, by initial conditions prescribed in Q for $t = t_0$ and by boundary conditions prescribed on an appropriate remaining part of the boundary of Q. Thus, for instance, if $Q = \mathbb{R}^+ \times [t_0, \infty)$ then the boundary conditions are prescribed for $x = 0$, $t \geq t_0$ and the initial conditions for $t = t_0$, $x \geq 0$; if $Q = [0, \ell] \times [t_0, \infty)$ we prescribe the boundary conditions for $x = 0$, $t \geq t_0$ and $x = \ell$, $t \geq t_0$ and the initial conditions for $x \in [0, \ell]$, $t = t_0$. However, the number and the form of the boundary conditions is not arbitrary. As we shall see in the next simple example, they are closely connected with the shape of the domain Q and of the characteristics.

Let us investigate the system

$$\begin{aligned} u_{1t} \quad + \lambda_1 u_{1x} \quad\quad\quad\quad &= 0, \\ u_{2t} \quad\quad\quad\quad - \lambda_2 u_{2x} &= 0, \\ u_{3t} \quad\quad\quad\quad\quad\quad\quad\quad &= 0, \end{aligned}$$

in $Q = [0, \ell] \times \mathbb{R}^+$ with λ_1, λ_2 positive numbers and with initial conditions $u_j(x, 0) = \varphi_j(x)$, $j = 1, 2, 3$, $x \in [0, \ell]$.

The characteristics are

$$\xi = \lambda_1 \tau + c_1, \qquad \xi = -\lambda_2 \tau + c_2, \qquad \text{and} \qquad \xi = c_3$$

on which the components u_1, u_2 and u_3, respectively, are constant. Let us show why the boundary conditions may be prescribed only on appropriate parts of the boundary. For instance, since the component u_1 is constant on the characteristics $\xi = \lambda_1 \tau + c_1$ with a positive slope it must not be prescribed on the right-hand part of the boundary, $x = \ell$, $t \in [0, \ell\lambda_1^{-1}]$ since it is already defined there by the initial condition:

$$u_1(x, t) = \varphi_1 (x - \lambda_1 t) \qquad \text{for} \quad x \in [0, \ell],\, 0 \leq t \leq x\lambda_1^{-1};$$

if we prescribe

$$u_1(0, t) = h_1(t), \qquad t \geq t_0 \tag{2.119}$$

(which is usually most natural) then u_1 must not be given for $x = \ell$ even for $t > \ell\lambda_1^{-1}$ since it is fully determined by (2.119):

$$u_1(x, t) = h_1 \left(t - x\lambda_1^{-1}\right) \qquad \text{for} \quad x \in [0, \ell],\, t \geq x\lambda_1^{-1}.$$

Wishing to have a uniquely determined solution we must require that

$$\varphi_1(0) = h_1(0).$$

If $\varphi_1 \in C[0,\ell]$, $h_1 \in C(\mathbf{R}^+)$ and the last condition is satisfied then the component u_1 is continuous in Q. Requiring the solution to be a classical one we must add further conditions, namely

$$\varphi_1'(0) = -\frac{1}{\lambda_1} h_1'(0) \text{ and } \varphi_1 \in C^1[0,\ell],\ h_1 \in C^1(\mathbf{R}^+).$$

Replacing the roles of the left-hand and right-hand boundary parts of ∂Q we may state quite analogous assertions for the component u_2. As concerns the component u_3, it is easy to see that besides the initial condition no boundary condition can be prescribed.

2.4.10 The propagation of a signal

Let us investigate the initial-boundary value problem for the linearized system of acoustics

$$\begin{aligned} u_t + \frac{c^2}{\rho_0}\rho_x &= 0, \\ \rho_t + \rho_0 u_x &= 0, \end{aligned} \qquad (x,t) \in Q = \mathbf{R}^+ \times \mathbf{R}^+, \tag{2.120}$$

$$u(x,0) = 0, \qquad \rho(x,0) = \rho_0, \qquad x \geq 0, \tag{2.121}$$

$$\rho(0,t) = h(t), \qquad t \geq 0, \tag{2.122}$$

which describes the propagation of a signal along the positive x-half-axis. Transforming the variables according to the relations

$$\begin{aligned} v &= \rho_0 u + c\rho, \\ \sigma &= \rho_0 u - c\rho, \end{aligned}$$

we bring the system (2.120) to the diagonal form

$$\begin{aligned} v_t + c v_x &= 0, \\ \sigma_t - c\sigma_x &= 0, \end{aligned}$$

with the general solution

$$v(x,t) = \varphi(x - ct), \quad \sigma(x,t) = \psi(x + ct),$$

where $\varphi \in C^1(\mathbf{R})$, $\psi \in C^1(\mathbf{R}^+)$. By the inverse transformation we get

$$\begin{aligned} u(x,t) &= \frac{1}{2\rho_0}\left(v(x,t) + \sigma(x,t)\right) = \frac{1}{2\rho_0}\left(\varphi(x-ct) + \psi(x+ct)\right), \\ \rho(x,t) &= \frac{1}{2c}\left(v(x,t) - \sigma(x,t)\right) = \frac{1}{2c}\left(\varphi(x-ct) - \psi(x+ct)\right). \end{aligned}$$

By (2.121)

$$\begin{aligned} 0 &= \varphi(x) + \psi(x), \\ 2c\rho_0 &= \varphi(x) - \psi(x), \end{aligned}$$

which yields $\varphi(x) = c\rho_0$; $\psi(x) = -c\rho_0$ for $x \geq 0$. Whereas ψ is already fully defined we have still to determine φ for $x < 0$. To this end let us make use of the boundary condition (2.122). We have

$$2c\,h(t) = \varphi(-ct) - \psi(ct) = \varphi(-ct) + c\rho_0.$$

Thus $\varphi(x) = 2c\,h\,(-x/c) - c\rho_0$ for $x < 0$.

Hence, finally,

$$\begin{aligned} u(x,t) &= 0 && \text{for} \quad x - ct \geq 0, \\ u(x,t) &= -c + \frac{c}{\rho_0} h\left(t - \frac{x}{c}\right) && \text{for} \quad x - ct < 0, \\ \rho(x,t) &= \rho_0 && \text{for} \quad x - ct \geq 0, \\ \rho(x,t) &= h\left(t - \frac{x}{c}\right) && \text{for} \quad x - ct < 0. \end{aligned}$$

We see that the signal propagates with velocity c. We will not determine the conditions of continuity of this solution since usually there is no reason to require that the signal starts smoothly from the null state.

2.5 A SYSTEM IN SEVERAL SPACE VARIABLES

The hyperbolicity of systems in several space variables is defined. The existence of travelling waves for the Maxwell equations is studied. The equation of characteristic surfaces is derived and the result is applied to equations of an electromagnetic field, to equations of elasticity and to equations of hydrodynamics.

2.5.1 Hyperbolic systems

Let us investigate a linear or a quasilinear system in n space variables which has the form

$$u_t + \mathcal{A}_1(x,t)u_{x_1} + \cdots + \mathcal{A}_n(x,t)u_{x_n} + \mathcal{B}(x,t)u = f(x,t) \tag{2.123}$$

or

$$u_t + \mathcal{A}_1(x,t,u)\,u_{x_1} + \cdots + \mathcal{A}_n(x,t,u)\,u_{x_n} = f(x,t,u) \tag{2.124}$$

respectively, where $x = (x_1, \ldots, x_n)$, $u = (u_1, \ldots, u_m)^T, f = (f_1, \ldots, f_m)^T$ and $\mathcal{A}_j, \mathcal{B}$ are $m \times m$ square matrices. Let us suppose that $\mathcal{A}_j, \mathcal{B}$ and f are as smooth as necessary.

We say that the system (2.123) is *hyperbolic (strictly hyperbolic)* in a domain $Q \subset \mathbf{R}^{n+1}$ if all roots $\lambda = \lambda(x,t,\eta)$ of the algebraic equation of m-th degree

$$\det\,(\lambda\mathcal{I} + \sum_{j=1}^{n} \mathcal{A}_j(x,t)\eta_j) = 0 \tag{2.125}$$

are real (real and distinct) for any $\eta_1, \ldots, \eta_n$ real such that $\sum_{j=1}^{n} \eta_j^2 \neq 0$ and for any $(x,t) \in Q$.

We say that the system (2.124) is *hyperbolic (strictly hyperbolic)* in a domain $Q \subset \mathbf{R}^{n+1}$ if all roots $\lambda = \lambda(x,t,u,\eta)$ of the algebraic equation of m-th degree

$$\det(\lambda\mathcal{I} + \sum_{j=1}^{n} A_j(x,t,u)\eta_j) = 0 \tag{2.126}$$

are real (real and distinct) for any $\eta_1, \ldots, \eta_n$, $u_1, \ldots, u_m$ real such that $\sum_{j=1}^{n} \eta_j^2 \neq 0$ and for any $(x,t) \in Q$.

(Note that the notion of hyperbolicity corresponds to the notion of weak hyperbolicity for systems in one space variable.)

2.5.2 Exercises

Let us note that the following three exercises are of special importance, since if we put $\lambda = \partial v/\partial t$, $\eta_j = \partial v/\partial x_j$, $j = 1,2,3$, they allow us to derive immediately the equations of characteristic surfaces for the three problems investigated in §§ 2.5.6, 2.5.7, 2.5.8.

2.5.2.1 Show that the system of the Maxwell equations

$$\frac{\partial E}{\partial t} - \frac{1}{\varepsilon\mu}\operatorname{curl} B + \frac{\gamma}{\varepsilon}E = -\frac{1}{\varepsilon}J,$$
$$\frac{\partial B}{\partial t} + \operatorname{curl} E = 0,$$

for unknowns $E_1, E_2, E_3, B_1, B_2, B_3$ is hyperbolic in $\mathbf{R}^4$.

Answer.

The equation $\lambda^2(\lambda^2 - \frac{1}{\varepsilon\mu}(\eta_1^2 + \eta_2^2 + \eta_3^2))^2 = 0$ has the roots $\lambda_1 = \lambda_2 = 0$, $\lambda_3 = \lambda_4 = (\frac{1}{\varepsilon\mu}(\eta_1^2 + \eta_2^2 + \eta_3^2))^{1/2}$, $\lambda_5 = \lambda_6 = -(\frac{1}{\varepsilon\mu}(\eta_1^2 + \eta_2^2 + \eta_3^2))^{1/2}$.

2.5.2.2 Show that the system of equations of elasticity

$$\frac{\partial v_j}{\partial t} - \frac{1}{\rho}\sum_{k=1}^{3}\frac{\partial \tau_{kj}}{\partial x_k} = F_j, \quad j = 1,2,3, \tag{2.127}$$

$$\frac{\partial \tau_{kj}}{\partial t} - \lambda_0\delta_{kj}\sum_{r=1}^{3}\frac{\partial v_r}{\partial x_r} - \mu_0(\frac{\partial v_k}{\partial x_j} + \frac{\partial v_j}{\partial x_k}) = 0, \quad j,k = 1,2,3,\ k \leq j, \tag{2.128}$$

$\tau_{kj} = \tau_{jk}$ with λ_0, μ_0 and ρ non-vanishing constants is hyperbolic in $\mathbf{R}^4$.

Answer.

The equation $\lambda^3(\mu_0(\eta_1^2 + \eta_2^2 + \eta_3^2) - \rho\lambda^2)^2((\lambda_0 + 2\mu_0)(\eta_1^2 + \eta_2^2 + \eta_3^2) - \rho\lambda^2) = 0$ has the roots $\lambda_1 = \lambda_2 = \lambda_3 = 0$, $\lambda_4 = \lambda_5 = \sqrt{\mu_0/\rho}(\eta_1^2 + \eta_2^2 + \eta_3^2)^{1/2}$, $\lambda_6 = \lambda_7 = -\sqrt{\mu_0/\rho}(\eta_1^2 + \eta_2^2 + \eta_3^2)^{1/2}$, $\lambda_{8,9} = \pm\sqrt{(\lambda_0 + 2\mu_0)/\rho}(\eta_1^2 + \eta_2^2 + \eta_3^2)^{1/2}$.

2.5.2.3 Show that the Euler equations

$$\frac{\partial u_j}{\partial t} + \sum_{k=1}^{3} u_k \frac{\partial u_j}{\partial x_k} + \frac{1}{\rho}\frac{\partial p}{\partial x_j} = F_j, \quad j = 1,2,3,$$

$$\frac{\partial \rho}{\partial t} + \sum_{k=1}^{3} \frac{\partial \rho}{\partial x_k} u_k + \rho \sum_{k=1}^{3} \frac{\partial u_k}{\partial x_k} = 0,$$

with $p = p(\rho)$, $p' \geq 0$ for the unknowns u_1, u_2, u_3, ρ is hyperbolic in $\mathbf{R}^4$.
Answer.
The equation $(\lambda + \sum_{k=1}^{3} u_k\eta_k)^2[(\lambda + \sum_{k=1}^{3} u_k\eta_k)^2 - p'(\rho)\sum_{k=1}^{3}\eta_k^2] = 0$ has the roots $\lambda_1 = \lambda_2 = -\sum_{k=1}^{3} u_k\eta_k$, $\lambda_{3,4} = -\sum_{k=1}^{3} u_k\eta_k \pm (p'(\rho)\sum_{k=1}^{3}\eta_k^2)^{1/2}$.

2.5.2.4 Show that the system

$$v_{1t} + \alpha_{11}v_{1x_1} + \alpha_{12}v_{2x_1} + \beta_{11}v_{1x_2} + \beta_{12}v_{2x_2} + \gamma_{11}v_1 + \gamma_{12}v_2 = f_1(x_1, x_2, t),$$
$$v_{2t} + \alpha_{21}v_{1x_1} + \alpha_{22}v_{2x_1} + \beta_{21}v_{1x_2} + \beta_{22}v_{2x_2} + \gamma_{21}v_1 + \gamma_{22}v_2 = f_2(x_1, x_2, t),$$

is strictly hyperbolic if and only if

$$\det \mathcal{D} > 0, \; d_{11} > 0,$$

where $\mathcal{D} = (d_{jk})_{j,k=1}^2$, $d_{11} = (\alpha_{11} - \alpha_{22})^2 + 4\alpha_{12}\alpha_{21}$, $d_{22} = (\beta_{11} - \beta_{22})^2 + 4\beta_{12}\beta_{21}$, $d_{12} = d_{21} = \alpha_{11}\beta_{11} - \alpha_{22}\beta_{11} - \alpha_{11}\beta_{22} + \alpha_{22}\beta_{22} + 2\alpha_{21}\beta_{12} + 2\alpha_{12}\beta_{21}$.
Answer.
The quadratic form in the discriminant of the equation

$$\lambda^2 + \lambda((\alpha_{11} + \alpha_{22})\eta_1 + (\beta_{11} + \beta_{22})\eta_2) + (\alpha_{11}\eta_1 + \beta_{11}\eta_2)(\alpha_{22}\eta_1 + \beta_{22}\eta_2)$$
$$- (\alpha_{12}\eta_1 + \beta_{12}\eta_2)(\alpha_{21}\eta_1 + \beta_{21}\eta_2) = 0$$

is positive definite under the above conditions.

2.5.3 Remark

In this section we do not establish any existence theorem for the initial-value problem for systems with several space variables. We have already mentioned above that there is a method by which such an existence theorem can be proved with the help of Riemann functions, but that it needs the theory of distributions which we wish to avoid, see Courant (1962), pp. 727–744. Another method often used to prove such an existence theorem is the method of Fourier transform, see Bers, John and Schechter (1979), pp. 62–93, Mizohata (1973), pp. 320–338, 348–361, but again the detailed elaboration of this method is beyond the scope of this book. In the sequel we deal with two other problems, namely with the existence of travelling waves and the existence of characteristic surfaces for some standard equations of mathematical physics [here we follow roughly the approach of Smirnov (1981)]. We touch the former problem only very briefly since for instance in the case of a *travelling plane wave* it consists in looking for a solution in the form $\varphi(\sum_{j=1}^{n} k_j x_j - \omega t)$. This leads to the resolution of a system of ordinary differential equations, which is beyond the framework of partial differential equations. As an illustration we will deal in §2.5.4 with the existence of travelling waves for the Maxwell equations. The latter problem is more important since seeking a characteristic surface in the form of a conoid (or of a cone in some special cases) with the vertex at the point $(a_1, \dots, a_n, t_0)$ we get the only surface along which a weak discontinuity starting at the time t_0 at the point $(a_1, \dots, a_n)$ can propagate, see Courant (1962), p. 416.

2.5.4 Travelling plane waves for the Maxwell equations

We will investigate plane waves in a dielectric, which are described by the Maxwell equations with $\gamma = 0$, $J = 0$, that is

$$\varepsilon \frac{\partial E}{\partial t} - \frac{1}{\mu} \operatorname{curl} B = 0,$$
$$\frac{\partial B}{\partial t} + \operatorname{curl} E = 0.$$

Looking for E and B in the form $E_j(x,t) = \varphi_j(\sum_{l=1}^{3} k_l x_l - \omega t)$, $B_j(x,t) = \psi_j(\sum_{l=1}^{3} k_l x_l - \omega t)$, with $\sum_{l=1}^{3} k_l^2 = 1$, $j = 1,2,3$ we get the system

$$\begin{aligned} \omega\varphi' + \frac{1}{\varepsilon\mu} k \times \psi' &= 0, \\ \omega\psi' - k \times \varphi' &= 0. \end{aligned} \tag{2.129}$$

Forming the scalar product of these equations with k and taking into account the relation $k \cdot (k \times a) = 0$, we get the following necessary conditions for the existence of plane waves in a dielectric

$$k \cdot \varphi' = 0, \quad k \cdot \psi' = 0. \tag{2.130}$$

These conditions are always satisfied in a dielectric because they are equivalent to the equations

$$\operatorname{div} E = 0, \quad \operatorname{div} B = 0.$$

By virtue of the equalities $\|k\|^2 = 1$ and $a \times (b \times c) = (a \cdot c)b - (a \cdot b)c$, (2.129) and (2.130) imply that

$$(\varepsilon\mu\omega^2 - 1)\varphi' = 0. \tag{2.131}$$

If $\varepsilon\mu\omega^2 - 1 \neq 0$ then the only plane waves are the constant ones, which is quite uninteresting. On the other hand, if

$$\varepsilon\mu\omega^2 - 1 = 0, \tag{2.132}$$

an arbitrary function φ satisfying the first condition from (2.130) together with ψ evaluated from the relation $\psi' = \frac{1}{\omega} k \times \varphi'$ gives a solution to the Maxwell equations in the form of a plane wave. Hence in any direction (k_1, k_2, k_3) there exist infinitely many plane waves restricted only by the relation (2.130). However, for physical reasons we usually require that they remain bounded in the whole space and at any moment. Since the processes studied in this way are usually of oscillatory character, the most natural way is to require that the vectors φ and ψ be periodic.

As the velocity V_0 of propagation of this wave is concerned, for each component E_j or B_j we have $\|V_0\| = \omega$, confer (2.145), and according to (2.132)

$$\|V_0\| = \frac{1}{\sqrt{\varepsilon\mu}},$$

which is the known formula for the velocity of light.

2.5.5 Characteristic surfaces

We shall investigate generalized solutions to (2.124) in a domain $Q \subset \mathbb{R}^{n+1}$ with $\mathcal{A}_k \in C^1(Q \times \mathbb{R}^m)$. Let Q be divided by a surface Σ into two parts Q_1, Q_2. Let Σ be given by the equation

$$v(x_1, \ldots, x_n, t) = 0 \text{ where } v \in C^1(\mathbb{R}^{n+1}).$$

We look for generalized solutions to (2.124) which are continuous in Q and have continuous first derivatives separately in Q_1, Q_2 , with possible jumps on Σ. Let us denote by $u_t^+, u_{x_1}^+, \ldots, u_{x_n}^+$ and $u_t^-, u_{x_1}^-, \ldots, u_{x_n}^-$ the limits of the corresponding derivatives from that part of Q where $v > 0$ or $v < 0$, respectively. Passing in (2.124) to the limits from both sides of Σ, subtracting these two limits and denoting $[u_t] = u_t^+ - u_t^-$, $[u_{x_1}] = u_{x_1}^+ - u_{x_1}^-, \ldots, [u_{x_n}] = u_{x_n}^+ - u_{x_n}^-$, we obtain

$$[u_t] + \sum_{k=1}^{n} \mathcal{A}_k(x, t, u)[u_{x_k}] = 0. \tag{2.133}$$

It turns out that the surface Σ cannot be an arbitrary one. Surfaces on which the first derivatives of a solution may have jumps are called *characteristic surfaces.* Let us derive their equation. On one side of the surface Σ let us choose a curve

$$x = x(s), \quad t = t(s).$$

Evidently, every component of u satisfies

$$\frac{du_j(x(s), t(s))}{ds} = \sum_{k=1}^{n} \frac{\partial u_j(x(s), t(s))}{\partial x_k} \frac{dx_k(s)}{ds} + \frac{\partial u_j(x(s), t(s))}{\partial t} \frac{dt(s)}{ds}$$

and thus

$$\begin{aligned} &u_j(x(s), t(s)) - u_j(x(s_0), t(s_0)) \\ &= \int_{s_0}^{s} \{\sum_{k=1}^{n} \frac{\partial u_j(x(r), t(r))}{\partial x_k} x_k'(r) + \frac{\partial u_j(x(r), t(r))}{\partial t} t'(r)\} dr. \end{aligned} \tag{2.134}$$

Let us choose a curve C on Σ and shift it by a small distance δ in both directions from Σ. Since the formula (2.134) is valid on both sides of Σ, passing to the limits for $\delta \to 0$ and subtracting the two limits we get, by virtue of the continuity of u_j and the arbitrariness of s_0 and s, the relation

$$\sum_{k=1}^{n} [\frac{\partial u_j}{\partial x_k}] \frac{dx_k}{ds} + [\frac{\partial u_j}{\partial t}] \frac{dt}{ds} = 0. \tag{2.135}$$

Since the curve C was chosen arbitrarily and the vector $(dx_1/ds, \ldots, dx_n/ds, dt/ds)$ lies in the tangent plane to Σ, the relation (2.135) says that the vector $([\partial u_j/\partial x_1], \ldots,$ $[\partial u_j/\partial x_n], [\partial u_j/\partial t])$ has the direction of the normal to Σ. Thus there exists a function $h_j = h_j(x,t)$, $(x,t) \in \Sigma$ such that

$$([\frac{\partial u_j}{\partial x_1}], \ldots, [\frac{\partial u_j}{\partial x_n}], [\frac{\partial u_j}{\partial t}]) = h_j(\frac{\partial v}{\partial x_1}, \ldots, \frac{\partial v}{\partial x_n}, \frac{\partial v}{\partial t}). \tag{2.136}$$

It can be shown that $h_j \in C(\Sigma)$. Substituting (2.136) for $j = 1, \ldots, m$ into (2.133) we get

$$(\frac{\partial v}{\partial t} \mathcal{I} + \sum_{k=1}^{n} \mathcal{A}_k(x,t,u) \frac{\partial v}{\partial x_k}) h = 0. \tag{2.137}$$

If jumps actually exist, at least one of the h_j's ($j = 1, \ldots, m$) must be different from zero. Hence necessarily

$$\det\,(\frac{\partial v}{\partial t} \mathcal{I} + \sum_{k=1}^{n} \mathcal{A}_k(x,t,u) \frac{\partial v}{\partial x_k}) = 0, \tag{2.138}$$

which represents the desired *equation of a characteristic surface*. Obviously, in the quasilinear case the equation of a characteristic surface depends on the solution u to (2.124).

2.5.6 Characteristic surfaces for the Maxwell equations

Le us derive the equation of characteristic surfaces for the Maxwell equations

$$\begin{aligned} &\frac{\partial E}{\partial t} - \frac{1}{\varepsilon\mu} \operatorname{curl} B + \frac{\gamma}{\varepsilon} E = -\frac{1}{\varepsilon} J, \\ &\frac{\partial B}{\partial t} + \operatorname{curl} E = 0. \end{aligned}$$

By (2.138) the equation of a characteristic surface is

$$\det \begin{pmatrix} p_0 & 0 & 0 & 0 & \frac{1}{\varepsilon\mu}p_3 & -\frac{1}{\varepsilon\mu}p_2 \\ 0 & p_0 & 0 & -\frac{1}{\varepsilon\mu}p_3 & 0 & \frac{1}{\varepsilon\mu}p_1 \\ 0 & 0 & p_0 & \frac{1}{\varepsilon\mu}p_2 & -\frac{1}{\varepsilon\mu}p_1 & 0 \\ 0 & -p_3 & p_2 & p_0 & 0 & 0 \\ p_3 & 0 & -p_1 & 0 & p_0 & 0 \\ -p_2 & p_1 & 0 & 0 & 0 & p_0 \end{pmatrix} = 0,$$

where $p_0 = \partial v/\partial t$, $p_j = \partial v/\partial x_j$, $j = 1,2,3$. The evaluation of the determinant yields the equation

$$[p_0^2 - \frac{1}{\varepsilon\mu}(p_1^2 + p_2^2 + p_3^2)]^2 p_0^2 = 0.$$

The characteristic surface defined by the equation $p_0 = 0$ is of little interest since it determines only a standing wave $v(x_1, x_2, x_3, t) = \psi(x_1, x_2, x_3)$. On the other hand, the equation

$$p_0^2 - \frac{1}{\varepsilon\mu}(p_1^2 + p_2^2 + p_3^2) = 0$$

is already of physical significance. Its general solution was investigated in §§ 2.3.5, 2.3.6.1 and its solution in the form of a cone was studied in § 2.3.9.3. Putting $c = 1/\sqrt{\varepsilon\mu}$ in Example 2.3.9.3 we find that the characteristic surface in the form of a cone with the vertex at the point (a_1, a_2, a_3, t_0) is given by

$$\frac{1}{\sqrt{\varepsilon\mu}}(t - t_0) \pm \sqrt{\sum_{j=1}^{3}(x_j - a_j)^2} = 0.$$

This means that at time $t = t_1$ the discontinuity may occur on the sphere $\sum_{j=1}^{3}(x_j - a_j)^2 = 1/\sqrt{\varepsilon\mu}\,(t_1 - t_0)^2$. Since $\sqrt{\sum_{j=1}^{3}(x_j - a_j)^2}$ is the length of the path of the signal and $t - t_0$ is the time needed to cover this distance, we see that $\|V_0\| = 1/\sqrt{\varepsilon\mu}$ is again the velocity of propagation of the wave.

2.5.7 Characteristic surfaces for the equations of elasticity

Let us derive the equation of characteristic surfaces for the equations of elasticity (2.127), (2.128). Denoting $u_1 = v_1$, $u_2 = v_2$, $u_3 = v_3$, $u_4 = \tau_{11}$, $u_5 = \tau_{12}$, $u_6 = \tau_{13}$, $u_7 = \tau_{22}$, $u_8 = \tau_{23}$, $u_9 = \tau_{33}$, in (2.136), the system (2.137) reads

$$\mathcal{A}\, h = 0, \tag{2.139}$$

where

$$\mathcal{A} = \begin{pmatrix} p_0 & 0 & 0 & -\frac{p_1}{\rho} & -\frac{p_2}{\rho} & -\frac{p_3}{\rho} & 0 & 0 & 0 \\ 0 & p_0 & 0 & 0 & -\frac{p_1}{\rho} & 0 & -\frac{p_2}{\rho} & -\frac{p_3}{\rho} & 0 \\ 0 & 0 & p_0 & 0 & 0 & -\frac{p_1}{\rho} & 0 & -\frac{p_2}{\rho} & -\frac{p_3}{\rho} \\ -kp_1 & -\lambda_0 p_2 & -\lambda_0 p_3 & p_0 & 0 & 0 & 0 & 0 & 0 \\ -\mu_0 p_2 & -\mu_0 p_1 & 0 & 0 & p_0 & 0 & 0 & 0 & 0 \\ -\mu_0 p_3 & 0 & -\mu_0 p_1 & 0 & 0 & p_0 & 0 & 0 & 0 \\ -\lambda_0 p_1 & -kp_2 & -\lambda_0 p_3 & 0 & 0 & 0 & p_0 & 0 & 0 \\ 0 & -\mu_0 p_3 & -\mu_0 p_2 & 0 & 0 & 0 & 0 & p_0 & 0 \\ -\lambda_0 p_1 & -\lambda_0 p_2 & -kp_3 & 0 & 0 & 0 & 0 & 0 & p_0 \end{pmatrix},$$

$k = \lambda_0 + 2\mu_0, h = (h_1, \dots, h_9)^T$. This system has a non-trivial solution if and only if its determinant vanishes (compare (2.138)), that is, if

$$p_0^3(\mu_0(p_1^2 + p_2^2 + p_3^2) - \rho p_0^2)^2((\lambda_0 + 2\mu_0)(p_1^2 + p_2^2 + p_3^2) - \rho p_0^2) = 0.$$

The first three roots define standing waves of the form $v(x_1, x_2, x_3, t) = \psi(x_1, x_2, x_3)$. By § 2.3.9.3 the second and the third expressions determine two characteristic cones $\sqrt{\frac{\mu_0}{\rho}}(t - t_0) = \pm\sqrt{\sum_{j=1}^3 (x_j - a_j)^2}$ and $\sqrt{\frac{\lambda_0 + 2\mu_0}{\rho}}(t - t_0) = \pm\sqrt{\sum_{j=1}^3 (x_j - a_j)^2}$, along which the waves propagate with velocities $\sqrt{\mu_0/\rho}$ and $\sqrt{(\lambda_0 + 2\mu_0)/\rho}$, respectively. (Here we have neglected the deviation of molecules since they are inessential.)

To determine the character of the wave in more detail let us return to the system (2.139). Calculating $h_4, \dots, h_9$ from the last six equations and inserting for them into the first three equations we obtain the relation

$$\left[\mu_0(p_1^2 + p_2^2 + p_3^2) - \rho p_0^2\right](h_1, h_2, h_3) + (\lambda_0 + \mu_0)\Big(\sum_{j=1}^3 p_j h_j\Big)(p_1, p_2, p_3) = 0.$$

Denoting by $\nu = (p_1, p_2, p_3)/\sqrt{p_1^2 + p_2^2 + p_3^2} = \dfrac{\operatorname{grad}_x v}{\|\operatorname{grad}_x v\|}$ the unit vector in the direction of the normal to the characteristic surface and by h'_ν the projection of $h' = (h_1, h_2, h_3)$ onto ν we can rewrite the last equation as

$$[\mu_0(p_1^2 + p_2^2 + p_3^2) - \rho p_0^2]h' + (\lambda_0 + \mu_0)(p_1^2 + p_2^2 + p_3^2)h'_\nu = 0.$$

The expression $[\mu_0(p_1^2 + p_2^2 + p_3^2) - \rho p_0^2]$ being zero on the first cone we have $h'_\nu = 0$ which means that the vector (h_1, h_2, h_3) lies in the tangent plane to the characteristic surface and thus the wave is transversal. On the other hand, on the second cone we have $h' = h'_\nu$, that is, h has the direction of the normal and the wave is longitudinal.

2.5.8 Characteristic surfaces for the Euler equations

Let us derive the equation of characteristic surfaces for the Euler equations for a barotropic fluid (that is with $p = p(\rho)$, $p' \geq 0$)

$$\frac{\partial u_j}{\partial t} + \sum_{k=1}^3 u_k \frac{\partial u_j}{\partial x_k} + \frac{1}{\rho}\frac{\partial p}{\partial x_j} = F_j, \quad j = 1, 2, 3, \tag{2.140}$$

$$\frac{\partial \rho}{\partial t} + \sum_{k=1}^3 \frac{\partial \rho}{\partial x_k} u_k + \rho \sum_{k=1}^3 \frac{\partial u_k}{\partial x_k} = 0. \tag{2.141}$$

Denoting $u_4 = \rho$ in (2.136), the equations (2.137) assume the form

$$(p_0 + \sum_{k=1}^3 p_k u_k)h_j + \frac{1}{\rho}p'(\rho)p_j h_4 = 0, \quad j = 1, 2, 3, \tag{2.142}$$

$$\rho \sum_{k=1}^3 p_k h_k + (p_0 + \sum_{k=1}^3 p_k u_k)h_4 = 0. \tag{2.143}$$

This system has a non-trivial solution $h = (h_1, \ldots, h_4)$ if and only if

$$(p_0 + \sum_{k=1}^{3} p_k u_k)^2 \{(p_0 + \sum_{k=1}^{3} p_k u_k)^2 - p'(\rho) \sum_{k=1}^{3} p_k^2\} = 0. \tag{2.144}$$

Let us start with the determination of the velocity of the wave which is given as a difference of the propagation velocity of the characteristic surface V_0 and the normal velocity u_ν of the fluid particles, $V = V_0 - u_\nu$. The magnitude $\|V_0\|$ of the velocity V_0 can be found as follows. At the times t and $t + \Delta t$ let us fix two characteristic surfaces. Let us choose an arbitrary point P on the first surface and let the normal at P to this surface intersect the second surface at the point Q. The magnitude of the velocity V_0 is defined as $\|V_0\| = \lim\limits_{\Delta t \to 0} \dfrac{\|PQ\|}{\Delta t}$. Evidently $Q = P + s\,\mathrm{grad}_x v(P, t)$ and $v(Q, t + \Delta t) = 0$, that is

$$v(P + s\,\mathrm{grad}_x v, t + \Delta t) = 0.$$

Differentiating this relation we get

$$s\,\|\mathrm{grad}_x v(P, t)\|^2 + \frac{\partial v(P, t)}{\partial t} \Delta t = 0.$$

Since $\|PQ\| = |s|\,\|\mathrm{grad}_x v(P, t)\|$ we have

$$\|V_0\| = |\frac{\partial v}{\partial t}| / \|\mathrm{grad}_x v\|. \tag{2.145}$$

Further $\|u_\nu\| = |u \cdot \mathrm{grad}_x v| / \|\mathrm{grad}_x v\|$. Taking into account all possible mutual orientations of V_0, u_ν and ν we get

$$\|V\| = |\frac{\partial v}{\partial t} + u \cdot \mathrm{grad}_x v| / \|\mathrm{grad}_x v\|. \tag{2.146}$$

Returning to the equation (2.144) we see that it is equivalent to the relations

$$\|V\|^2 = 0 \quad \text{or} \quad \|V\|^2 = p'(\rho).$$

The first of them determines a standing wave, while the other one defines the velocity of propagation of sound

$$\|V\| = \sqrt{p'(\rho)}. \tag{2.147}$$

Now let us show that the waves are longitudinal ones. Indeed, from (2.142) we easily find that

$$\begin{aligned} h_j &= -\frac{1}{\rho} p'(\rho) p_j h_4 \frac{1}{p_0 + u \cdot \mathrm{grad}_x v} \\ &= -\frac{1}{\rho} \|V\|^2 \frac{p_j}{\|\mathrm{grad}_x v\|} h_4 \frac{\|\mathrm{grad}_x v\|}{|p_0 + u \cdot \mathrm{grad}_x v|} \mathrm{sgn}(p_0 + u \cdot \mathrm{grad}_x v) \\ &= -\mathrm{sgn}(p_0 + u \cdot \mathrm{grad}_x v) \frac{\|V\| h_4}{\rho} \nu_j, \quad j = 1, 2, 3, \end{aligned}$$

which implies that h is parallel to the normal. Obviously, for an arbitrary solution u, ρ of the Euler equation we are not able to find the function v explicitly. It can be found in some special cases, for instance for u and ρ constant a procedure similar to that in Example 2.3.9.1 may be applied.

CHAPTER 3

Second order hyperbolic equations

3.1 THE WAVE EQUATION IN ONE SPACE VARIABLE

This chapter deals with equations and systems of equations of the second order often arising from the study of processes for which the propagation of waves in the underlying medium is characteristic (typical representatives: the wave equation, the telegraph equation).

Second order equations can be rewritten in an equivalent form as systems of first order equations (see for example § 2.4.5). In doing so, however, we get only very special types of systems and in turn this suggests that special methods might be developed for the solution of the second order equations thereby obtaining more detailed results. Although this chapter is naturally linked with the preceding one it is written in such a way to be largely independent.

In Section 3.1 we examine the *wave equation in one space variable* or alternately the *one-dimensional wave equation* [on a region $Q \subset \mathbb{R}^2$], that is,

$$u_{tt}(x,t) - c^2 u_{xx}(x,t) = g(x,t)$$

where c is a positive constant and the point (x,t) runs over the region Q. In physical terms, t stands for the time variable, x for the (single) space variable, the unknown function u measures some sort of "displacement", the function g represents external forces and is supposed to be known. In the case $g \equiv 0$ we speak of the homogeneous (one-dimensional) wave equation.

The main goal of this section is to derive explicit formulae for classical solutions of some initial-boundary value problems. In deriving these formulae we shall always tacitly assume that the classical solution possesses continuous partial derivatives up to order two up to the boundary, that is,

$$u \in C^2(\bar{Q}).$$

It simplifies the argument and, as to the problems discussed below, causes no loss of generality.

As to the literature, the topic is covered by a wide variety of textbooks, monographs, etc., for example, Courant (1962), Baldock and Bridgeman (1987).

3.1.1 Free longitudinal vibration of a rod

We begin with an example of a process which can be described by the wave equation. Consider a rod of homogeneous elastic material of constant (for instance, circular) cross-section which, for a certain time, has been subjected to an external force acting solely along the axis of the rod and being constant at each cross-section. (The gravity is neglected in our considerations.) After this axial force is removed the points of the rod start moving along the axial direction, in other words, the rod undergoes a free longitudinal vibration.

We are interested in the mathematical description of this movement. To this end, let the coordinate system be such that the axis of the rod in its rest position is coincident with the x-axis. Under our assumptions each point of the axis moves solely along the x-axis and hence the vibration of the rod may be described by a single function u of two variables, the space variable x and the time t, while $u(x,t)$ means the displacement at the time t of the point having been placed at x in the rest position. [If, at rest, a point is situated at x, its position at the time t is $x + u(x,t)$.]

Consider now a typical element Σ of the rod extending from x to $x+h$ (as drawn schematically in Fig. 4; for simplicity we assume that the rod has been at rest at $t = 0$).

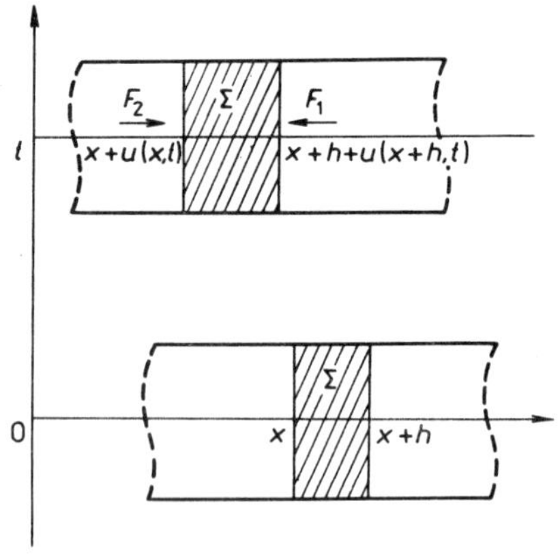

Fig. 4

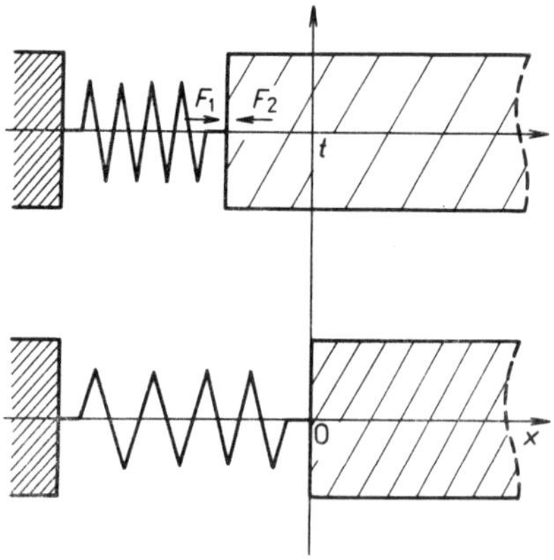

Fig. 5

By Newton's second law of motion the time rate of the momentum of the element Σ (at time t) is equal to the resultant of all external forces acting on Σ (at the same time), momentum being defined as the product of the mass and velocity. The momentum of the element Σ at t is given by

$$\rho_0 A \int_x^{x+h} u_t(\xi, t)\, d\xi,$$

where ρ_0 is the mass density of the material of the rod at rest and A is the cross-sectional area. If at the time t no more external forces are acting on the rod then the only forces acting on the element Σ are due to the remaining portions of the rod.

By Hooke's law (under the assumptions of the perfect elasticity of the material) these forces are given by

$$F_1 = EA\,u_x(x+h,t)$$

for the force acting "from the right" and

$$F_2 = -EA\,u_x(x,t)$$

for the force acting "from the left", where E is the modulus of elasticity of the material. The Newton law yields the equality

$$\rho_0 A \int_x^{x+h} u_{tt}(\xi,t)\,d\xi = EA\left[u_x(x+h,t) - u_x(x,t)\right],$$

from which it is deduced upon dividing by $\rho_0 Ah$ and passing to the limit $h \to 0+$ an equation to be satisfied by the function u,

$$u_{tt}(x,t) - \frac{E}{\rho_0}\,u_{xx}(x,t) = 0,$$

that is the homogeneous wave equation with the constant c given by

$$c = \left(\frac{E}{\rho_0}\right)^{\frac{1}{2}}.$$

The reader will have noticed that in the above procedure some relations (e.g. Hooke's law) have been used the validity of which is limited by certain restrictions on the magnitude of the function u or of its derivatives. Because of this fact the linear wave equation indeed fits in with only a relatively small vibration of the rod. □

Next, let us ask about what boundary conditions for the wave equation can be supplied. For instance, assume that one end of the rod (for definiteness we consider the left-hand end) is located at $x = 0$ at the time $t = 0$ (position of rest) and it is attached to a spring which is also at rest at $t = 0$ (see Fig. 5). At the time t the force F_1 exerted by the spring upon the end of the rod is proportional to the displacement: $F_1 = -k\,u(0,t)$ [k is the so-called stiffness constant of the spring], and the force F_2 exerted by the rod upon the spring is given (on account of Hooke's law) by the relation $F_2 = EA\,u_x(0,t)$. The two forces F_1 and F_2 constitute the action and reaction and according to Newton's third law $-F_1 = F_2$ which amounts to

$$-u_x(0,t) + \frac{k}{EA}\,u(0,t) = 0.$$

We have arrived at the Newton boundary condition (see § 1.1.4) which corresponds to the so-called *springy* end of the rod. The Dirichlet and Neumann boundary conditions have a good physical meaning as well. Namely, the Dirichlet boundary condition is obtained from the latter condition by letting $k \to \infty$ (the spring becomes rigid) and the Neumann condition is obtained by setting $k = 0$ (the end is free from stress, the spring has no resistance). In the former case we speak of a *fixed (clamped)* end, in the latter one of a *free* end of the rod. □

In concluding this subsection let us add that the wave equation (with the above mentioned types of boundary conditions) occurs often by the description of physical systems of mechanical or electrodynamical character with one degree of freedom (for instance, torsional vibrations of rods, transverse vibrations of strings, and so forth).

3.1.2 General solution of the homogeneous wave equation

Let us now turn our attention to the mathematical investigation of the wave equation. To begin with, we show a simple yet important fact that, for certain regions, a formula for the general solution of the homogeneous wave equation can be furnished.

Let us assume first that u is the classical solution of the equation

$$u_{tt} - c^2 u_{xx} = 0 \tag{3.1}$$

in the whole plane $\mathbf{R}^2$ of variables (x,t). Make the transformation of independent variables

$$x \to \xi = x + ct, \qquad t \to \tau = x - ct$$

and look at which equation is satisfied by the function u of the new variables or more precisely by the function v defined by $v(\xi,\tau) = u(x,t)$. The chain rule gives

$$\begin{aligned} u_x &= v_\xi \xi_x + v_\tau \tau_\xi = v_\xi + v_\tau \\ u_{xx} &= v_{\xi\xi} + 2v_{\xi\tau} + v_{\tau\tau} \\ u_t &= cv_\xi - cv_\tau \\ u_{tt} &= c^2 v_{\xi\xi} - 2c^2 v_{\xi\tau} + c^2 v_{\tau\tau} \\ u_{tt} - c^2 u_{xx} &= -4c^2 v_{\xi\tau} \end{aligned}$$

and thus the function v is a solution of the equation

$$v_{\xi\tau}(\xi,\tau) = 0. \tag{3.2}$$

From this equation it follows that the function v_τ is independent of ξ and hence there exists a function s of single variable τ such that

$$v_\tau(\xi,\tau) = s(\tau), \qquad (\xi,\tau) \in \mathbf{R}^2.$$

This implies an expression for the function v given by

$$v(\xi,\tau) = v(\xi,0) + \int_0^\tau v_\tau(\xi,\eta)\,d\eta = v(\xi,0) + \int_0^\tau s(\eta)\,d\eta,$$

that is to say,

$$v(\xi,\tau) = \sigma(\xi) + \varrho(\tau), \qquad (\xi,\tau) \in \mathbf{R}^2, \tag{3.3}$$

where we have denoted

$$\sigma(\xi) = v(\xi,0), \qquad \varrho(\tau) = \int_0^\tau s(\eta)\,d\eta.$$

It may be easily shown that both these functions σ and ϱ are twice continuously differentiable. Relation (3.3) says nothing other than that the function u has the form

$$u(x,t) = \sigma(x+ct) + \varrho(x-ct), \qquad (x,t) \in \mathbf{R}^2. \tag{3.4}$$

In like manner we can show that every solution u of the wave equation (3.1) on a region $Q \subset \mathbf{R}^2$ takes on the form (3.4) if (and only if) the intersection of the region Q with any straight line $x + ct = k$ ($k \in \mathbf{R}$) is a connected set and the same is true for the straight lines $x - ct = k$ ($k \in \mathbf{R}$). In other words, if Q enjoys the following property:

$$\begin{aligned}&\text{For any } k \in \mathbf{R} \text{ the set of all } x \in \mathbf{R} \text{ for which}\\&\text{there exists } t \in \mathbf{R} \text{ such that } (x,t) \in Q \text{ and}\\&x + ct = k \quad (x - ct = k) \text{ is an interval.}\end{aligned} \tag{3.5}$$

On the other hand, by a direct computation we can verify that any function of the form (3.4) is a solution of the wave equation. Our results can be gathered as follows:

Let the region $Q \subset \mathbf{R}^2$ possess the property (3.5). *Then a function u in Q is the classical solution of the wave equation* (3.1) *if and only if it assumes the form* (3.4) *where σ and ϱ are twice continuously differentiable functions.*

In what follows we encounter mostly regions of the form $Q = (a, b) \times (T_1, T_2)$ (where $-\infty \le a < b \le \infty, -\infty \le T_1 < T_2 \le \infty$) which do obey (3.5). Note, too, that for such regions relation (3.4) defines the solution of Eq. (3.1) as long as the function σ is defined on the interval $(a + cT_1,\ b + cT_2)$ and the function ϱ on the interval $(a - cT_2,\ b - cT_1)$. □

Relation (3.4) for the solution of the wave equation admits of a simple physical interpretation. Namely, look at the part of the solution which corresponds to the function $\sigma(x + ct)$. At any fixed time the shape of this function is the same and as time evolves it "moves to the left" with velocity c. Hence this part of the solution describes a *progressive wave* which spreads out in the negative direction of the x-axis. Similarly we can speak of the part of the solution represented by the function $\varrho(x - ct)$ as a progressive wave which propagates in the positive x-direction. Going back to the interpretation of relation (3.4) we can say that any process described by the homogeneous wave equation is the superposition of two progressive waves travelling without change of shape in opposite directions each with velocity c. Accordingly the constant c in the equation has a definite physical meaning, for instance, the *velocity of propagation* of disturbance, or signal, in a medium if processes of mechanical or electrodynamical character are under consideration.

On the basis of relation (3.4) we can carry through an analysis of a number of effects (especially impact effects) arising in processes described by the homogeneous wave equation. As an illustration of how this result might be used we explain, through a simple example, the effect of the reflection of waves.

3.1.3 Example: the reflection of waves from a fixed end

For definiteness we consider free longitudinal vibrations of a semi-infinite rod (placed along the positive part of the x-axis) with end held fixed (at $x = 0$). Further, let us assume that for small values of time $t > 0$ the free vibration of the rod has the form of a localized wave travelling toward the clamped end of the rod. We are interested in what happens when the initial disturbance arrives at the clamped end and what we can expect the free vibration of the rod to look like.

Formulated mathematically the problem is to find a solution u of the homogeneous wave equation on the region $(0, \infty) \times (0, \infty)$ satisfying the (boundary) condition

$u(0,t) = 0$ for $t \geq 0$ and having the form $u(x,t) = \beta(x+ct)$ for small $t > 0$ where β is a function on the interval $(0, \infty)$ vanishing everywhere except for a small neighbourhood of a point $x_0 > 0$ (localization of the wave).

By the preceding subsection the solution must look like $u(x,t) = \sigma(x+ct)+\varrho(x-ct)$ where σ and ϱ are twice continuously differentiable functions on the interval $(0, \infty)$ and $(-\infty, \infty)$, respectively. In order to meet the condition $u(0,t) = 0$ it is necessary that $\sigma(ct) + \varrho(-ct) = u(0,t) = 0$, that is to say

$$\varrho(\xi) = -\sigma(-\xi), \qquad \xi \leq 0$$

should be satisfied. Since $u(x,t) = \beta(x+ct)$ for small values of t (and for all $x \geq 0$) it holds $\sigma'(x) + \varrho'(x) = u_x(x,0) = \beta'(x)$ and $c\sigma'(x) - c\varrho'(x) = u_t(x,0) = c\beta'(x)$, whence

$$\varrho'(\xi) = 0, \qquad \sigma'(\xi) = \beta'(\xi), \qquad \xi > 0.$$

Thus, there exist constants k_1 and k_2 such that $\varrho(\xi) = k_1$ and $\sigma(\xi) = \beta(\xi) + k_2$ for $\xi \geq 0$. Besides, from the relation $\varrho(0) = -\beta(0)$ it follows $k_1 = -k_2$. We can put $k_1 = k_2 = 0$. (This choice is permissible due to the observation that the functions σ and ϱ in relation (3.4) are determined uniquely up to a constant in the sense that we are at liberty to add any constant to either function and to subtract it from the other.)

Because of these facts the given conditions determine the solution uniquely for all $t > 0$ and one has

$$u(x,t) = \beta(x+ct) - \tilde{\beta}(-x+ct), \tag{3.6}$$

where $\tilde{\beta}$ has arisen from β by extending the latter function as zero for negative values of argument, i.e.

$$\tilde{\beta}(\xi) = \begin{cases} \beta(\xi) & \text{for } \xi \geq 0 \\ 0 & \text{for } \xi < 0. \end{cases} \tag{3.7}$$

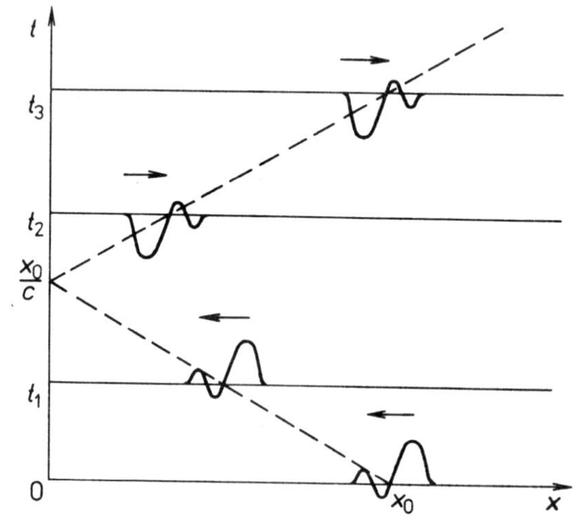

Fig. 6

Let us look at conclusions which can be drawn at this stage (see a sketch in Fig. 6). For times satisfying the crude upper bound $c^{-1}x_0$ (more precisely, until the front of the disturbance reaches the clamped end of the rod) the second term in (3.6) equals zero and the process within the rod is reduced to the propagation of the original disturbance down the rod towards its rigidly terminated end. On the other hand, for times greater than "$c^{-1}x_0$" (how much it depends on the "width" of the original disturbance) the first term in (3.6) vanishes and the solution has the form of a wave travelling to the right which differs from the original (incoming) one only by the "reversed" shape and "opposite" amplitude. It is natural to call this (outcoming) wave the *reflected wave*. In a neighbourhood of the fixed end of the rod and for times near to the value $c^{-1}x_0$ the form of the solution becomes more complicated – this is due to the effect of superposition of the original and reflected waves. □

In what follows we focus our attention on the study of initial-boundary value problems for the wave equation (in general inhomogeneous). One such concrete case has already been solved in this subsection. The approach we have chosen appears to be applicable to the investigations in general cases. Namely, as already mentioned in § 1.1.5 the general solution of any linear inhomogeneous equation is the sum of the general solution of the corresponding homogeneous equation and one particular solution of the inhomogeneous one. Therefore, every solution of the wave equation (on simple regions) may be put into the form

$$u(x,t) = \sigma(x+ct) + \varrho(x-ct) + w(x,t),$$

where σ, ϱ are arbitrary (twice continuously differentiable) functions and w is one special solution. The initial and boundary conditions to be met might now be ensured by a suitable choice of functions σ and ϱ. However, in what follows we apply a different procedure based on certain relations ensuing from Green's theorem. One of the reasons justifying this approach is that we can proceed quite similarly when dealing with partial differential equations for which in contrast to the homogeneous wave equation no amenable form of the general solution is available.

3.1.4 The initial-value problem

We shall now examine the initial-value problem for the wave equation, that is, we shall be concerned in obtaining results on existence and uniqueness of the classical solution on the half-plane

$$Q = (-\infty,\infty) \times (0,\infty)$$

satisfying the initial conditions

$$u(x,0) = \varphi(x), \quad u_t(0,x) = \psi(x), \quad -\infty < x < \infty \tag{3.8}$$

where φ and ψ are given functions. Our aim is to find the value of the solution at an arbitrarily chosen (but fixed) point (x,t) of the region Q. Accordingly we shall denote a generic point of Q by (ξ,τ) and write the wave equation in the form

$$u_{\tau\tau}(\xi,\tau) - c^2 u_{\xi\xi}(\xi,\tau) = g(\xi,\tau). \tag{3.9}$$

Let us repeat that the initial-value problem for the wave equation might be taken up with the aid of results of Chapter 2 upon having recourse to a system of first order equations as shown in § 2.4.5. A decisive role was played by straight lines $\xi \pm c\tau = \text{const.}$, the so-called characteristics of the corresponding system. As may be surmised these lines keep their importance even if a different method of investigation is exposed as it is in the present case. We retain the terminology calling them the *characteristics* of the wave equation. (The counterparts of characteristics for general hyperbolic equations of the second order are the so-called characteristic surfaces we encounter in § 3.5.3.)

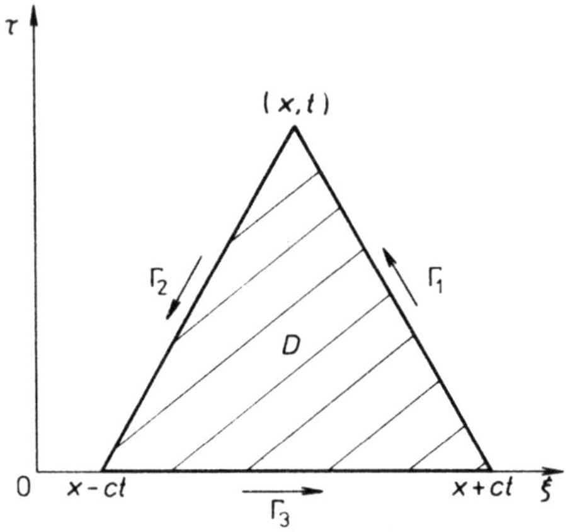

Fig. 7

We choose now a fixed point (x,t) in the region Q. For any twice continuously differentiable function u by virtue of the Green's theorem (see § 1.2.3) we have the equality

$$\int_D (u_{\tau\tau} - c^2 u_{\xi\xi})\, d\xi d\tau = -\int_\Gamma u_\tau d\xi + c^2 u_\xi d\tau, \tag{3.10}$$

where D is a region (triangle) bounded by the initial line (ξ-axis) and by the characteristics $\xi + c\tau = x + ct$ and $\xi - c\tau = x - ct$ (see Fig. 7). The line integral in (3.10) over the boundary Γ of the triangle D can be broken up into the sum of integrals over the sides Γ_1, Γ_2 and Γ_3. The curve Γ_1 may be parametrically described by functions $\xi = x + ct - cs$, $\tau = s$, $s \in [0,t]$ and this implies the relation

$$-\int_{\Gamma_1} u_\tau d\xi + c^2 u_\xi d\tau = \int_0^t [c\, u_\tau(x+ct-cs, s) - c^2 u_\xi(x+ct-cs,s)]\, ds. \tag{3.11}$$

If we put

$$f(s) = c\, u(x+ct-cs, s), \qquad s \in [0,t],$$

then by the chain rule

$$f'(s) = c\, u_\tau(x+ct-cs,s) - c^2 u_\xi(x+ct-cs,s)$$

which shows that the integral on the right-hand side of (3.11) is equal to

$$\int_0^t f'(s)\,ds = f(t) - f(0) = c[u(x,t) - u(x+ct,0)].$$

Hence the equality (3.11) takes the form

$$-\int_{\Gamma_1} u_\tau d\xi + c^2 u_\xi d\tau = c[u(x,t) - u(x+ct,0)].$$

Quite analogously (after the parametrization of the portion Γ_2 of the boundary Γ) we obtain

$$-\int_{\Gamma_2} u_\tau d\xi + c^2 u_\xi d\tau = c[u(x,t) - u(x-ct,0)].$$

To evaluate the line integral over Γ_3 we use the parametrization $\xi = s$, $\tau = 0$, $s \in [x-ct, x+ct]$ which leads to

$$-\int_{\Gamma_3} u_\tau d\xi + c^2 u_\xi d\tau = -\int_{x-ct}^{x+ct} u_\tau(s,0)\,ds.$$

Upon inserting the obtained relations into (3.10) and making simple arrangements we get for any $u \in C^2(\bar{Q})$ and for any $(x,t) \in \bar{Q}$

$$\begin{aligned} u(x,t) &= \frac{1}{2}[u(x+ct,0) + u(x-ct,0)] \\ &+ \frac{1}{2c}\int_{x-ct}^{x+ct} u_\tau(\xi,0)\,d\xi + \frac{1}{2c}\int_0^t \int_{x-c(t-\tau)}^{x+c(t-\tau)} [u_{\tau\tau}(\xi,\tau) - c^2 u_{\xi\xi}(\xi,\tau)]\,d\xi d\tau, \end{aligned} \tag{3.12}$$

where the latter integral corresponds to the double integral on the left-hand side of (3.10) written down, by well-known rules, in the form of an iterated integral.

Now, let us assume that u is the classical solution of the wave equation (3.9) satisfying the initial conditions (3.8). Then by (3.12) u can be written in terms of known quantities as

$$\begin{aligned} u(x,t) = \frac{1}{2}[\varphi(x+ct) + \varphi(x-ct)] &+ \frac{1}{2c}\int_{x-ct}^{x+ct} \psi(\xi)\,d\xi \\ &+ \frac{1}{2c}\int_0^t \int_{x-c(t-\tau)}^{x-c(t+\tau)} g(\xi,\tau)\,d\xi d\tau. \end{aligned} \tag{3.13}$$

An immediate consequence is that the classical solution of the initial-value problem for the wave equation is determined in a unique manner.

It remains to clarify the conditions on functions φ, ψ and g under which the function u defined by formula (3.13) is the solution of the wave equation. The function u can be written in the form

$$u(x,t) = \sigma(x+ct) + \varrho(x-ct) + w(x,t), \tag{3.14}$$

where

$$\sigma(s) = \frac{1}{2}\varphi(s) + \frac{1}{2c}\int_0^s \psi(\xi)\,d\xi, \qquad s \in \mathbf{R},$$

$$\varrho(s) = \frac{1}{2}\varphi(s) - \frac{1}{2c}\int_0^s \psi(\xi)\,d\xi, \qquad s \in \mathbf{R},$$

$$w(x,t) = \frac{1}{2c}\int_0^t \int_{x-c(t-\tau)}^{x+c(t-\tau)} g(\xi,\tau)\,d\xi d\tau, \qquad (x,t) \in Q.$$

If $\varphi \in C^2(\mathbf{R})$ and $\psi \in C^1(\mathbf{R})$ then σ and ϱ are twice continuously differentiable functions and by § 3.1.2 the first two terms on the right-hand side of (3.14) solve the homogeneous wave equation. Assume further that the function g is continuous on $\bar{Q}$ together with its first partial derivative with respect to the first variable. Then the function w is twice continuously differentiable and by § 1.2.5 its derivatives can be evaluated as follows:

$$w_x(x,t) = \frac{1}{2c}\int_0^t [g(x + c(t-\tau),\tau) - g(x - c(t-\tau),\tau)]d\tau,$$

$$w_{xx}(x,t) = \frac{1}{2c}\int_0^t [g_x(x + c(t-\tau),\tau) - g_x(x - c(t-\tau),\tau)]d\tau,$$

$$w_t(x,t) = \frac{1}{2}\int_0^t [g(x + c(t-\tau),\tau) + g(x - c(t-\tau),\tau)]d\tau,$$

$$w_{tt}(x,t) = \frac{c}{2}\int_0^t [g_x(x + c(t-\tau),\tau) - g_x(x - c(t-\tau),\tau)]d\tau + g(x,t).$$

It is readily seen that the function w is the classical solution of the inhomogeneous wave equation (3.9) [with vanishing initial data]. Likewise, the same is true if g is continuous together with its first derivative with respect to the second variable. (To prove the latter assertion it appears to be more convenient to express the function w in the form

$$w(x,t) = \frac{1}{2c}\int_{x-ct}^{x}\int_0^{t+(\xi-x)/c} g(\xi,\tau)\,d\tau d\xi + \frac{1}{2c}\int_x^{x+ct}\int_0^{t-(\xi-x)/c} g(\xi,\tau)\,d\tau d\xi$$

arising from the original formula by reversing the order of integration.) It may be well to point out here that the mere continuity of the function g is in general insufficient to ensure that the function w be twice continuously differentiable.

Since it is apparent from (3.13) that the function u obeys the initial conditions (3.8) we can state the following theorem.

Let $\varphi \in C^2(\mathbf{R})$, $\psi \in C^2(\mathbf{R})$ and g be continuous on $\bar{Q}$ together with its first derivative with respect to either variable, x or t. Then there exists exactly one classical solution of the problem (3.8), (3.9). *This solution is given by formula* (3.13).

3.1.5 The initial-boundary value problem for the quarter plane: the Dirichlet boundary condition

In § 3.1.4 we studied wave processes in systems unlimited in the space. As a rule, however, we encounter bounded systems (such as rods, strings, etc.). For them the foregoing results are yet of importance. Namely, we have shown that the disturbances described by the wave equation propagate with finite velocity and therefore the effect of the boundary at points which are far-away is not perceivable for a certain period [this means that the solution is still given by formula (3.13) for this time interval]. The solution of the wave equation in a neighbourhood of the boundary has been discussed in a very special case in § 3.1.3. Now, we shall pursue this subject more thoroughly. (For simplicity of exposition we keep on treating a system the boundary of which consists of just one point.)

So let us turn our attention to the question on existence and uniqueness of the classical solution of the wave equation on the quarter plane

$$Q = (0, \infty) \times (0, \infty)$$

satisfying the initial conditions

$$u(x,0) = \varphi(x), \qquad u_t(x,0) = \psi(x), \qquad x \geq 0 \tag{3.15}$$

and the Dirichlet boundary condition

$$u(0,t) = h(t), \qquad t \geq 0. \tag{3.16}$$

As previously, we derive first the form of the solution. To this end, assume that u is the solution of Eq. (3.9) satisfying (3.15) and (3.16). Provided that $x \geq ct$ the triangle D (see Fig. 7) lies in the quarter plane Q and thus the value of $u(x,t)$ is given again by (3.13).

Further, let us consider a point $(x,t) \in Q$ for which $x < ct$. In this case we cannot proceed as in § 3.1.4 since the triangle D does not belong to the quarter plane Q. However, by inspection we find out that the success of the method yielding ultimately the value of the solution $u(x,t)$ results from integrating over a suitable region $D \subset Q$ with the boundary formed by a portion of the initial line (where the function u together with its first derivative with respect to the second variable is known) and by segments of characteristics the end points of which either coincide with (x,t) (at which the function u is to be evaluated) or the value of u is known therein. In the present case such a region D which meets these requirements is sketched in Fig. 8. Upon integrating Eq. (3.9) over D we obtain analogously as in § 3.1.4

$$u(x,t) = h\left(t - \frac{x}{c}\right) + \frac{1}{2}[\varphi(x+ct) - \varphi(-x+ct)] + \frac{1}{2c}\int_{-x+ct}^{x+ct} \psi(\xi)\,d\xi$$
$$+ \frac{1}{2c}\int_0^{t-x/c}\int_{-x+c(t-\tau)}^{x+c(t-\tau)} g(\xi,\tau)\,d\xi d\tau + \frac{1}{2c}\int_{t-x/c}^{t}\int_{x-c(t-\tau)}^{x+c(t-\tau)} g(\xi,\tau)\,d\xi d\tau. \tag{3.17}$$

From the preceding it is evident that the solution of our initial-boundary value problem is determined uniquely. Further, assume that $\varphi \in C^2(\mathbf{R}^+)$, $\psi \in C^2(\mathbf{R}^+)$, $h \in C^2(\mathbf{R}^+)$, g is continuous on $\bar{Q}$ together with its first derivative with respect to the variable x (or t) and denote by u^+ the function defined by formula (3.17) for $x \leq ct$ and by u^- the function defined by formula (3.13) for $x \geq ct$. By a straightforward computation we can then check that u^+ is the classical solution of the wave equation

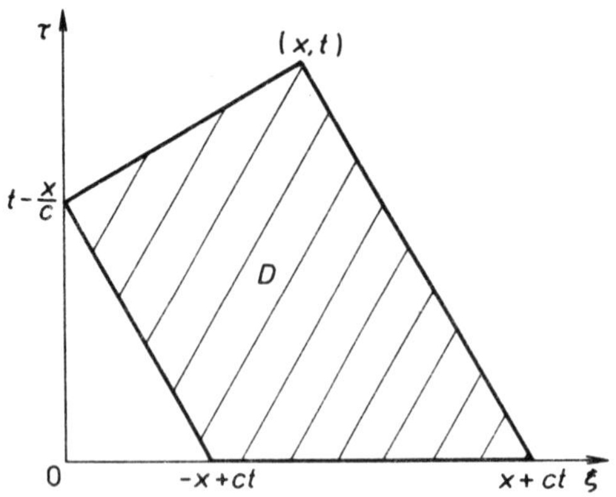

Fig. 8

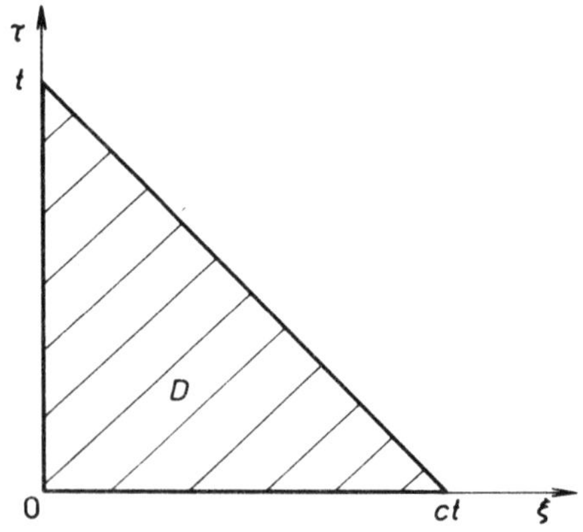

Fig. 9

(3.9) on the set $x \le ct$ (more precisely on the intersection of the quarter plane Q with the half-plane $x \le ct$) satisfying the boundary condition (3.16) and the function u^- is the classical solution of (3.9) for $x \ge ct$ satisfying the initial conditions (3.15). Hence, the function

$$u = \begin{cases} u^+ & \text{for } x \le ct \\ u^- & \text{for } x > ct \end{cases} \tag{3.18}$$

is the classical solution of our problem if and only if the values of u^+, u^- and their partial derivatives up to the second order match on the straight line $x = ct$. By a direct computation we arrive at the following relations which hold on the line $x = ct$:

$$\begin{aligned} u^+ - u^- &= h(0) - \varphi(0), \\ u_t^+ - u_t^- &= h'(0) - \psi(0), \\ u_x^+ - u_x^- &= -\frac{1}{c}[h'(0) - \psi(0)], \\ u_{tt}^+ - u_{tt}^- &= h''(0) - c^2\varphi''(0) - g(0,0), \\ u_{tx}^+ - u_{tx}^- &= -\frac{1}{c}[h''(0) - c^2\varphi''(0) - g(0,0)], \\ u_{xx}^+ - u_{xx}^- &= \frac{1}{c^2}[h''(0) - c^2\varphi''(0) - g(0,0)] \end{aligned}$$

and the coincidence of derivatives up to the second order of u^+ and u^- for $x = ct$ is tantamount to the conditions

$$h(0) = \varphi(0), \quad h'(0) = \psi(0), \quad h''(0) = c^2\varphi''(0) + g(0,0). \tag{3.19}$$

In summary, we have proved the following theorem.

Let $\varphi \in C^2(\mathbf{R}^+)$, $\psi \in C^2(\mathbf{R}^+)$, $h \in C^2(\mathbf{R}^+)$ *and* g *be a function continuous on* $\bar{Q}$ *together with the first derivative with respect to the first or second variable. Then there exists the classical solution of the problem* (3.9), (3.15), (3.16) *if and only if the compatibility conditions* (3.19) *are fulfilled. Moreover, the solution is determined uniquely and is given by* (3.18).

3.1.6 The initial-value problem for the quarter plane: interconnections among values of the solution on the time axis

In § 3.1.5 we have shown that the solution of the wave equation for the quarter plane $Q = (0,\infty)\times(0,\infty)$ satisfying the initial conditions (3.15) is determined uniquely once its values on the τ-axis are known. From this it follows that there is necessarily a relation between the values on the τ-axis of the function u (or u_t) and those of the function u_x. Before proceeding to the investigation of other initial-boundary value problems it proves useful to set down this dependence explicitly.

Hence assume that u is the classical solution of the wave equation (3.9) on Q satisfying the initial conditions (3.15). Further, let us pick out an arbitrary number $t > 0$ and integrate Eq. (3.9) over the region D which is bounded by the axes ξ and τ and by the characteristic $\xi + c\tau = ct$ (see Fig. 9). Using similar manipulations as in § 3.1.4 (based on Green's theorem) we arrive at the sought relation

$$u(0,t) + c\int_0^t u_x(0,\tau)\,d\tau = \varphi(ct) + \frac{1}{c}\int_0^{ct}\psi(\xi)\,d\xi + \frac{1}{c}\int_0^t\int_0^{c(t-\tau)} g(\xi,\tau)\,d\xi d\tau \quad (3.20)$$

which may be reduced further (by differentiating with respect to t) to a relation between u_t and u_x, namely

$$u_t(0,t) + c\,u_x(0,t) = c\varphi'(ct) + \psi(ct) + \int_0^t g(c(t-\tau),\tau)\,d\tau. \quad (3.21)$$

3.1.7 The initial-boundary value problem for the quarter plane: the Newton boundary condition

As mentioned in § 3.1.1 another boundary condition of importance is the Newton boundary condition

$$u_x(0,t) + \alpha(t)\,u(0,t) = h(t), \qquad t \geq 0 \quad (3.22)$$

of which a special case (for $\alpha \equiv 0$) is the Neumann boundary condition.

Let us assume that u is the classical solution of the wave equation in the quarter plane satisfying the initial conditions (3.15) and the boundary condition (3.22). Inserting $u_x(0,t)$ from (3.22) into (3.21) we obtain

$$u_t(0,t) - c\,\alpha(t)\,u(0,t) = c\varphi'(ct) + \psi(ct) + \int_0^t g(c(t-\tau),\tau)\,d\tau - c\,h(t). \quad (3.23)$$

This is an ordinary differential equation for the function $v(t) = u(0,t)$:

$$v'(t) - c\,\alpha(t)\,v(t) = c\varphi'(ct) + \psi(ct) + \int_0^t g(c(t-\tau),\tau)\,d\tau - c\,h(t). \quad (3.24)$$

In addition we know the initial condition for the function v since owing to (3.15)

$$v(0) = \varphi(0). \quad (3.25)$$

The function v is determined uniquely by Eq. (3.24) in conjunction with the condition (3.25).

On the other hand, if u is the classical solution of the wave equation satisfying the initial conditions (3.15) and the Dirichlet boundary condition

$$u(0,t) = v(t), \qquad t \geq 0, \tag{3.26}$$

where v is the solution of the problem (3.24), (3.25), then by combining relations (3.23) [which is true by virtue of (3.24) and (3.26)] and (3.21) we find that u obeys the Newton boundary condition (3.22).

In this way, we have reduced (by equivalent arrangements) the problem of existence and uniqueness of the classical solution of the initial-boundary value problem with the Newton boundary condition to the problem with the Dirichlet boundary condition studied in detail in § 3.1.5. By results of the mentioned subsection usual assumptions on the functions φ, ψ and g ensure that the function v (corresponding now to the function h in § 3.1.5) should be twice continuously differentiable [which is evidently satisfied if we add the assumptions $\alpha \in C^1(\mathbf{R}^+)$, $h \in C^1(\mathbf{R}^+)$] and that the compatibility conditions (3.19) which now take the form

$$v(0) = \varphi(0), \quad v'(0) = \psi(0), \quad v''(0) = c^2\varphi''(0) + g(0,0) \tag{3.27}$$

should be satisfied. The first of these conditions is true automatically on account of (3.25). Upon inserting $t = 0$ into Eq. (3.24) and into the same equation differentiated with respect to t we obtain the relations

$$v'(0) = c\,\alpha(0)\,\varphi(0) + c\,\varphi'(0) + \psi(0) - c\,h(0),$$
$$v''(0) = c\,\alpha'(0)\,\varphi(0) + c\,\alpha(0)\,\varphi'(0) + c^2\varphi''(0) + c\,\psi'(0) + g(0,0) - c\,h'(0).$$

Hence, the remaining two compatibility conditions in (3.27) are met if and only if

$$\begin{aligned} h(0) &= \alpha(0)\,\varphi(0) + \varphi'(0), \\ h'(0) &= \alpha'(0)\,\varphi(0) + \alpha(0)\,\varphi'(0) + \psi'(0). \end{aligned} \tag{3.28}$$

To conclude we state the following theorem.

Let $\varphi \in C^2(\mathbf{R}^+)$, $\psi \in C^1(\mathbf{R}^+)$, g be continuous on $\bar{Q}$ together with the first derivative with respect to either variable, $\alpha \in C^1(\mathbf{R}^+)$ and $h \in C^1(\mathbf{R}^+)$. Furthermore, let the conditions (3.28) *be satisfied. Then there exists exactly one classical solution of the wave equation on the quarter plane Q satisfying the initial conditions* (3.15) *and the Newton boundary condition* (3.22).

Let us remark that the differential equation (3.24) can be solved explicitly. Hence the solution u might be expressed explicitly, though a fairly intricate formula would be obtained. For concrete situations it is enough to keep in mind that the solution u is given by (3.18) where, for h, the solution of the differential equation (3.24) satisfying the initial condition (3.25) is to be substituted.

3.1.8 The initial-boundary value problems for the semi-infinite strip

As indicated above the effect of the reflection of the waves from the end makes the mathematical description of corresponding problems technically more complicated. *A fortiori* the same relates to the description of bounded systems where repeated reflections of the waves from both ends take place. Due to this fact we merely touch upon the salient aspects here. For more details on the subject we refer to Section 3.3 where the Fourier method will be introduced which allows us, in a relatively simple way, to cope with questions as regards existence and uniqueness of solutions of initial-boundary value problems even if their explicit representations are not available.

First let us look at the problem of what is the form of the solution of the wave equation for the semi-infinite strip with Dirichlet boundary conditions. Hence suppose that u is the classical solution of the wave equation on

$$Q = (0,\ell) \times (0,\infty)$$

satisfying the initial conditions

$$u(x,0) = \varphi(x), \quad u_t(x,0) = \psi(x), \qquad 0 \le x \le \ell \tag{3.29}$$

and the boundary conditions

$$u(0,t) = h_0(t), \quad u(\ell,t) = h_1(t), \qquad t \ge 0. \tag{3.30}$$

Provided $0 \le ct \le x \le \ell - ct$ the triangle D from Fig. 7 lies in the region Q and the value $u(x,t)$ is thus given by formula (3.13). Similarly for $0 \le x \le ct \le \ell - x$ the solution is given (since the region D from Fig. 8 lies in Q) by formula (3.17) where h is to be replaced by h_0. For $0 \le \ell - x \le ct \le x$ upon integrating the wave equation over the region D from Fig. 10 we get

$$u(x,t) = h_1\big(t - \frac{\ell - x}{c}\big) + \frac{1}{2}[\varphi(x-ct) - \varphi(2\ell - x - ct)] \\ + \frac{1}{2c}\int_{x-ct}^{2\ell-x-ct} \psi(\xi)\,d\xi + \frac{1}{2c}\int_D g(\xi,\tau)\,d\xi d\tau$$

and for $0 \le \ell - ct \le x \le ct$ upon integrating over the region D from Fig. 11 similarly

$$u(x,t) = h_0\big(t - \frac{x}{c}\big) + h_1\big(t - \frac{\ell - x}{c}\big) \\ - \frac{1}{2}[\varphi(ct - x) + \varphi(2\ell - x - ct)] + \frac{1}{2c}\int_{ct-x}^{2\ell-x-ct} \psi(\xi)\,d\xi + \frac{1}{2c}\int_D g(\xi,\tau)\,d\xi d\tau.$$

In this way the solution is determined (uniquely) for all x and $0 \le t \le \ell/c$. Now, consider a point $(x,t) \in Q$ such that $t > \ell/c$. Let $n \in \mathbb{N}$ be such that $0 < t - n\ell/c \le \ell/c$ and denote by $D_1, \ldots, D_n$ the regions sketched in Fig. 12 with boundaries consisting of segments of characteristics. Integrating the wave equation over the region D_1 and performing usual arrangements we obtain

$$u(x,t) = h_0\big(t - \frac{x}{c}\big) + h_1\big(t - \frac{\ell - x}{c}\big) + \frac{1}{2c}\int_{D_1} g(\xi,\tau)\,d\xi d\tau - u\big(\ell - x, t - \frac{\ell}{c}\big).$$

In an analogous manner (integration over D_2) we express the value $u(\ell - x, t - \ell/c)$ by means of known quantities and of the value $u(x, t - 2\ell/c)$. Continuing in this way, we eventually arrive at the relation

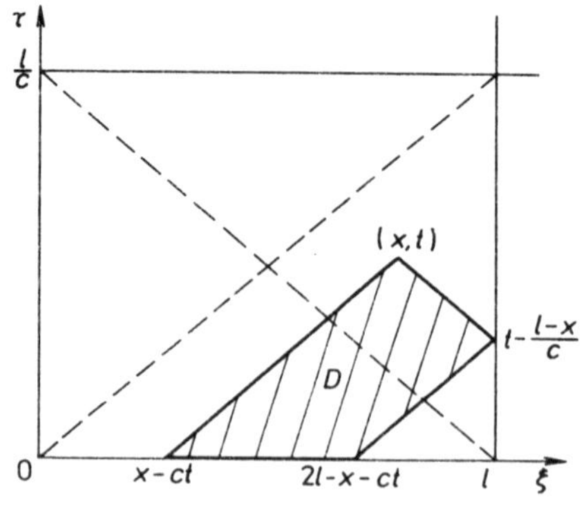

Fig. 10

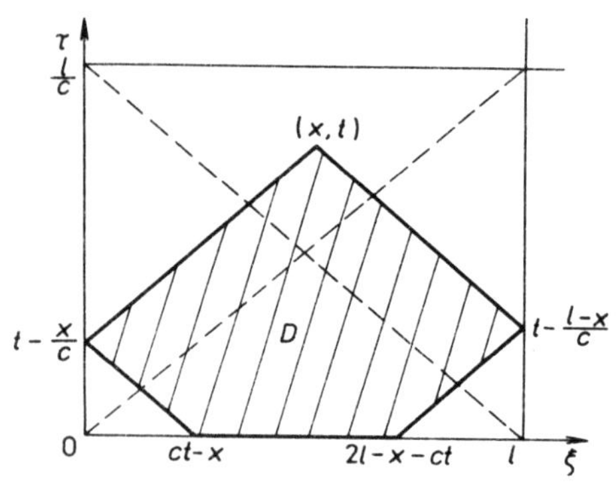

Fig. 11

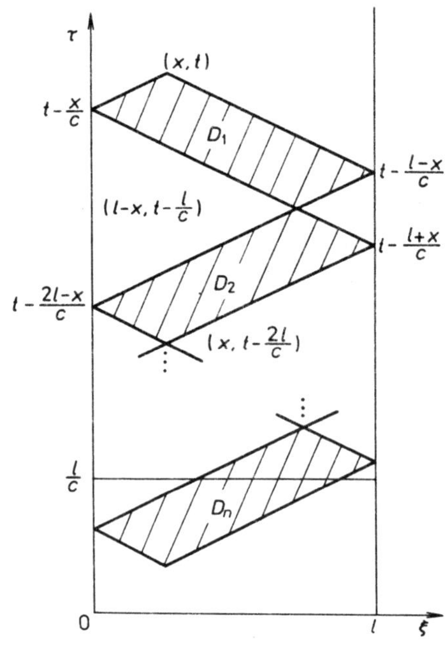

Fig. 12

$$u(x,t)=\sum_{j=1}^{n}(-1)^{j+1}\Big\{h_0\big(t-\frac{1}{c}\big[(j-\frac{1}{2})\ell+(-1)^j(\frac{\ell}{2}-x)\big]\big)$$

$$+h_1\big(t-\frac{1}{c}\big[(j-\frac{1}{2})\ell+(-1)^{j-1}(\frac{\ell}{2}-x)\big]\big)+\frac{1}{2c}\int_{D_j}g(\xi,\tau)\,d\xi d\tau\Big\}$$

$$+(-1)^n u\big(\frac{1-(-1)^n}{2}\ell+(-1)^n x,t-n\frac{\ell}{c}\big).$$

Now, the last term may be represented in terms of the known functions (since $t - n\ell/c \le \ell/c$) and consequently the solution u is determined uniquely by the initial and boundary conditions and can be written out explicitly. Looking through the above formulae we discover that in order to ensure the existence of the classical solution the following compatibility conditions apart from usual smoothness assumptions have to be satisfied:

$$\begin{aligned} h_0(0) &= \varphi(0), \quad h_0'(0) = \psi(0), \quad h_0''(0) = c^2\varphi''(0) + g(0,0), \\ h_1(0) &= \varphi(\ell), \quad h_1'(0) = \psi(\ell), \quad h_1''(0) = c^2\varphi''(\ell) + g(\ell,0). \end{aligned} \tag{3.31}$$

To sum up, we have the following theorem.

Let $\varphi \in C^2[0,\ell]$, $\psi \in C^2[0,\ell]$, $h_0 \in C^2(\mathbf{R}^+)$, $h_1 \in C^2(\mathbf{R}^+)$, g *be a function continuous in* $\bar{Q}$ *together with its first derivative with respect to either variable and the compatibility conditions* (3.31) *be satisfied. Then there exists exactly one classical solution of the wave equation satisfying the initial condition* (3.29) *and the Dirichlet boundary conditions* (3.30). □

The investigation of problems with other types of boundary conditions is still more complicated and we do not enter here into any details. We content ourselves with the single remark that such an investigation can be carried out by reducing the problem to the case of the Dirichlet boundary conditions as was done in § 3.1.7. This reduction is possible on account of the following relations which are valid for any solution of the wave equation on the semi-infinite strip satisfying the initial conditions (3.29) and which are analogous to those derived in § 3.1.6:

$$\begin{aligned} u_t(0,t) + c\,u_x(0,t) &= \psi(ct) + c\varphi'(ct) \\ &\quad + \int_0^t g(c(t-\tau),\tau)\,d\tau, \qquad 0 \le t \le \frac{\ell}{c}, \end{aligned}$$

$$\begin{aligned} u_t(0,t) + c\,u_x(0,t) &= u_t(\ell, t - \frac{\ell}{c}) + c\,u_x(\ell, t - \frac{\ell}{c}) \\ &\quad + \int_{t-\ell/c}^t g(c(t-\tau),\tau)\,d\tau, \qquad t > \frac{\ell}{c}, \end{aligned}$$

$$\begin{aligned} u_t(\ell,t) - c\,u_x(\ell,t) &= \psi(\ell - ct) - c\varphi'(\ell - ct) \\ &\quad + \int_0^t g(\ell - c(t-\tau),\tau)\,d\tau, \qquad 0 \le t \le \frac{\ell}{c}, \end{aligned}$$

$$\begin{aligned} u_t(\ell,t) - c\,u_x(\ell,t) &= u_t(0, t - \frac{\ell}{c}) - c\,u_x(0, t - \frac{\ell}{c}) \\ &\quad + \int_{t-\ell/c}^t g(\ell - c(t-\tau),\tau)\,d\tau, \qquad t > \frac{\ell}{c}. \end{aligned}$$

3.1.9 Remarks

In the preceding subsections we have studied some initial-boundary value problems for the wave equation on "rectangular" regions. The solutions of these problems have been found by means of a method which can be called the *method of characteristics* (which relates very closely to the method of the same name from Chapter 2).

The method of characteristics is suitable even when dealing with the wave equation considered on more general regions. As an illustration, consider a region Q bounded by the curve

$$\tau = a(\xi) \tag{3.32}$$

where

$$|a'(\xi)| \leq \text{const.} < \frac{1}{c}, \qquad \xi \in \mathbf{R}. \tag{3.33}$$

For any $(x, t) \in Q$ either characteristic passing through (x, t) intersects the curve (3.32) in a single point. For example, this point of intersection, say (ξ_1, τ_1), corresponding to the characteristic $\xi - c\tau = x - ct$ is defined by the relation

$$\tau_1 = a(\xi_1) = \frac{\xi_1}{c} - \frac{x}{c} + t,$$

equivalently

$$\xi_1 - c\,a(\xi_1) = x - ct. \tag{3.34}$$

Owing to (3.33), the equation (3.34) for ξ_1 has exactly one solution, x and t being arbitrary. To visualize the dependence of ξ_1 on x and t we write $\xi_1 = \xi_1(x, t)$. It turns out that analogously there exists a function $\xi_2 = \xi_2(x, t)$ such that the characteristic $\xi + c\tau = x + ct$ cuts the curve (3.32) only once at a point with coordinates $(\xi_2(x, t),\ a(\xi_2(x, t)))$. Now, assume that u is the classical solution of the wave equation and that we know the values of the functions u and u_t on the curve (3.32), specifically,

$$u(\xi, a(\xi)) = \varphi(\xi), \quad u_t(\xi, a(\xi)) = \psi(\xi), \qquad \xi \in \mathbf{R}. \tag{3.35}$$

Integration of the wave equation over the region D from Fig. 13 and the usual devices yield

$$\begin{aligned} u(x, t) &= \frac{1}{2}\left[\varphi(\xi_1(x, t)) + \varphi(\xi_2(x, t))\right] \\ &\quad + \frac{1}{2c}\int_{\xi_1(x,t)}^{\xi_2(x,t)} \left[\psi(\xi) - c^2\varphi'(\xi)\,a'(\xi)\right] d\xi + \frac{1}{2c}\int_D g(\xi, \tau)\,d\xi d\tau. \end{aligned} \tag{3.36}$$

If, moreover, the function a is twice continuously differentiable then the functions ξ_1 and ξ_2 are also twice continuously differentiable and relation (3.36) under the usual assumptions on the functions φ, ψ and g determines a (unique) solution of the wave equation on the region Q satisfying "the initial" conditions (3.35). (This formula, however, is as much explicit as are the functions ξ_1 and ξ_2.) So, there is practically no difference between this case and the initial-value problem. An analogous situation arises for a region Q bounded by the curve (3.32) and by the curve

$$\xi = b(\tau), \quad \text{where} \quad |b'(\tau)| \leq \text{ const. } < c,$$

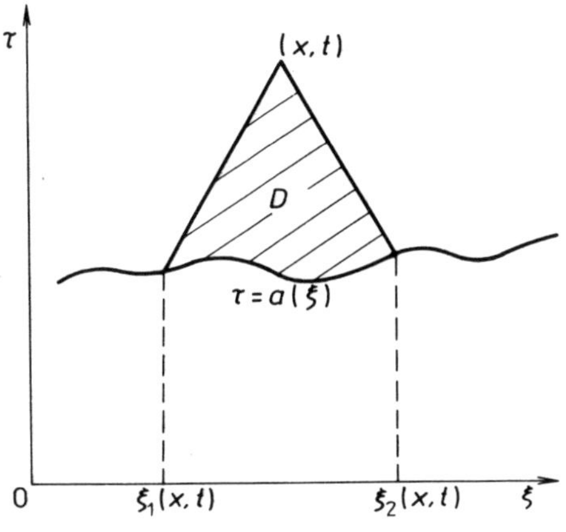

Fig. 13

or else by two such curves, when a similar procedure as in the case of the quarter plane or the semi-infinite strip may be applied. Let us add that by the same method we can examine even the problems on a region Q which lies "under" the curve (3.32), this means the history of the process. Incidentally, this expresses the fact that the wave equation describes reversible processes. More general regions than those mentioned above are of marginal interest in terms of usefulness in applications and therefore we omit any further discussion of these topics. □

It is noteworthy that the knowledge of one set of the data (functions from initial (boundary) conditions) for which the solution exists and is determined uniquely, yields a natural method which enables us to approach other problems – it consists in determining the data in such a way that the newly-imposed conditions be satisfied. This approach was applied in § 3.1.7 when reducing the problem with the Newton boundary condition to the problem with the Dirichlet one. As a matter of fact, the investigation of initial-boundary value problems for the quarter plane and for the semi-infinite strip could be entirely based on the knowledge of the solution of the initial-value problem: the initial conditions are to be extended to the whole line so that the corresponding solution of the initial-value problem satisfies the imposed conditions. (In this case the original problem is "reduced" to a problem on an "extended" region; this can be done even for other equations and we speak in this context of the *method of extension*.) □

To conclude we mention another physically important problem which consists in finding periodic solutions (solutions satisfying

$$u(x, t + \omega) = u(x, t) \qquad \text{for all } x \text{ and } t,$$

where ω denotes the period). The problem can be solved either directly by means of the method of characteristics or by using the preceding results and looking for initial functions φ and ψ for which the corresponding solution satisfies the relation

$$u(x, \omega) = \varphi(x), \qquad u_t(x, \omega) = \psi(x).$$

[For more details on periodic solutions we refer the reader to Vejvoda *et al.* (1982).]

3.1.10 Exercises

In the following exercises the problem is to find classical solutions of the wave equation

$$u_{tt} - c^2 u_{xx} = g(x,t) \tag{3.37}$$

with specified data.

3.1.10.1 Find all solutions of Eq. (3.37) with $g \equiv 0$ on the plane $\mathbf{R}^2$ satisfying
a) $u(0,t) = u(\ell,t), \quad t \in \mathbf{R}$;
b) $u(0,t) = u(x,0) = 0, \quad t, x \in \mathbf{R}$.
Answer.

$$u(x,t) = \sigma(x+ct) - \sigma(-x+ct),$$

where $\sigma \in C^2(\mathbf{R})$ and either a) 2ℓ-periodic, i.e. $\sigma(x+2\ell) = \sigma(x)$, $x \in \mathbf{R}$;
or b) even, i.e. $\sigma(x) = \sigma(-x)$, $x \in \mathbf{R}$.

3.1.10.2 Describe the reflection of waves from the free end in case of Eq. (3.37) with $g \equiv 0$ if for small $t > 0$ the solution has the form of a localized wave travelling to the left.
Answer.
If $u(x,t) = \beta(x+ct)$ for small t then the solution of the homogeneous equation on the quarter plane with the boundary condition $u_x(0,t) = 0$ for $t \geq 0$ has the form $u(x,t) = \beta(x+ct) + \tilde{\beta}(-x+ct)$ where $\tilde{\beta}$ is a function on $\mathbf{R}$ defined in (3.7). In this case the reflected wave has the "reversed" shape and equal amplitude.

3.1.10.3 Determine the solution of the initial-value problem for Eq. (3.37) if
a) $\varphi(x) = \sin x, \ \psi \equiv 0, \ g \equiv 0$;
b) $\varphi \equiv 0, \ \psi(x) = \sin x, \ g \equiv 0$;
c) $\varphi \equiv 0, \ \psi \equiv 0, \ g(x,t) = \sin x$.
Answer.

a)
$$u(x,t) = \sin x \, \cos(ct);$$

b)
$$u(x,t) = \frac{1}{c} \sin x \, \sin(ct);$$

c)
$$u(x,t) = \frac{1}{c^2}\left[\sin x - \frac{1}{2}\sin(x+ct) - \frac{1}{2}\sin(x-ct)\right].$$

3.1.10.4 Verify the compatibility conditions and write down the solution of Eq. (3.37) on the quarter plane with the Dirichlet boundary condition if all given functions but one indicated below vanish identically.
a) $\varphi(x) = x$; b) $\psi(x) = x$; c) $h(t) = t^3$; d) $g(x,t) = t$.
Answer.

a)
$$u(x,t) = x;$$

b)
$$u(x,t) = xt;$$

c)
$$u(x,t) = \begin{cases} 0 & \text{for } x \geq ct, \\ \left(t - \frac{x}{c}\right)^3 & \text{for } x < ct; \end{cases}$$

d)
$$u(x,t) = \begin{cases} \frac{1}{6}t^3 & \text{for } x \geq ct, \\ \frac{1}{6}\left(t^3 - \left(t - \frac{x}{c}\right)^3\right) & \text{for } x < ct. \end{cases}$$

3.1.10.5 Find the solution of the homogeneous equation (3.37) on the semi-infinite strip $(0,\ell) \times (0,\infty)$ with the homogeneous Dirichlet boundary conditions if for some $n \in \mathbf{N}$

a)
$$\varphi(x) = \sin\left(\frac{\pi n}{\ell}x\right), \qquad \psi \equiv 0;$$

b)
$$\varphi \equiv 0, \qquad \psi(x) = \sin\left(\frac{\pi n}{\ell}x\right).$$

Hint. Take advantage of the method of extension. By Exercise 3.1.10.1 a) look for a solution in the form $u(x,t) = \sigma(x+ct) - \sigma(-x+ct)$ where σ is a 2ℓ-periodic function.
Answer.

a)
$$\sigma(x) = \frac{1}{2}\sin\left(\frac{\pi n}{\ell}x\right) \text{ and hence}$$

$$u(x,t) = \frac{1}{2}\sin\left(\frac{\pi n}{\ell}(x+ct)\right) - \frac{1}{2}\sin\left(\frac{\pi n}{\ell}(-x+ct)\right) = \sin\left(\frac{\pi n}{\ell}x\right)\cos\left(\frac{c\pi n}{\ell}t\right);$$

b)
$$\sigma(x) = -\frac{\ell}{2c\pi n}\cos\left(\frac{\pi n}{\ell}x\right) \text{ and hence}$$

$$\begin{aligned} u(x,t) &= -\frac{\ell}{2c\pi n}\left\{\cos\left(\frac{\pi n}{\ell}(x+ct)\right) - \cos\left(\frac{\pi n}{\ell}(-x+ct)\right)\right\} \\ &= \frac{\ell}{c\pi n}\sin\left(\frac{\pi n}{\ell}x\right)\sin\left(\frac{c\pi n}{\ell}t\right). \end{aligned}$$

3.1.10.6 Find an expression for the solution of Eq. (3.37) on the region bounded by the curve $\tau = k\xi$ ($|k| < c^{-1}$) wherein the values of u and u_t are prescribed, i.e.

$$u(\xi, k\xi) = \varphi(\xi), \qquad u_t(\xi, k\xi) = \psi(\xi), \qquad \xi \in \mathbf{R}.$$

Answer.

$$u(x,t) = \frac{1}{2}\left\{\varphi\left(\frac{x-ct}{1-ck}\right) + \varphi\left(\frac{x+ct}{1+ck}\right)\right\} + \frac{1}{2c}\int_{\frac{x-ct}{1-ck}}^{\frac{x+ct}{1+ck}} \left[\psi(\xi) - c^2k\varphi'(\xi)\right] d\xi$$

$$+ \frac{1}{2c}\int_{\frac{x-ct}{1-ck}}^{x}\int_{k\xi}^{t+\frac{\xi-x}{c}} g(\xi,\tau)\, d\tau d\xi + \frac{1}{2c}\int_{x}^{\frac{x+ct}{1+ck}}\int_{k\xi}^{t-\frac{\xi-x}{c}} g(\xi,\tau)\, d\tau d\xi.$$

3.1.10.7 By means of the Fourier integral transform over the space coordinate find the solution of the homogeneous wave equation (3.37) satisfying the initial conditions

$$u(x,0) = 0, \;\; u_t(x,0) = \psi(x), \quad x \in \mathbf{R}.$$

Hint. See § 1.2.6 and § 4.2.3. The transformed problem is

$$\hat{u}_{tt}(y,t) - c^2y^2\hat{u}(y,t) = 0, \quad t > 0,$$

$$\hat{u}(y,0) = 0, \;\; \hat{u}_t(y,0) = \hat{\psi}(y)$$

and has the solution

$$\frac{\sin(cyt)}{cy}\,\hat{\psi}(y).$$

Derive the relation

$$\frac{\sin(cyt)}{cy}\exp(iyx) = \frac{1}{2c}\int_{x-ct}^{x+ct}\exp(i\sigma y)\, d\sigma.$$

Use the inversion formula (1.38) and interchange the order of integration.
Answer.

$$u(x,t) = \frac{1}{2c}\int_{x-ct}^{x+ct}\psi(\xi)\, d\xi.$$

3.1.10.8 *The Duhamel principle.* For any $\tau > 0$ determine the solution $v(x,t,\tau)$ of the problem

$$u_{tt} - c^2u_{xx} = 0, \;\; (x,t) \in \mathbf{R}\times(\tau,\infty), \quad u(x,\tau) = 0, \; u_t(x,\tau) = g(x,\tau), \;\; x \in \mathbf{R}$$

and verify that

$$\int_0^t v(x,t,\tau)\, d\tau$$

is a solution (at least formal) of the problem

$$u_{tt} - c^2u_{xx} = g(x,t), \;\; (x,t) \in \mathbf{R}\times(0,\infty), \quad u(x,0) = u_t(x,0) = 0, \;\; x \in \mathbf{R}.$$

Answer.

$$v(x,t,\tau) = \frac{1}{2c}\int_{x-c(t-\tau)}^{x+c(t-\tau)} g(\xi,\tau)\, d\xi.$$

3.1.10.9 By means of the Laplace integral transform over the time coordinate find the solution of the homogeneous wave equation (3.37) on the quarter plane satisfying the initial conditions (3.15) with $\varphi = \psi \equiv 0$ and the boundary condition (3.16).
Hint. See § 1.2.7 and § 4.3.7. The transformed problem is

$$\tilde{u}(x,s) - \frac{s^2}{c^2}\,\tilde{u}_{xx}(x,s) = 0, \quad 0 < x < \infty,$$

$$\tilde{u}(0,s) = \tilde{h}(s), \quad \tilde{u}(x,s) \text{ is bounded as } x \to \infty$$

and has the solution

$$\tilde{u}(x,s) = \tilde{h}(s)\,\exp(-\frac{sx}{c}).$$

Apply formula (1.47_2).
Answer.

$$u(x,t) = \begin{cases} h\left(t - \frac{x}{c}\right) & \text{for } 0 \le x \le ct, \\ 0 & \text{for } x > ct. \end{cases}$$

3.2 THE TELEGRAPH EQUATION IN ONE SPACE VARIABLE

The *telegraph equation* [on a region $Q \subset \mathbf{R}^2$] is an equation of the form

$$u_{tt} - c^2 u_{xx} + 2\alpha(x,t)\,u_t + 2\beta(x,t)\,u_x + \gamma(x,t)\,u = g(x,t),$$

which unlike the wave equation also contains at least one of the lower order derivatives of the function u with respect to either variable x and t. In this section we find solutions [again from $C^2(\bar{Q})$] of some initial-boundary value problems by means of the so-called Riemann method (which generalizes the approach previously used for the wave equation). An important role is played again by the straight lines $x \pm ct =$ const. still called *characteristics*.

3.2.1 The telegraph equation

To begin with, we mention briefly some situations where the telegraph equation can be encountered. Consider, for example, a system of equations

$$u_x + LI_t + RI = 0, \tag{3.38}$$
$$I_x + Cu_t + Gu = 0 \tag{3.39}$$

describing the flow of the electric current in a long-distance transmission line (see Exercise 2.4.3.5). By differentiating (3.38) with respect to x and (3.39) with respect to t we get the equations

$$u_{xx} + LI_{tx} + RI_x = 0, \tag{3.40}$$
$$I_{xt} + Cu_{tt} + Gu_t = 0. \tag{3.41}$$

After computing I_{tx} from (3.41) and I_x from (3.39) and inserting these relations into (3.40) we obtain an equation for the function u,

$$u_{xx} - LC\,u_{tt} - (LG + CR)\,u_t - GRu = 0$$

which becomes

$$u_{tt} - c^2 u_{xx} + 2\alpha u_t + \gamma u = 0 \tag{3.42}$$

upon denoting

$$c = \frac{1}{(LC)^{\frac{1}{2}}}, \qquad \alpha = \frac{LG + CR}{2LC}, \qquad \gamma = \frac{GR}{LC}.$$

It may be easily checked that the function I also satisfies an equation of the same form. The constants α and γ are non-negative and describe electric losses in the transmission line.

Furthermore, the equation of the form (3.42) also governs a number of mechanical processes, for example, longitudinal vibrations of rods, transverse vibrations of strings and so forth. In these cases the terms $2\alpha u_t$ and γu reflect the situation when a surrounding medium provides a restricting force to the motion proportional to the velocity (external friction) and an additional restoring force proportional to the displacement, respectively. The constants α and γ are therefore again usually non-negative.

From the above examples it is apparent that the telegraph equation describes processes of the wave character which are damped owing to the surroundings. By means of a very special case we illustrate how the presence of this damping manifests itself in mathematical terms. Consider an equation of the form (3.42) with $\gamma = \alpha$, that is

$$u_{tt} - c^2 u_{xx} + 2\alpha u_t + \alpha u = 0 \tag{3.43}$$

Introducing a new unknown function v by $v(x,t) = \exp(\alpha t)\,u(x,t)$ we discover easily that the function v satisfies the homogeneous wave equation $v_{tt} - c^2 v_{xx} = 0$. Invoking the results from § 3.1.2 we arrive at the conclusion that every solution of Eq. (3.43) assumes the form

$$u(x,t) = \exp(-\alpha t)\,[\sigma(x + ct) + \varrho(x - ct)]$$

and consequently it is the superposition of two waves whose amplitudes (in case $\alpha > 0$) are exponentially decaying as $t \to \infty$. □

For the telegraph equation the study of the same initial-boundary value problems as for the wave equation makes sense. However, if one tries to apply the same method some difficulties arise. To illustrate for the initial-value problem, upon integrating the telegraph equation over the triangle D from Fig. 7 and making some re-arrangements we get

$$\begin{aligned} u(x,t) = {} & \frac{1}{c}\int_\Gamma [\alpha(\xi,\tau)\,u(\xi,\tau)\,d\xi - \beta(\xi,\tau)\,u(\xi,\tau)\,d\tau] \\ & + \frac{1}{c}\int_D [\alpha_\tau(\xi,\tau) + \beta_\xi(\xi,\tau) - \frac{1}{2}\gamma(\xi,\tau)]\,u(\xi,\tau)\,d\xi d\tau \\ & + \frac{1}{2}[\varphi(x - ct) + \varphi(x + ct)] + \frac{1}{2c}\int_{x-ct}^{x+ct} \psi(\xi)\,d\xi \\ & + \frac{1}{2c}\int_D g(\xi,\tau)\,d\xi d\tau. \end{aligned} \tag{3.44}$$

Unlike the situation for the wave equation we have not obtained any explicit expression for the solution but only a further (equivalent) equation for u. To get around this difficulty we shall proceed in a somewhat modified way.

3.2.2 Riemann's method

Let us multiply the telegraph equation by any twice continuously differentiable function r and integrate the resulting equation

$$ru_{\tau\tau} - c^2 ru_{\xi\xi} + 2\alpha ru_\tau + 2\beta ru_\xi + \gamma ru = rg$$

over the region D sketched in Fig. 14 where the parts Γ_1 and Γ_2 of its boundary Γ are formed by the characteristics through (x, t).

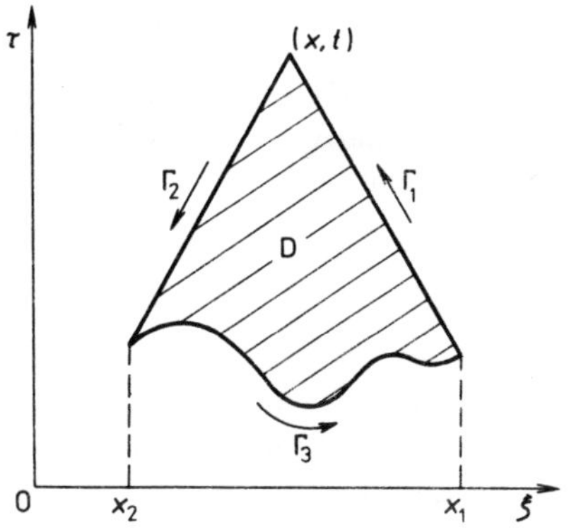

Fig. 14

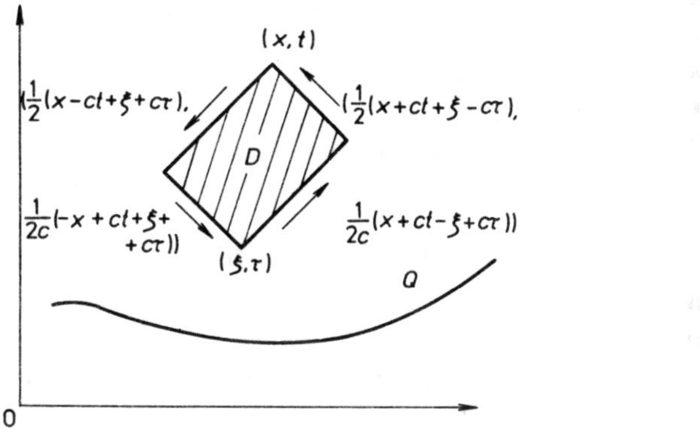

Fig. 15

Analogous manipulations as in § 3.1.4 (a key ingredient being Green's theorem) yield

$$\begin{aligned}
r(x,t)\,u(x,t) &= \frac{1}{2}\left[r(x_1,t_1)\,u(x_1,t_1) + r(x_2,t_2)\,u(x_2,t_2)\right] \\
&+ \frac{1}{2c}\int_{\Gamma_3}\left[(ru_\tau - r_\tau u + 2\alpha ru)\,d\xi + (c^2 ru_\xi - c^2 r_\xi u - 2\beta ru)\,d\tau\right] \\
&+ \frac{1}{c}\int_{-x_1}^{-x}\left[\left(-cr_\xi + r_\tau - (\alpha + \frac{1}{c}\beta)r\right)u\right]\left(-s, t + \frac{1}{c}(x+s)\right)ds \\
&+ \frac{1}{c}\int_{x}^{x_2}\left[\left(cr_\xi + r_\tau - (\alpha + \frac{1}{c}\beta)r\right)u\right]\left(-s, t - \frac{1}{c}(x+s)\right)ds \\
+ \frac{1}{2c}\int_D rg\,d\xi d\tau &- \frac{1}{2c}\int_D\left[r_{\tau\tau} - c^2 r_{\xi\xi} - 2(\alpha r)_\tau - 2(\beta r)_\xi + \gamma r\right]u\,d\xi d\tau. \quad (3.45)
\end{aligned}$$

Now, let us assume that the function r satisfies the following conditions

$$r(x,t) = 1, \tag{3.46$_1$}$$

$$-cr_\xi + r_\tau - \left(\alpha + \frac{1}{c}\beta\right) r = 0 \qquad \text{on } \Gamma_1, \tag{3.46$_2$}$$

$$cr_\xi + r_\tau - \left(\alpha - \frac{1}{c}\beta\right) r = 0 \qquad \text{on } \Gamma_2, \tag{3.46$_3$}$$

$$r_{\tau\tau} - c^2 r_{\xi\xi} - 2(\alpha r)_\tau - 2(\beta r)_\xi + \gamma r = 0 \qquad \text{in } D. \tag{3.46$_4$}$$

The relation (3.45) becomes considerably simplified:

$$\begin{aligned} u(x,t) = {} & \frac{1}{2}\left[r(x_1,t_1)\,u(x_1,t_1) + r(x_2,t_2)\,u(x_2,t_2)\right] \\ & + \frac{1}{2c}\int_{\Gamma_3} \left[(ru_\tau - r_\tau u + 2\alpha ru)\,d\xi + \left(c^2 ru_\xi - c^2 r_\xi u - 2\beta ru\right)d\tau\right] \\ & + \frac{1}{2c}\int_D rg\,d\xi d\tau. \end{aligned} \tag{3.47}$$

Once the values of functions u, u_τ and u_ξ are known on Γ_3, relation (3.47) determines values of the solution of the telegraph equation at the point (x,t); this expression is explicit according as the function r is known explicitly. First of all, however, a question arises whether such a function r exists at all. In the affirmative case r is then called the *Riemann function* for the corresponding telegraph equation.

3.2.3 The Riemann function

We start by clarifying the role of the first three conditions (3.46) for the existence of the function r. The expression $-cr_\xi + r_\tau$ is nothing other than the derivative of the function r along the characteristic Γ_1. Consequently, the function r must fulfil a certain ordinary differential equation on Γ_1 which can be solved. More precisely, if we put $p(s) = r(x + c(t-s), s)$ then in virtue of (3.46$_2$) the function p satisfies the differential equation

$$p'(s) - \left(\alpha + \frac{1}{c}\beta\right)(x + c(t-s), s)\,p(s) = 0$$

which together with (3.46$_1$) implies

$$r(x + c(t-s), s) = \exp\left(\int_0^s \left(\alpha + \frac{1}{c}\beta\right)(x + c(t-\lambda), \lambda)\,d\lambda\right).$$

Quite analogously we derive a relation to be satisfied by r on the characteristic Γ_2 [see (3.49$_2$)].

Further, let us note that the Riemann function depends not only on the variables ξ and τ but also on the coordinates of the point (x,t); we write accordingly $r = r(x,t,\xi,\tau)$. Hence by the Riemann function is to be understood a (twice continuously differentiable) function of four variables which satisfies (as a function of the latter two variables) the differential equation

$$r_{\tau\tau} - c^2 r_{\xi\xi} - 2(\alpha r)_\tau + \gamma r = 0 \tag{3.48}$$

and the conditions

$$r(x,t,x+c(t-s),s) = \exp\Big(\int_0^s (\alpha + \frac{1}{c}\beta)(x+c(t-\lambda),\lambda)\,d\lambda\Big), \tag{3.49$_1$}$$

$$r(x,t,x-c(t-s),s) = \exp\Big(\int_0^s (\alpha - \frac{1}{c}\beta)(x+c(t-\lambda),\lambda)\,d\lambda\Big). \tag{3.49$_2$}$$

We proceed to outline the proof of the existence of the Riemann function in the general case. By integrating Eq. (3.48) over the region D which is bounded by the characteristics through (x,t) and (ξ,τ) (see Fig. 15) we get for r the equation

$$\begin{aligned} r(\xi,\tau) = &-2\int_\Gamma \alpha r\,d\xi + 2\int_\Gamma \beta r\,d\tau - \int_D \gamma r\,d\xi d\tau \\ &+ r\left(\frac{1}{2}(x-ct+\xi+c\tau), \frac{1}{2c}(-x+ct+\xi+c\tau)\right) \\ &+ r\left(\frac{1}{2}(x+ct+\xi-c\tau), \frac{1}{2c}(x+ct-\xi+c\tau)\right) - r(x,t). \end{aligned}$$

The last three terms are determined by conditions (3.49). The existence and uniqueness of the solution of this equation can be established if, for example, the functions α and β are twice continuously differentiable being, of course, together with γ defined on the whole region D. If the latter assumption is not met the function r can be found after making a suitable extension of the functions α, β and γ; since this extension is not unambiguous the Riemann function (for those values of x, t, ξ and τ) is not defined in a unique manner.

Hence finding the Riemann function does not appear to be an easy matter. We abandon any further general investigation in favour of the equation with constant coefficients.

3.2.4 The Riemann function for the equation with constant coefficients

Let us take up first of all the special case $\alpha = \beta = 0$ when the equation is reduced to

$$u_{tt} - c^2 u_{xx} + \gamma u = g(x,t) \tag{3.50}$$

The Riemann function must fulfil the equation

$$r_{\tau\tau} - c^2 r_{\xi\xi} + \gamma r = 0 \tag{3.51}$$

together with the conditions resulting from (3.49)

$$r(x,t,x \pm c(t-s),s) = 1. \tag{3.52}$$

The conditions (3.52) are fulfilled automatically if the function r has the form

$$r(x,t,\xi,\tau) = q\left(c^2(t-\tau)^2 - (x-\xi)^2\right), \tag{3.53}$$

where a function q is such that

$$q(0) = 1. \tag{3.54}$$

Indeed, in this case $r(x,t,x \pm c(t-s),s) = q\left(c^2(t-s)^2 - c^2(t-s)^2\right) = q(0)$. A necessary and sufficient condition for a function r of the form (3.53) to satisfy Eq. (3.51) turns out to be

$$sq''(s) + q'(s) + \frac{\gamma}{4c^2}\, q(s) = 0. \tag{3.55}$$

The solution of Eq. (3.55) which meets condition (3.54) is given by the formula

$$q(s) = \begin{cases} \mathrm{J}_0\big(\frac{1}{c}(\gamma s)^{\frac{1}{2}}\big) & \text{for } \gamma s \geq 0, \\ \mathrm{I}_0\big(\frac{1}{c}(-\gamma s)^{\frac{1}{2}}\big) & \text{for } \gamma s < 0, \end{cases} \tag{3.56}$$

where J_0 and I_0 are Bessel functions of order zero (see § 1.2.8). □

Let us now turn our attention to a general telegraph equation with constant coefficients

$$u_{tt} - c^2 u_{xx} + 2\alpha u_t + 2\beta u_x + \gamma u = g(x,t). \tag{3.57}$$

After making the change of the dependent variable

$$u(x,t) = \exp\big(-\alpha t + \frac{\beta}{c^2}x\big)\, v(x,t) \tag{3.58}$$

it is easily seen that the function u is a solution of Eq. (3.57) if and only if v is a solution of the equation

$$v_{tt} - c^2 v_{xx} + \delta v = \exp\big(\alpha t - \frac{\beta}{c^2}x\big)\, g(x,t) \tag{3.59}$$

where

$$\delta = -\alpha^2 + \frac{\beta^2}{c^2} + \gamma. \tag{3.60}$$

In this way it is possible (and it may be sometimes convenient) to reduce the study of initial-boundary value problems for Eq. (3.57) to the study of analogous problems (with changed data, though) for Eq. (3.59) the Riemann function of which is known. Further, the transformation (3.58) makes it possible to show that the function r defined by

$$r(x,t,\xi,\tau) = \exp\big(\alpha(\tau - t) - \frac{\beta}{c^2}(\xi - x)\big)\, q\left(c^2(t-\tau)^2 - (x-\xi)^2\right)$$

[where δ is given by (3.60) and q by (3.56) with γ replaced by δ] is the Riemann function for Eq. (3.57).

3.2.5 The initial-value problem

We are now in a position to write down easily a formula for the solution to the problem given by Eq. (3.57) on the half-plane $Q = (-\infty, \infty) \times (0, \infty)$ in conjunction with the initial conditions (3.8). If the Riemann function r exists, hence if the functions α, β and γ are sufficiently smooth, then relation (3.47) [where for the region D we take the triangle from Fig. 7] yields

$$\begin{aligned} u(x,t) &= \frac{1}{2}[r(x,t,x-ct,0)\,\varphi(x-ct) + r(x,t,x+ct,0)\,\varphi(x+ct)] \\ &+ \frac{1}{2c}\int_{x-ct}^{x+ct} \{r(x,t,\xi,0)\,\psi(\xi) + [2\alpha(\xi,0)\,r(x,t,\xi,0) - r_\tau(x,t,\xi,0)]\}\,d\xi \\ &\qquad + \frac{1}{2c}\int_0^t \left(\int_{x-c(t-\tau)}^{x+c(t-\tau)} r(x,t,\xi,\tau)\,g(\xi,\tau)\,d\xi\right)d\tau. \qquad (3.61) \end{aligned}$$

Conversely, it can be proved that the function u defined by formula (3.61) is indeed a solution of (3.57), (3.8) provided $\varphi \in C^2(\mathbf{R})$ and $\psi \in C^1(\mathbf{R})$.

Let us mention some conclusions to be drawn from the latter formula. First of all, the disturbances in processes described by the telegraph equation spread out with velocity c. Hence compared with the wave equation the velocity remains unaffected by the presence of lower order derivatives of u in the equation. Further, relation (3.61) is amenable to the study of damping questions concerning the processes described by the telegraph equation. For example, the asymptotic properties of the Riemann function for the equation with constant coefficients are well-known and this allows us to perform with relative ease the relevant estimates for an arbitrary solution of the initial-value problem.

3.2.6 Initial-boundary value problems for the quarter plane

As seen earlier, the study of initial-boundary value problems for the wave equation affords a considerable simplification in the case of the Dirichlet boundary condition. This situation, however, does not apply to the telegraph equation: relation (3.47) where D is taken from Fig. 8 does not yield directly the value of the solution at the point (x,t).

Hence, let us assume first that u is the classical solution of the telegraph equation on the quarter plane $(0,\infty) \times (0,\infty)$ satisfying the initial conditions (3.15). Assume further that there exists the Riemann function r. For $x \geq ct$ the value of $u(x,t)$ is given again by relation (3.61). Let $x < ct$. Relation (3.47) with D taken from Fig. 16

implies

$$u(x,t) = \frac{1}{2} r\left(x,t,0,t-\frac{x}{c}\right) u\left(0,t-\frac{x}{c}\right)$$
$$+ \frac{1}{2c}\int_0^{t-x/c} \left\{ \left[c^2 r_\xi(x,t,0,\tau) + 2\beta(0,\tau)\, r(x,t,0,\tau)\right] u(0,\tau) - c^2 r(x,t,0,\tau)\, u_\xi(0,\tau)\right\} d\tau$$
$$+ \frac{1}{2c}\int_0^{x+ct} \left\{ r(x,t,\xi,0)\,\psi(\xi) + \left[2\alpha(\xi,0)\, r(x,t,\xi,0) - r_\tau(x,t,\xi,0)\right]\varphi(\xi)\right\} d\xi$$
$$+ \frac{1}{2} r(x,t,x+ct,0)\,\varphi(x+ct) + \frac{1}{2c}\int_D r(x,t,\xi,\tau)\, g(\xi,\tau)\, d\xi d\tau. \quad (3.62)$$

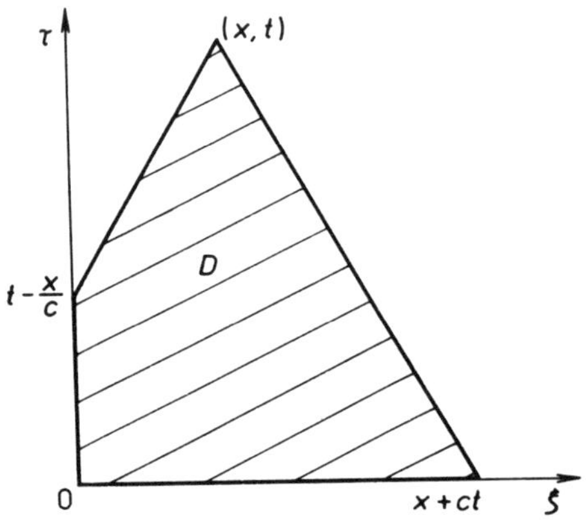

Fig. 16

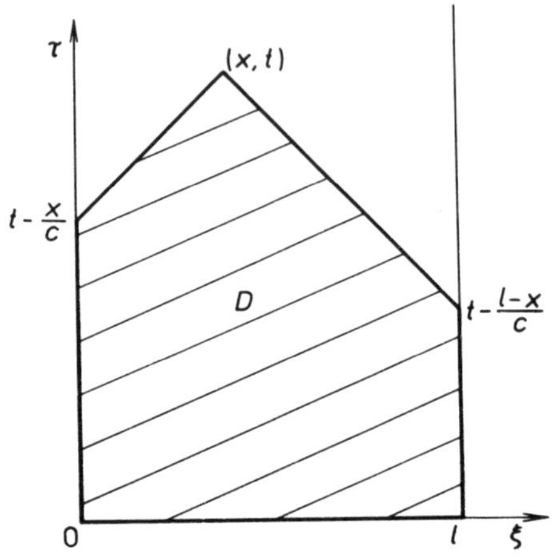

Fig. 17

Of course, the boundary values of u and u_ξ on the τ-axis are not known simultaneously. The boundary condition determines either of these functions (in case of the Dirichlet or Neumann boundary conditions) or even only a relation between them (in case of the Newton ones). A further equation to be added is obtained by inserting $x = 0$ into (3.62):

$$u(0,t) + \frac{1}{c}\int_0^{t} \left\{ c^2 r(0,t,0,\tau)\, u_\xi(0,\tau) - \left[c^2 r_\xi(0,t,0,\tau) \right.\right.$$
$$\left.\left. + \ 2\beta(0,\tau)\, r(0,t,0,\tau)\right] u(0,\tau)\right\} d\tau$$
$$= \frac{1}{c}\int_0^{ct} \left\{ r(0,t,\xi,0)\,\psi(\xi) + \left[2\alpha(\xi,0)\, r(0,t,\xi,0) - r_\tau(0,t,\xi,0)\right]\varphi(\xi)\right\} d\xi$$
$$+ r(0,t,ct,0)\,\varphi(ct) + \frac{1}{c}\int_0^{t}\int_0^{c(t-\tau)} r(0,t,\xi,\tau)\, g(\xi,\tau)\, d\xi d\tau. \quad (3.63)$$

The latter relation together with a (usual) boundary condition determines uniquely the values of u and u_ξ on the boundary. Their explicit expressions, however, are far from being obvious in the general case.

The above considerations already suggest a method of obtaining the solution of the initial-boundary value problem: for $x \geq ct$ the solution is given by (3.61). By means of the boundary condition and relation (3.63) we get the values of u and u_ξ on the τ-axis. For $x < ct$ the solution is then given by (3.62).

On the basis of this procedure we can also show that the solution of initial-boundary value problems for the telegraph equation on the quarter plane exists and is unique assuming that the data are sufficiently smooth and the compatibility conditions (analogous to those indicated for the wave equation) are satisfied.

3.2.7 Example

We look for a solution of the telegraph equation

$$u_{tt} - c^2 u_{xx} + \gamma u = 0 \qquad \text{on } (0,\infty) \times (0,\infty)$$

(γ is a constant) satisfying the initial conditions

$$u(x,0) = u_t(x,0) = 0, \qquad x \geq 0$$

and the Neumann boundary condition

$$u_x(0,t) = h(t), \qquad t \geq 0.$$

For definiteness assume $\gamma > 0$. By § 3.2.4 the Riemann function has the form $\left[\text{for } (x-\xi)^2 \leq c^2(t-\tau)^2\right]$

$$r(x,t,\xi,\tau) = \mathrm{J}_0\Big(\frac{1}{c}\left[\gamma c^2(t-\tau)^2 - \gamma(x-\xi)^2\right]^{\frac{1}{2}}\Big).$$

From this relation it is clear that $r_\xi(0,t,0,\tau) \equiv 0$ and by (3.63) we find that

$$u(0,t) = -c\int_0^t \mathrm{J}_0\Big(\frac{\gamma^{\frac{1}{2}}}{c}(t-\tau)\Big)\,h(\tau)\,d\tau$$

must hold. Upon inserting into (3.62) we get for $x < ct$

$$\begin{aligned} u(x,t) = -\frac{c}{2}\int_0^{t-x/c} &\Big\{\mathrm{J}_0\Big(\frac{\gamma^{\frac{1}{2}}}{c}(t-\frac{x}{c}-\tau)\Big) \\ &+ \mathrm{J}_0\Big(\frac{1}{c}\left[\gamma c^2(t-\tau)^2 - \gamma x^2\right]^{\frac{1}{2}}\Big)\Big\}\,h(\tau)\,d\tau - \frac{c\gamma^{\frac{1}{2}}}{2}x\int_0^{t-x/c} \mathrm{J}_0'\Big(\frac{1}{c}\left[\gamma c^2(t-\tau)^2 - \gamma x^2\right]^{\frac{1}{2}}\Big) \\ &\cdot\left[c^2(t-\tau)^2 - x^2\right]^{-\frac{1}{2}}\left[\int_0^\tau \mathrm{J}_0\Big(\frac{\gamma^{\frac{1}{2}}}{c}(\tau-\lambda)\Big)h(\lambda)\,d\lambda\right]d\tau. \end{aligned}$$

For $x \geq ct$ clearly $u(x,t) = 0$. The compatibility conditions in this case take the form

$$h(0) = h'(0) = 0.$$

3.2.8 Initial-boundary value problems for the semi-infinite strip

Again, the solution of the usual initial-boundary value problems for the telegraph equation on the semi-infinite strip exists and is determined uniquely if the compatibility conditions are satisfied and the data are sufficiently smooth. As we have seen in Section 3.1 the study of this case is rather intricate even for the wave equation. Therefore, we content ourselves here with giving only a brief outline of how to proceed.

Let us assume that u is the classical solution of the telegraph equation on the semi-infinite strip $(0,\ell)\times(0,\infty)$ (for which the Riemann function exists) satisfying the initial conditions (3.29). Fix a point $(x,t)\in Q$. From relation (3.47) where D is the region bounded by the characteristics through (x,t) and by the straight lines $\xi=0$, $\xi=\ell$ and $\tau=0$, we obtain an expression for the solution u involving the values of u and u_ξ on the boundary. For instance, in the situation such as that in Fig. 17 (that is, for $t\geq \ell/c$) we get

$$\begin{aligned}
u(x,t) = \frac{1}{2}\Big[& r\big(x,t,0,t-\frac{x}{c}\big)\,u\big(0,t-\frac{x}{c}\big) \\
& + r\big(x,t,\ell,t-\frac{\ell-x}{c}\big)\,u\big(\ell,t-\frac{\ell-x}{c}\big)\Big] \\
& + \frac{1}{2c}\int_0^{t-x/c}\big\{\big[c^2 r_\xi(x,t,0,\tau)+2\beta(0,\tau)\,r(x,t,0,\tau)\big]\;u(0,\tau) \\
& - c^2 r(x,t,0,\tau)\,u_\xi(0,\tau)\big\}\,d\tau + \frac{1}{2c}\int_0^{t-(\ell-x)/c}\big\{c^2 r(x,t,\ell,\tau)\,u_\xi(\ell,\tau) \\
& - \big[c^2 r_\xi(x,t,\ell,\tau)+2\beta(\ell,\tau)\,r(x,t,\ell,\tau)\big]\;u(\ell,\tau)\big\}\,d\tau \\
& + \frac{1}{2c}\int_0^{\ell}\{r(x,t,\xi,0)\,\psi(\xi)+[2\alpha(\xi,0)\,r(x,t,\xi,0)-r_\tau(x,t,\xi,0)]\,\varphi(\xi)\}\,d\xi \\
& + \frac{1}{2c}\int_D r(x,t,\xi,\tau)\,g(\xi,\tau)\,d\xi d\tau.
\end{aligned} \tag{3.64}$$

Upon inserting $x=0$ and $x=\ell$ into this expression and using the boundary conditions we obtain a system of four equations for four unknown functions $u(0,t)$, $u_x(0,t)$, $u(\ell,t)$ and $u_x(\ell,t)$ which can be shown to have exactly one solution. After inserting these functions into (3.64) (or into analogous relations for $t<\ell/c$) the solution of the corresponding initial-boundary value problems is expressed in terms of known functions and the discussion of existence and uniqueness can be carried out.

Let us note that the above procedure for finding solutions of concrete problems is fairly cumbersome and that the solutions are often obtainable by a simpler one as it is, for example, the method of reducing the given problem to the initial-value problem sketched in Section 3.1.

3.2.9 Exercises

The problems concern the telegraph equation (3.57) with constant coefficients.

3.2.9.1 Determine the Riemann function in case a) $\alpha = \beta = \gamma = 0$; b) $\beta = 0$, $\alpha = \gamma = 1$.

Answer.

a) $r = 1$;

b) $r = \exp(t - \tau)$.

3.2.9.2 Find the solution of the initial-value problem in case $\alpha = \gamma = 1$, $\beta = 0$ if

a) $\varphi(x) = \sin x$, $\psi \equiv 0$, $g \equiv 0$;

b) $\varphi \equiv 0$, $\psi(x) = \sin x$, $g \equiv 0$;

c) $\varphi \equiv 0$, $\psi \equiv 0$, $g(x,t) = \sin x$.

Answer.

a) $u(x,t) = \exp(-t)\left[\cos(ct) + \frac{1}{c}\sin(ct)\right]\sin x$;

b) $u(x,t) = \frac{1}{c}\exp(-t)\,\sin(ct)\,\sin x$;

c) $u(x,t) = \frac{1}{1+c^2}\left\{1 - \exp(-t)\left[\cos(ct) + \frac{1}{c}\sin(ct)\right]\right\}\sin x$.

3.3 THE FOURIER METHOD

In the preceding subsections we have seen that the use of the Riemann method to the investigation of initial-boundary value problems for the telegraph and wave equations on a semi-infinite strip is feasible but technically fairly difficult. Now we introduce the so-called Fourier method which enables us to explore the existence, uniqueness and other important properties of solutions of these problems with relative ease even though it does not provide solutions in a closed form. The Fourier method extends successfully to far more general equations and therefore it belongs by right to modern techniques of solution of partial differential equations.

We commence our discussion with the co-called method of separation of variables, an essential ingredient of the Fourier method.

3.3.1 Separation of variables

The solution of initial-boundary value problems for partial differential equations appears to be usually the more difficult the more (independent) variables are involved. For special data we often succeed in reducing the given problem to a simpler one (involving less variables) by means of the ***separation-of-variables method***. This method consists (roughly speaking) in splitting the variables up into two or more groups (the separation of variables) and looking for a solution in the form of the product of functions each depending only upon the variables of the respective group.

We elucidate the use of the separation-of-variables method in the case of the telegraph equation

$$u_{tt} - c^2 u_{xx} + 2\alpha u_t = g(x,t) \tag{3.65}$$

(α a constant). Assume that the right-hand side takes a special form

$$g(x,t) = e(x)\,h(t) \tag{3.66}$$

and look for a solution in the form

$$u(x,t) = e(x)\, v(t). \tag{3.67}$$

If such a function is a solution of Eq. (3.65) then substitution into (3.65) yields that, for all x and t, the following equality takes place

$$e(x)[v''(t) + 2\alpha v'(t) - h(t)] - c^2 e''(x)\, v(t) = 0. \tag{3.68}$$

As a matter of fact, we are interested in non-trivial solutions (non-vanishing identically) and thus we can assume that there exist x_0 and t_0 such that $e(x_0) \neq 0$ and $v(t_0) \neq 0$. For $x = x_0$ the equality (3.68) says that the function v satisfies the ordinary differential equation

$$v''(t) + 2\alpha v'(t) + \lambda\, c^2 v(t) = h(t), \tag{3.69}$$

where

$$\lambda = -\frac{e''(x_0)}{e(x_0)}.$$

Further, inserting $t = t_0$ into (3.68) we get that e is a solution of the equation

$$e''(x) + \lambda\, e(x) = 0. \tag{3.70}$$

Conversely, if the functions v and e satisfy the differential equations (3.69) and (3.70) respectively, where λ is an arbitrary real number, then it is easily to see that the function u of the form (3.67) is a solution of the telegraph equation (3.65) with the right-hand side given by (3.66). □

The above described procedure can be also used in solving some special initial-boundary value problems for Eq. (3.65). Consider, for instance, Eq. (3.65) on the semi-infinite strip $Q = (0, \ell) \times (0, \infty)$ and first look at what initial conditions are satisfied by the solutions of the form (3.67). Evidently

$$u(x,0) = a\, e(x), \quad u_t(x,0) = b\, e(x), \tag{3.71}$$

where the constants a and b are determined by

$$v(0) = a, \quad v'(0) = b. \tag{3.72}$$

The point is that there exists a solution satisfying the conditions (3.71) if a and b are any constants whatever. This is evidently equivalent to the claim that Eq. (3.69) possesses a solution satisfying the initial conditions (3.72) for any a and b. In addition, such a solution can be written down explicitly:
putting

$$\varrho = |\lambda\, c^2 - \alpha^2|^{\frac{1}{2}}$$

then

a) $$v(t) = a\, \exp(-\alpha t)\, \cos(\varrho t) + \frac{1}{\varrho}(b + \alpha a) \exp(-\alpha t) \sin(\varrho t) + \frac{1}{\varrho} \int_0^t \sin(\varrho(t-s)) \exp(-\alpha(t-s))\, h(s)\, ds$$

$$\text{if} \qquad \lambda c^2 - \alpha^2 > 0;$$

b) $$v(t) = a\exp(-\alpha t) + (b + \alpha a)t\exp(-\alpha t) + \int_0^t (t-s)\exp(-\alpha(t-s))\,h(s)\,ds$$

$$\text{if} \qquad \lambda c^2 - \alpha^2 = 0;$$

c) $$v(t) = \frac{1}{2\varrho}[(\varrho + \alpha)a + b]\exp((-\alpha + \varrho)t) + \frac{1}{2\varrho}[(\varrho - \alpha)a - b]\exp(-(\alpha + \varrho)t)$$
$$+ \frac{1}{2\varrho}\int_0^t \exp(-\alpha(t-s))[\exp(\varrho(t-s)) - \exp(-\varrho(t-s))]\,h(s)\,ds$$

$$\text{if} \qquad \lambda c^2 - \alpha^2 < 0.$$

Let us now inquire whether the boundary conditions might also be fulfilled. For the sake of definiteness we consider the Dirichlet boundary conditions

$$u(0,t) = u(\ell,t) = 0, \qquad t > 0. \tag{3.73}$$

A non-trivial function u of the form (3.67) obeys (3.73) if and only if

$$e(0) = e(\ell) = 0. \tag{3.74}$$

The latter relations induce restrictions as regards to possible values of the constant λ which appear in Eqs. (3.69) and (3.70). Namely, as we shall see below, a non-trivial solution of Eq. (3.70) satisfying the conditions (3.74) exists only for

$$\lambda = \frac{k^2\pi^2}{\ell^2},$$

where $k \in \mathbf{N}$, and is given (up to a multiplicative constant) by

$$e(x) = \sin\frac{k\pi x}{\ell}, \qquad 0 < x < \ell.$$

These facts can be verified in the following way:

- For $\lambda = 0$ the general solution of (3.70) is

$$e(x) = d_1 x + d_2,$$

where d_1 and d_2 are arbitrary constants. Substitution into (3.74) gives

$$d_2 = 0, \quad d_1\ell + d_2 = 0$$
$$\text{whence} \quad d_1 = d_2 = 0 \quad \text{and} \quad e \equiv 0.$$

- For λ negative, say $\lambda = -\mu^2$, every solution of Eq. (3.70) can be written in the form

$$e(x) = d_1\exp(\mu x) + d_2\exp(-\mu x).$$

Upon inserting into (3.74) we get

$$d_1 + d_2 = 0, \qquad d_1\exp(\mu\ell) + d_2\exp(-\mu\ell) = 0$$

and again $d_1 = d_2 = 0$ can be deduced.

- For $\lambda = \mu^2$, $\mu > 0$ for definiteness, we have

$$e(x) = d_1 \sin(\mu x) + d_2 \cos(\mu x)$$

and (3.74) yields

$$d_2 = 0, \qquad d_1 \sin(\mu \ell) = 0.$$

For $\mu\ell \neq k\pi$, $\sin(\mu\ell) \neq 0$ and the foregoing relations imply $d_1 = d_2 = 0$. For $\mu\ell = k\pi$, d_1 may be arbitrary and hence functions of the form

$$e(x) = d_1 \sin \frac{k\pi x}{\ell}$$

are the only solutions of Eq. (3.70) satisfying the conditions (3.74) as predicted. □

As could have been surmised from above the same methodology applies to seeking some particular solutions of initial-boundary value problems for wider classes of equations. The problems encountered mostly in this chapter can be written in the following abstract setting:

$$u_{tt} + 2\alpha u_t + L\, u = g(x,t), \tag{3.75}$$

where the time variable t runs over an interval $(0,T)$ [$T = \infty$ admissible], the space variable $x = (x_1, \ldots, x_n)$ belongs to a region $\Omega \subset \mathbf{R}^n$ and L is a differential expression (operator) to be specified in the variables $x_1, \ldots, x_n$, with boundary conditions

$$B_j\, u = 0 \quad \text{on} \quad \partial\Omega \times (0,T), \quad j = 1, \ldots, m, \tag{3.76}$$

where B_j are differential operators again only in variables $x_1, \ldots, x_n$.

The solution is sought in the form

$$u(x,t) = e(x)\, v(t), \qquad x = (x_1, \ldots, x_n).$$

If a function e solves the (so-called) *eigenvalue problem*

$$\begin{cases} L\, e = \lambda e & \text{in} \quad \Omega \\ B_j e = 0 & \text{on} \quad \partial\Omega, \quad j = 1, \ldots, m, \end{cases} \tag{3.77}$$

where λ is a constant, and a function v satisfies the equation

$$v''(t) + 2\alpha v'(t) + \lambda\, v(t) = h(t)$$

and the conditions

$$v(0) = a, \qquad v'(0) = b,$$

then the function u is a solution of Eq. (3.75) with the right-hand side

$$g(x,t) = e(x)\, h(t)$$

and satisfies the boundary conditions (3.76) and the initial conditions

$$u(x,0) = a\, e(x), \quad u_t(x,0) = b\, e(x).$$

The resulting problem for the function v is solvable as we have already shown in the case of the telegraph equation. The main task consists in finding non-trivial functions e and numbers λ satisfying (3.77). [Such functions are called the *eigenfunctions* and the numbers λ are called the *eigenvalues* of the problem (3.77).] □

In some cases we manage to solve the problem (3.77) again by means of the separation of variables. To illustrate we take the Laplace operator Δ given in (1.11) for $-L$ together with the Dirichlet boundary condition. In other words, we want to find a non-trivial function e satisfying the equation

$$\Delta e \equiv \sum_{j=1}^{n} \frac{\partial^2 e}{\partial x_j^2} = -\lambda e \qquad \text{in} \quad \Omega \tag{3.78}$$

subject to the condition

$$u = 0 \qquad \text{on} \quad \partial\Omega. \tag{3.79}$$

Provided that Ω is a rectangular parallelepiped

$$\Omega = (0, \ell_1) \times \cdots \times (0, \ell_n) \tag{3.80}$$

the function e can be sought in the form

$$e(x) = e_1(x_1) \cdots e_n(x_n). \tag{3.81}$$

The condition (3.79) will be met if

$$e_j(0) = e_j(\ell_j) = 0, \qquad j = 1, \ldots, n. \tag{3.82}$$

Upon inserting (3.81) into (3.78) we conclude in a way similar to the derivation of Eqs. (3.69) and (3.70) that the functions e_j must solve the differential equations

$$e_j'' + \lambda_j e_j = 0 \qquad \text{in} \quad (0, \ell_j), \quad j = 1, \ldots, n, \tag{3.83}$$

where λ_j are constants. We already know that non-trivial solutions of these equations satisfying the conditions (3.82) exist if and only if

$$\lambda_j = \frac{k_j^2 \pi^2}{\ell_j^2},$$

where $k_j \in \mathbb{N}$ and the form of these solutions (up to a multiplicative constant) is

$$e_j(x) = \sin \frac{k_j \pi x}{\ell_j}, \qquad x \in (0, \ell_j).$$

To summarize, the problem (3.78), (3.79) on region (3.80) is solved by functions of the form

$$e(x) = \sin\left(\frac{k_1 \pi x_1}{\ell_1}\right) \cdots \sin\left(\frac{k_n \pi x_n}{\ell_n}\right),$$

where $k_1, \ldots, k_n$ are arbitrary positive integers; the corresponding eigenvalues are

$$\lambda = \frac{k_1^2 \pi^2}{\ell_1^2} + \cdots + \frac{k_n^2 \pi^2}{\ell_n^2}. \ \Box$$

We mention yet another important example of a region for which the eigenvalue problem with the Laplace operator and the Dirichlet boundary condition can be reduced to simpler problems in one variable. For convenience we restrict ourselves to the case of only two spatial variables and we shall look for (non-trivial) functions e which, for some λ, solve the equation

$$\Delta e \equiv \frac{\partial^2 e}{\partial x^2} + \frac{\partial^2 e}{\partial y^2} = -\lambda e \tag{3.84}$$

in the circle $\Omega = \{(x, y) \mid x^2 + y^2 < R^2\}$ and satisfy again the Dirichlet boundary condition

$$e(x, y) = 0 \qquad \text{for} \quad x^2 + y^2 = R^2. \tag{3.85}$$

We introduce the polar coordinates r and ϑ where

$$x = r\cos\vartheta, \qquad y = r\sin\vartheta$$

and Eq. (3.84) takes the form (see § 1.2.8)

$$\frac{1}{r}\frac{\partial}{\partial r}\Big(r\frac{\partial e}{\partial r}\Big) + \frac{1}{r^2}\frac{\partial^2 e}{\partial \vartheta^2} = -\lambda e. \tag{3.86}$$

The variables (r, ϑ) run over the rectangle $(0, R) \times (0, 2\pi)$ and this suggests the separation of the form

$$e(r, \vartheta) = f(r)\, w(\vartheta).$$

Substituting into Eq. (3.84) we find that such a function is a solution if and only if the functions w and f satisfy the equations

$$w'' + \mu\, w = 0, \tag{3.87}$$

$$\frac{1}{r}\frac{\partial}{\partial r}\Big(r\frac{\partial f}{\partial r}\Big) + \Big(\lambda - \frac{\mu}{r^2}\Big)f = 0, \tag{3.88}$$

where μ is a constant. Further conditions on the functions w and f are furnished by appealing to (3.85) and to the fact that the function e in the original variables should be twice continuously differentiable. The function w, as well as its derivatives, should assume the same values for $\vartheta = 0$ and $\vartheta = 2\pi$ since these quantities represent the same points in the (x, y)-plane. A non-trivial function w satisfying these requirements and Eq. (3.87) exists only for $\mu = n^2$, where $n \in \mathbb{N} \cup \{0\}$, and then has the form

$$w(\vartheta) = a\cos(n\vartheta) + b\sin(n\vartheta).$$

The solution of Eq. (3.88) can be expressed in terms of the Bessel functions (see § 1.2.8). For $\lambda > 0$ and $\mu = n^2$ the solution f which ultimately yields a twice continuously differentiable function e takes the form

$$f(r) = \mathrm{J}_n(\lambda^{\frac{1}{2}} r). \tag{3.89}$$

One other condition is imposed by (3.85) which leads to

$$f(R) = 0. \tag{3.90}$$

By inserting the function (3.89) into (3.90) we find out that the eigenvalues λ are of the form

$$\lambda = R^{-2}\mu_{nk}^2,$$

where $(\mu_{nk})_{k\in\mathbb{N}}$ is a sequence of positive zeros of the function J_n. So, we have found a double sequence of eigenfunctions and corresponding eigenvalues for the problem (3.84), (3.85). One particular case deserves special mention: we inquire often into solutions having no angular dependence (circular symmetric case). Such solutions evidently correspond to the case $n = 0$ and have the form

$$e_k(x,y) = J_0(R^{-1}\,\mu_k(x^2+y^2)^{\frac{1}{2}}),$$

where $(\mu_k)_{k\in\mathbb{N}}$ is a sequence of positive zeros of the function J_0.

Up to now we have confined ourselves to finding only positive eigenvalues and corresponding eigenfunctions. It turns out, however, that for $\lambda \le 0$ the problem (3.84), (3.85) cannot have non-trivial solutions. To see this let us multiply Eq. (3.84) by the function e and integrate over the region Ω. Applying the Green's theorem 1.2.3 and taking into account the condition (3.85) we arrive eventually at the relation

$$\int_\Omega \left[\left(\frac{\partial e}{\partial x}\right)^2 + \left(\frac{\partial e}{\partial y}\right)^2\right] d\Omega = \lambda \int_\Omega e^2\, d\Omega,$$

from which it follows that $e \equiv 0$ provided $\lambda \le 0$. In general, however, there are problems (for instance, Eq. (3.84) with a different type of boundary conditions) possessing also negative eigenvalues. We return to these questions in Chapter 4.

3.3.2 Elements of the Fourier method

We shall now examine in detail the question of existence of the solution to the initial-boundary value problem given by the telegraph equation

$$u_{tt} - c^2 u_{xx} + 2\alpha u_t = g(x,t) \tag{3.91}$$

on the semi-infinite strip

$$Q = (0,\ell)\times(0,\infty) \tag{3.92}$$

and by the initial and boundary conditions

$$u(x,0) = \varphi(x), \quad u_t(x,0) = \psi(x), \tag{3.93}$$

$$u(0,t) = u(\ell,t) = 0. \tag{3.94}$$

By § 3.3.1 a function of the form

$$u(x,t) = v_k(t)\sin\frac{k\pi x}{\ell}$$

solves this problem (for special right-hand sides and initial functions) if the function v_k obeys the relations

$$\begin{aligned} v_k''(t) + 2\alpha v_k'(t) + \frac{c^2k^2\pi^2}{\ell^2}\, v_k(t) &= h_k(t), \\ v_k(0) = a_k, \quad v_k'(0) &= b_k. \end{aligned} \tag{3.95}$$

Here k is an arbitrary positive integer, h_k is an arbitrary (continuous) function on $\mathbf{R}^+$ and a_k, b_k are arbitrary constants. Further solutions are obtained upon superposing these elementary solutions: the function

$$u(x,t) = \sum_{k=1}^{n} v_k(t) \sin \frac{k\pi x}{\ell}$$

solves the problem (3.91) through (3.94) for the data

$$\begin{aligned} g(x,t) &= \sum_{k=1}^{n} h_k(t) \sin \frac{k\pi x}{\ell}, \\ \varphi(x) &= \sum_{k=1}^{n} a_k \sin \frac{k\pi x}{\ell}, \\ \psi(x) &= \sum_{k=1}^{n} b_k \sin \frac{k\pi x}{\ell}. \end{aligned}$$

With increasing n we obtain an increasing number of right-hand sides and initial functions for which the solution can be furnished. A natural question arises whether a solution for arbitrary data might be eventually obtained after passing to the limit $n \to \infty$.

Having this idea in mind we shall look for a solution of the problem (3.91) through (3.94) in the form of an infinite series

$$u(x,t) = \sum_{k=1}^{\infty} v_k(t) \sin \frac{k\pi x}{\ell}. \tag{3.96}$$

The constants a_k, b_k and the functions h_k are to be chosen in such a way that the relations

$$\varphi(x) = \sum_{k=1}^{\infty} a_k \sin \frac{k\pi x}{\ell}, \tag{3.97}$$

$$\psi(x) = \sum_{k=1}^{\infty} b_k \sin \frac{k\pi x}{\ell}, \tag{3.98}$$

$$g(x,t) = \sum_{k=1}^{\infty} h_k(t) \sin \frac{k\pi x}{\ell} \tag{3.99}$$

hold. At this point we use substantially the fact that

$$\left\{\sin\frac{k\pi x}{\ell}\right\}_{k=1}^{\infty} \tag{3.100}$$

forms the so-called *orthogonal system* of functions on the interval $(0,\ell)$, that is

$$\int_0^\ell \sin\frac{k\pi x}{\ell}\,\sin\frac{j\pi x}{\ell}\,dx = 0, \quad k \neq j.$$

Upon multiplying equality (3.97) by the function $\sin\frac{j\pi x}{\ell}$ and integrating over $(0,\ell)$ we get

$$\int_0^\ell \varphi(x)\,\sin\frac{j\pi x}{\ell}\,dx = \sum_{k=1}^{\infty} a_k \int_0^\ell \sin\frac{k\pi x}{\ell}\,\sin\frac{j\pi x}{\ell}\,dx = \frac{\ell}{2}a_j, \quad j \in \mathbb{N}.$$

Meanwhile the relation

$$\int_0^\ell \sin^2\frac{j\pi x}{\ell}\,dx = \frac{\ell}{2}, \quad j \in \mathbb{N}$$

has been used. Hence we must put

$$a_k = \frac{2}{\ell}\int_0^\ell \varphi(x)\,\sin\frac{k\pi x}{\ell}\,dx, \quad k \in \mathbb{N}, \tag{3.101}$$

which represents a necessary condition for the validity of (3.97). This condition, however, turns out to be sufficient also since (3.100) is the so-called *complete system*, which means that the following implication is true:

$$\int_0^\ell f(x)\,\sin\frac{k\pi x}{\ell}\,dx = 0, \quad k \in \mathbb{N} \quad \Longrightarrow \quad f \equiv 0.$$

Setting

$$f(x) = \varphi(x) - \sum_{k=1}^{\infty} a_k \sin\frac{k\pi x}{\ell}$$

we see that the premise in the above implication is satisfied if and only if a_k are determined by relations (3.101).

Likewise, we can check that the equalities (3.98) and (3.99) hold if and only if we put

$$b_k = \frac{2}{\ell}\int_0^\ell \psi(x)\,\sin\frac{k\pi x}{\ell}\,dx, \quad k \in \mathbb{N}, \tag{3.102}$$

$$h_k(t) = \frac{2}{\ell}\int_0^\ell g(x,t)\,\sin\frac{k\pi x}{\ell}\,dx, \quad t > 0,\ k \in \mathbb{N}. \tag{3.103}$$

The previous considerations suggest a way how to proceed in solving the initial-boundary value problem (3.91) through (3.94). First, by relations (3.101) through (3.103) we determine the constants a_k, b_k and the functions h_k. Second, we find functions v_k as solutions of the problem (3.95). The solution of our initial-boundary value problem is then given by the series (3.96) (on condition that this series is convergent in some appropriate sense). This procedure is called the *Fourier method* for solving the initial-boundary value problem for the telegraph equation. The convergence of the series on the right-hand side of (3.96) will be tackled in § 3.3.4. (In this context we speak also of the convergence of the Fourier method.) □

We can proceed quite similarly in the case of many other equations and so we are entitled to speak of the Fourier method for solving initial-boundary value problems. A crucial role is always played by orthogonal systems of functions.

Let us recall [see, for example, B. Friedmann (1956), Stakgold (1967)] that a system $\{e_k\}_{k\in M}$ of non-trivial functions from $L_2(\Omega)$ where Ω is a region in $\mathbf{R}^n$, is called *orthogonal (on Ω)* [more precisely, *orthogonal in $L_2(\Omega)$*] if

$$\int_\Omega e_j\, e_k\, d\Omega = 0, \qquad j \neq k.$$

We speak here of a system rather than of a sequence since occasionally it is preferable to use the n-tuple-index notation instead of the single-index one commonly used for usual sequences.

Sometimes it is also convenient (to achieve a particularly simple fashion of the Parseval identity mentioned below, for example) that the functions e_k be normalized, that is to say,

$$\int_\Omega e_k^2\, d\Omega = 1, \quad k \in M.$$

The system is then said to be *orthonormal*. Every orthogonal system can be made orthonormal upon multiplying each function by a suitable constant. For instance, the orthogonal system (3.100) on the interval $(0,\ell)$ becomes orthonormal after multiplying every function by $(2/\ell)^{\frac{1}{2}}$, that is

$$\left\{ \left(\frac{2}{\ell}\right)^{\frac{1}{2}} \sin \frac{k\pi x}{\ell} \right\}_{k=1}^{\infty}$$

is an orthonormal system on $(0,\ell)$. We mention in passing that in the Fourier method for the telegraph equation instead of functions $\sin(k\pi x/\ell)$ we could have used their arbitrary (non-zero) multiples with the only difference that the right-hand sides of relations (3.101) through (3.103) would have been multiplied by certain constants.

For any function $f \in L_2(\Omega)$ the numbers

$$f_k = \int_\Omega f\, e_k\, d\Omega, \qquad k \in M$$

are called the *Fourier coefficients* of the function f (with respect to a given system). If $\{e_k\}_{k\in M}$ is an orthonormal system then for every $f \in L_2(\Omega)$ we have the so-called *Bessel inequality*

$$\sum_{k\in M} f_k^2 \leq \int_\Omega f^2\, d\Omega. \tag{3.104}$$

The symbol $\sum\limits_{k\in M}$ denotes (here and elsewhere) the series $\sum\limits_{k=1}^{\infty}$ if M coincides with the set $\mathbf{N}$ of positive integers (or a multiple series if the index set M consists of n-tuples of positive integers).

A system $\{e_k\}_{k\in M}$ is called *complete* if the zero function is the only function from $L_2(\Omega)$ whose Fourier coefficients are all zero, in other terms:

$$\int_\Omega f\, e_k\, d\Omega = 0, \quad k \in M \quad \Longrightarrow \quad f \equiv 0.$$

If $\{e_k\}_{k\in M}$ is a complete orthonormal system then for every $f, g \in L_2(\Omega)$ the so-called *Parseval identity*

$$\int_\Omega fg\,d\Omega = \sum_{k\in M} f_k\,g_k, \tag{3.105}$$

holds, where f_k and g_k are the Fourier coefficients of the function f and g respectively. By the Parseval identity we also mean the relation

$$\int_\Omega f^2\,d\Omega = \sum_{k\in M} f_k^2 \tag{3.106}$$

which follows from (3.105) by setting $f = g$. □

In the remainder of this section we return to the question of how to apply the Fourier method for solving initial-boundary value problems placed in the general framework having been set up in § 3.3.2, viz.

$$\begin{gathered} u_{tt} + 2\alpha u_t + L\,u = g(x,t), \\ B_j u = 0 \qquad \text{on} \quad \partial\Omega\times(0,T), \quad j = 1,\dots,m \\ u(x) = \varphi(x), \quad u_t(x,0) = \psi(x), \end{gathered} \tag{3.107}$$

where L and B_j are differential operators involving only space variables $x_1,\dots,x_n$ which run over a region $\Omega \subset \mathbf{R}^n$ and the time variable runs over an interval $(0,T)$. The method relies on the existence of a complete orthonormal system of (sufficiently smooth) eigenfunctions of the problem (3.77). So let us adopt the following assumption:

$$\begin{gathered} \text{there exists a complete orthonormal system } \{e_k\}_{k\in M} \text{ on } \Omega \text{ such that} \\ L\,e_k = \lambda_k e_k, \quad B_j\,e_k = 0 \quad \text{on } \partial\Omega \quad \text{for} \quad k\in M \quad \text{and} \quad j = 1,\dots,m. \end{gathered} \tag{3.108}$$

The Fourier method consists in looking for a solution in the form of the expansion in terms of the system $\{e_k\}_{k\in M}$

$$u(x,t) = \sum_{k\in M} v_k(t)\,e_k(x). \tag{3.109}$$

Quite analogously as we did for the telegraph equation we easily check that such a function u (on condition that the series on the right-hand side is convergent) is a solution provided that the functions v_k solve the problem [cf. (3.95)]

$$\begin{gathered} v_k''(t) + 2\alpha v_k'(t) + \lambda_k v_k(t) = h_k(t), \\ v_k(0) = a_k, \quad v_k'(0) = b_k, \end{gathered} \tag{3.110}$$

where a_k, b_k and $h_k(t)$ are the Fourier coefficients of the function φ, ψ and g respectively, that is,

$$\begin{gathered} a_k = \int_\Omega \varphi\,e_k\,d\Omega, \quad b_k = \int_\Omega \psi\,e_k\,d\Omega, \\ h_k(t) = \int_\Omega g(\cdot,t)\,e_k\,d\Omega. \end{gathered}$$

We proceed to the discussion of when the fundamental assumption (3.108) takes place.

3.3.3 Complete orthonormal systems of eigenfunctions

A detailed investigation of eigenvalue problems ensuring the existence of complete orthonormal systems of eigenfunctions requires a fuller knowledge of functional analysis and therefore we restrict ourselves only to stating some basic assertions without proofs.

We start by considering a fairly general differential expression L in one space variable of the form

$$\begin{aligned} L\,u(x) \equiv (-1)^m \frac{d^m}{dx^m}\Big(a_m(x)\frac{d^m u(x)}{dx^m}\Big) \\ + \sum_{r=0}^{m-1} \frac{d^r}{dx^r}\Big(a_r(x)\frac{d^r u(x)}{dx^r}\Big) \end{aligned} \tag{3.111}$$

(a differential operator of order $2m$), where the variable x runs over a bounded interval $(0,\ell)$ and a_r are sufficiently smooth functions on $[0,\ell]$. Let the boundary conditions be of the form

$$\sum_{s=0}^{2m-1} \Big(b_{js}\frac{d^2u(0)}{dx^s} + d_{js}\frac{d^2u(\ell)}{dx^s}\Big) = 0, \quad j = 1,\ldots,2m, \tag{3.112}$$

where b_{js} and d_{js} are real constants.

For brevity in writing we denote by D_L the set of all functions from $C^{2m}[0,\ell]$ which satisfy the boundary conditions (3.112). The eigenvalue problem in which we are interested may be written formally as

$$L\,e = \lambda e, \qquad e \in D_L. \tag{3.113}$$

The following assertion holds.

Let $a_r \in C^r[0,\ell]$ for $r = 0,\ldots,m$ and $a_m(x) > 0$, $x \in [0,\ell]$. Further, let the relation

$$\int_0^\ell (L\,u)v\,dx = \int_0^\ell u(L\,v)\,dx \tag{3.114}$$

be satisfied for a function $v \in C^{2m}[0,\ell]$ and for all $u \in D_L$ if and only if $v \in D_L$.

Then there exists a complete orthogonal (orthonormal, respectively) system $\{e_k\}_{k=1}^\infty$ on $(0,\ell)$ of eigenfunctions of the problem (3.113). If λ_k are the corresponding eigenvalues then there is only a finite number of them in every interval $(-\infty, B)$ where $B < \infty$ (in other words: *the eigenvalues tend to ∞*). □

In this text we need to solve eigenvalue problems for differential operators of the second order (i.e. for $m = 1$); an exception is Chapter 5 where we meet a differential operator of the fourth order. For most of this chapter we deal with the operator

$$L\,u(x) \equiv -u''(x) \tag{3.115}$$

on the interval $(0,\ell)$ [or $L\,u \equiv -c^2u''$ which has the same eigenfunctions and eigenvalues which differ only by the constant factor c^2]. For any twice continuously differentiable functions u and v on the interval $[0,\ell]$ we obtain upon integrating by parts twice

$$\int_0^\ell (L\,u)v\,dx = \int_0^\ell (-u'')v\,dx = -u'(\ell)\,v(\ell) + u'(0)\,v(0) + u(\ell)\,v'(\ell) - u(0)\,v'(0) + \int_0^\ell u(L\,v)\,dx.$$

The assumption (3.114) will be satisfied [for a definite set of boundary conditions] if the identity

$$-u'(\ell)\,v(\ell) + u'(0)\,v(0) + u(\ell)\,v'(\ell) - v(0)\,v'(0) = 0 \tag{3.116}$$

holds for any u satisfying the boundary conditions if and *only if* v satisfies the same boundary conditions.

Let us take, for example, the homogeneous Dirichlet boundary conditions

$$u(0) = u(\ell) = 0. \tag{3.117}$$

If also $v(0) = v(\ell) = 0$ then (3.116) is true. Conversely, there exist evidently such functions u satisfying the conditions (3.117) for which $u'(0) = 1$, $u'(\ell) = 0$ [and $u'(0) = 0$, $u'(\ell) = 1$, respectively] and substituting them into (3.116) we find that v must fulfil the conditions (3.117). Hence, in this case the assumption (3.114) is satisfied and consequently there exists a complete orthogonal system of eigenfunctions of the problem (3.113) with L given by (3.115) and boundary conditions specified by (3.117). In § 3.3.1 we showed that every eigenfunction e necessarily assumes the form

$$e(x) = \sin\frac{k\pi x}{\ell}$$

(up to a multiplicative constant) and this yields that (3.100) is a complete orthogonal system on $(0,\ell)$, a fact which constitutes a key ingredient of the analysis in § 3.3.2.

Eigenvalue problems for the operator (3.115) with other types of boundary conditions can be handled in a similar way. For instance, for the boundary conditions

$$\begin{aligned} b_1u'(0) + b_0u(0) &= 0,\\ d_1u'(\ell) + d_0u(\ell) &= 0 \end{aligned} \tag{3.118}$$

the assumption (3.114) is satisfied if the constants b_0, b_1 and d_0, d_1 do not vanish simultaneously, that is, if

$$b_0^2 + b_1^2 > 0, \qquad d_0^2 + d_1^2 > 0. \tag{3.119}$$

Thus, these conditions ensure the existence of a complete orthogonal system of eigenfunctions corresponding to (3.115) and (3.118). Similarly, as in the case of the Dirichlet boundary conditions, we can establish their analytical representations as well as the equations for eigenvalues (not so easily solvable, though). To this end we use again the fact that for an arbitrary λ we know the general solution of the equation

$$-u''(x) = \lambda\,u(x).$$

Substitution into (3.118) then gives conditions for possible values of λ, that is, the equations for the eigenvalues. We shall not enter into technical details but rather quote the conclusions which can be deduced in this case (some other details are surveyed in §4.4.1).

1. There exist at most two non-positive eigenvalues. To each eigenvalue there corresponds only one eigenfunction (up to a constant factor).
2. The number $\lambda = -\mu^2$, where $\mu > 0$, is an eigenvalue if and only if

$$(b_1 d_0 - b_0 d_1) \cosh(\mu\ell) + (b_1 d_1 \mu - b_0 d_0 \mu^{-1}) \sinh(\mu\ell) = 0.$$

The corresponding eigenfunction $e(x)$ has the form (up to a constant factor)

$$e(x) = b_1 \cosh(\mu x) - b_0 \mu^{-1} \sinh(\mu x).$$

3. The number $\lambda = 0$ is an eigenvalue if and only if

$$b_0 d_0 \ell - b_1 d_0 + b_0 d_1 = 0$$

and the corresponding eigenfunction has the form (up to a constant factor)

$$e(x) = b_0 x - b_1.$$

4. The number $\lambda = \mu^2$, where $\mu > 0$, is an eigenvalue if

$$(b_0 d_0 - b_0 d_1) \cos(\mu\ell) = (b_1 d_1 \mu + b_0 d_0 \mu^{-1}) \sin(\mu\ell). \tag{3.120}$$

The corresponding eigenfunction has the form (up to a constant factor)

$$e(x) = b_1 \cos(\mu x) - b_0 \mu^{-1} \sin(\mu x).$$

A fuller discussion yields that Eq. (3.120) possesses virtually infinitely many solutions which tend to ∞. To illustrate, taking $b_0 = 1$, $b_1 = d_0 = d_1 = -1$ and $\ell = 1$ we can rewrite this equation in the form

$$\operatorname{cotg} \mu = \frac{1}{2}(\mu - \frac{1}{\mu})$$

and it is apparent from Fig. 18 that for any $n \in \mathbb{N}$ there is exactly one solution in the interval $((n-1)\pi, n\pi]$. $\square$

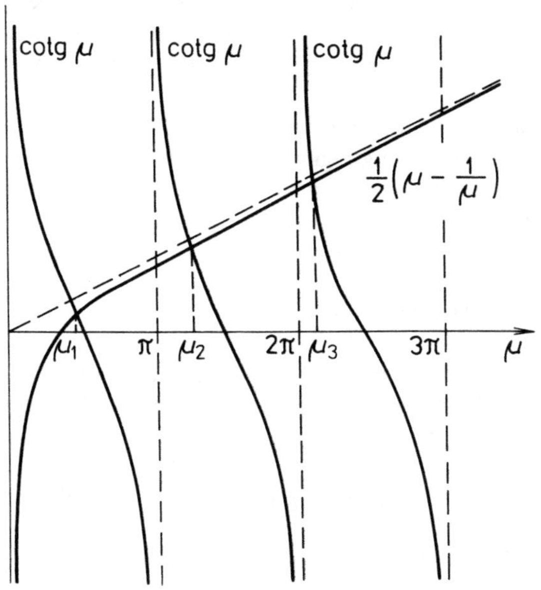

Fig. 18

The existence of a complete orthogonal (and hence orthonormal) system of eigenfunctions corresponding to a differential operator in several space variables can be ensured in some special cases by means of the following general assertion (stated without proof). *Let* $\{f_{j,k_j}\}_{k_j=1}^{\infty}$ *be a complete orthogonal (orthonormal, respectively) system of functions on the interval* $(0, \ell_j)$ *for* $j = 1, \ldots, n$. *Set*

$$e_{k_1,\ldots,k_n}(x) = f_{1,k_1}(x_1) \cdots f_{n,k_n}(x_n).$$

Then $\{e_{k_1,\ldots,k_n}\}_{k_1,\ldots,k_n=1}^{\infty}$ *is a complete orthogonal (orthonormal, respectively) system of functions on the region*

$$\Omega = (0, \ell_1) \times \cdots \times (0, \ell_n) \subset \mathbf{R}^n$$

[Let us add that the assertion remains true if the intervals $(0, \ell_j)$ are replaced by regions $\Omega_j \subset \mathbf{R}^{d_j}$. The resulting system will be orthogonal (orthonormal, respectively) and complete on the region $\Omega = \Omega_1 \times \cdots \times \Omega_n \subset \mathbf{R}^{d_1+\cdots+d_n}$.]

For example,

$$\left\{ \left(\frac{2}{\ell_j} \right)^{\frac{1}{2}} \sin \frac{k_j \pi x}{\ell_j} \right\}_{k_j=1}^{\infty}$$

is a complete orthonormal system on $(0, \ell_j)$ and in virtue of the above assertion

$$\left\{ \left(\frac{2^n}{\ell_1 \cdots \ell_n} \right)^{\frac{1}{2}} \sin \left(\frac{k_1 \pi x_1}{\ell_1} \right) \cdots \sin \left(\frac{k_n \pi x_n}{\ell_n} \right) \right\}_{k_1,\ldots,k_n=1}^{\infty}$$

is a complete orthonormal system on the region $\Omega = (0, \ell_1) \times \cdots \times (0, \ell_n)$. In § 3.3.1 we have shown that these functions are eigenfunctions of the problem (3.78), (3.79) and now the completeness of this system of functions is furnished. (We take benefit of this fact in § 3.4.5.)

In § 3.3.1 we have found yet another system of eigenfunctions corresponding to (3.84), (3.85) where the underlying region is a circle in $\mathbf{R}^2$. Again it can be shown that this system is complete orthogonal even though this fact does not follow directly from the above assertions.

3.3.4 Convergence of the Fourier method

We have shown that solutions of some initial-boundary value problems can be expressed by means of certain series of functions as long as these series are convergent in a convenient fashion, that is, as long as they determine functions with smoothness properties as required. Since the detailed investigation of conditions for the convergence of these series is usually rather lengthy we shall exemplify the procedure for the telegraph equation.

By § 3.3.2 the solution of the problem (3.91) through (3.94) is given by the series

$$u(x,t) = \left(\frac{2}{\ell}\right)^{\frac{1}{2}} \sum_{k=1}^{\infty} v_k(t) \sin\frac{k\pi x}{\ell}, \tag{3.121}$$

where the functions v_k solve the problem (3.95) with

$$\begin{aligned} a_k &= \left(\frac{2}{\ell}\right)^{\frac{1}{2}} \int_0^\ell \varphi(x) \sin\frac{k\pi x}{\ell}\, dx \\ b_k &= \left(\frac{2}{\ell}\right)^{\frac{1}{2}} \int_0^\ell \psi(x) \sin\frac{k\pi x}{\ell}\, dx, \\ h_k(t) &= \left(\frac{2}{\ell}\right)^{\frac{1}{2}} \int_0^\ell g(x,t) \sin\frac{k\pi x}{\ell}\, dx. \end{aligned} \tag{3.122}$$

(The series (3.121) differs somewhat from the series (3.96) since now we use the expansion in the series with respect to the orthonormal system of the corresponding eigenfunctions; note that the relations for a_k, b_k and $h_k(t)$ differ accordingly from those introduced in (3.101) through (3.103).)

The function (3.121) is actually a classical solution of the problem in question if it belongs for example to $C^2(\bar{Q})$. This property is ensured most easily by establishing the uniform convergence on every bounded subset of $\bar{Q}$ of the series on the right-hand side of (3.121) together with term-by-term differentiated series up to the second order. Simple estimates provide a sufficient condition:

$$\text{the series } \sum_{k=1}^{\infty} \left(k^2|v_k(t)| + k|v_k'(t)| + |v_k''(t)|\right) \tag{3.123}$$

is uniformly convergent on every bounded interval.

The function v_k solve the problem (3.95) and their explicit representations are known (see § 3.3.1). Rather lengthy estimates yield further that (3.123) holds provided

- $$\sum_{k=1}^{\infty} k^2 |a_k| < \infty \qquad (3.124_1)$$
- $$\sum_{k=1}^{\infty} k|b_k| < \infty \qquad (3.124_2)$$
- the series $\sum_{k=1}^{\infty} k|h_k(t)|$
  is uniformly convergent on bounded intervals. (3.124_3)

We claim that (3.124_1) is satisfied if

$$\varphi \in C^3[0,\ell], \quad \varphi(0) = \varphi(\ell) = \varphi''(0) = \varphi''(\ell) = 0. \qquad (3.125)$$

Indeed, integrating by parts (and using the boundary conditions imposed on φ) we get for any $k \in \mathbb{N}$

$$\begin{aligned} a_k &= \left(\frac{2}{\ell}\right)^{\frac{1}{2}} \int_0^\ell \varphi(x) \sin\frac{k\pi x}{\ell}\,dx = \left(\frac{2}{\ell}\right)^{\frac{1}{2}} \frac{\ell}{k\pi} \int_0^\ell \varphi'(x) \cos\frac{k\pi x}{\ell}\,dx \\ &= -\left(\frac{2}{\ell}\right)^{\frac{1}{2}} \frac{\ell^2}{k^2\pi^2} \int_0^\ell \varphi''(x) \sin\frac{k\pi x}{\ell}\,dx = -\left(\frac{2}{\ell}\right)^{\frac{1}{2}} \frac{\ell^3}{k^3\pi^3} \int_0^\ell \varphi'''(x) \cos\frac{k\pi x}{\ell}\,dx. \end{aligned}$$

Hence

$$k^3 a_k = -\frac{\ell^3}{\pi^3}\sigma_k,$$

where σ_k are the Fourier coefficients of the function φ''' with respect to the orthonormal system $\left\{\left(\frac{2}{\ell}\right)^{\frac{1}{2}} \cos\frac{k\pi x}{\ell}\right\}_{k=1}^{\infty}$. According to the Bessel inequality

$$\sum_{k=1}^{\infty} k^6 a_k^2 = \frac{\ell^6}{\pi^6} \sum_{k=1}^{\infty} \sigma_k^2 \le \frac{\ell^6}{\pi^6} \int_0^\ell |\varphi'''(x)|^2\,dx < \infty.$$

Now we use a) the well-known Cauchy inequality

$$\sum_{k=1}^{\infty} |\alpha_k \beta_k| \le \left(\sum_{k=1}^{\infty} \alpha_k^2\right)^{\frac{1}{2}} \left(\sum_{k=1}^{\infty} \beta_k^2\right)^{\frac{1}{2}} \qquad (3.126)$$

which is valid for any two sequences of real numbers and b) the fact that

$$\sum_{k=1}^{\infty} \frac{1}{k^2} \ (= \frac{\pi^2}{6}) < \infty. \qquad (3.127)$$

Putting $\alpha_k = 1/k$ and $\beta_k = k^3 a_k$ we get in view of (3.126), (3.127)

$$\sum_{k=1}^{\infty} k^2 |a_k| = \sum_{k=1}^{\infty} \frac{1}{k} k^3 |a_k| \leq \left(\sum_{k=1}^{\infty} \frac{1}{k^2}\right)^{\frac{1}{2}} \left(\sum_{k=1}^{\infty} k^6 a_k^2\right)^{\frac{1}{2}} < \infty$$

as desired. Analogously we can verify that the remaining two conditions in (3.124) are satisfied if

$$\psi \in C^2[0,\ell], \quad \psi(0) = \psi(\ell) = 0, \tag{3.128}$$

$$g, g_x, g_{xx} \in C(\bar{Q}), \quad g(0,t) = g(\ell,t) = 0,\ t > 0. \tag{3.129}$$

The assumptions on φ, ψ and g we have obtained for the existence of the classical solution to the problem (3.91) through (3.94) are stronger than those required by the Riemann method. The point here is that it is difficult to characterize the continuity (continuous differentiability) of functions by means of the behaviour of their Fourier coefficients. Notwithstanding the Fourier method is very well adaptable for treating generalized solutions (of even more general equations). For illustration we refer the reader to Section 4.3.

3.3.5 The Fourier method and the problem of resonance

As pointed out already the Fourier method enables us to obtain further important information concerning the solutions of initial-boundary value problems. To illustrate we will determine the so-called resonance frequencies.

The problem of resonance (in its simplest form) arises when an undamped physical system is acted on by harmonic external forces. Then it is well-known that for certain impressed frequencies, which are said to be *resonant* (or *in resonance* with natural frequencies), the energy of the system can grow indefinitely while for the other frequencies the energy remains bounded. A great increase in the energy can bring about damage to the system and for this reason it is often desirable that non-resonance frequency of the given external force be achieved (by changing some parameters of the system). On the other hand we sometimes need (for example, in radio engineering) to strengthen the response of the system to an external signal as much as possible and in this case we must "tune up" the system to a given frequency. These examples suggest that the dependence of the resonance frequencies upon the parameters of the given system is of considerable interest in practice.

We limit ourselves here to the investigation of the systems that are described by a function u satisfying certain homogeneous boundary conditions and a linear differential equation of the form

$$u_{tt} + L\,u = g(x,t), \tag{3.130}$$

where L is a linear differential operator in a region Ω of the space variables $x_1, \ldots, x_n$ and g represents an external force. The energy of such a system in time t is usually given (up to a multiplicative constant) by the expression

$$E(t) = \int_\Omega u_t^2(x,t)\,d\Omega + \int_\Omega Lu(x,t)\,u(x,t)\,d\Omega, \qquad (3.131)$$

where the first term corresponds to the kinetic energy and the second one to the potential energy. Multiplying equation (3.130) by the function $2u_t$ and integrating over the domain Ω we obtain, for the rate of the energy of the system, the relation

$$E'(t) = 2\int_\Omega g(x,t)\,u_t(x,t)\,d\Omega$$

which says nothing other than that this rate is equal to the power of the external forces. If no external forces are acting the energy is conserved: this is a more precise definition that the system under consideration be undamped.

Let us assume further that we can apply the Fourier method to Eq. (3.130). This means above all that there exists a complete orthonormal system $\{e_k\}_{k\in M}$ of eigenfunctions of the eigenvalue problem (3.113). The corresponding eigenvalues $\{\lambda_k\}_{k\in M}$ are supposed to be positive and their number to be finite in every interval of finite length.

As we have already shown in previous subsections the solution u of Eq. (3.130) can be written in the form of a series as

$$u(x,t) = \sum_{k\in M} v_k(t)\,e_k(x)$$

where v_k solve the problem (3.110) with $\alpha = 0$, that is, referring to p. 132,

$$v_k(t) = a_k\cos(\mu_k t) + \frac{b_k}{\mu_k}\sin(\mu_k t) + \frac{1}{\mu_k}\int_0^t \sin\left(\mu_k(t-s)\right) g_k(s)\,ds. \qquad (3.132)$$

Here μ_k are the *natural angular frequencies*,

$$\mu_k = \lambda_k^{\frac{1}{2}}$$

and a_k, b_k and $g_k(t)$ are the Fourier coefficients of initial functions and of the function g respectively.

The Fourier coefficients of the functions u_t, Lu and u are given by $v_k'(t)$, $\lambda_k v_k(t)$ and $v_k(t)$ respectively, and according to the Parseval identity (3.106) we get

$$E(t) = \sum_{k\in M} E_k(t) \qquad (3.133)$$

where

$$E_k(t) = v_k'^2(t) + \lambda_k v_k^2(t), \quad k \in M. \qquad (3.134)$$

The relation (3.133) expresses the important fact that the energy corresponding to a solution is equal to the sum of the energies of individual Fourier components of this solution.

Let us investigate now the time evolution of the energy if the external force is of the form

$$g(x,t) = h(x)\,\sin(2\pi\omega t).$$

First of all we note that in the study of resonance problems it is sufficient to consider only the solutions with vanishing initial data. [Indeed, every solution may be written as a sum of two functions, the first of which is a solution of the given equation with vanishing initial functions and the second one solves the homogeneous equation; the resonance can be influenced only by the first part of the solution since the energy corresponding to the second part is constant as we have seen above.] Let $E^{\omega}(t)$ denote the energy corresponding to the solution and $E_k^{\omega}(t)$ be its components (in the decomposition (3.133)). The function E_k^{ω} is given by the expression on the right in (3.134) where v_k are given by (3.132) with $a_k = b_k = 0$ and $g_k(t) = h_k \sin(2\pi\omega t)$, h_k being the Fourier coefficients of the function h.

We find by an easy computation

$$
\begin{aligned}
E_k^{\omega}(t) = \frac{2\pi\omega h_k^2}{(2\pi\omega)^2 - \lambda_k} \Bigg[& \frac{1}{2\pi\omega + \mu_k} - \frac{\cos((2\pi\omega + \mu_k)t)}{2\pi\omega + \mu_k} - \frac{1}{2\pi\omega - \mu_k} \\
& \frac{+}{\cos((2\pi\omega - \mu_k)t)} 2\pi\omega - \mu_k - \frac{1}{4\pi\omega} + \frac{\cos(4\pi\omega t)}{4\pi\omega} \Bigg]
\end{aligned}
\tag{3.135}
$$

for $2\pi\omega \neq \mu_k$ and

$$
E_k^{\omega}(t) = h_k^2 \left(\frac{1}{4} t^2 - \frac{1}{4\pi\omega} t \sin(2\pi\omega t) \cos(2\pi\omega t) + \frac{1}{16\pi^2\omega^2} \sin^2(2\pi\omega t) \right) \tag{3.136}
$$

for $2\pi\omega = \mu_k$.

Let ω be such a frequency of the external force that $2\pi\omega \neq \mu_k$ for all $k \in M$. Then $|2\pi\omega - \mu_k| > d, k \in M$, where $d > 0$ is a suitable constant. Estimating (crudely for simplicity) in (3.135) we obtain

$$
E_k^{\omega}(t) \leq \frac{h_k^2}{\pi d \omega}, \quad k \in M, \quad t > 0.
$$

It follows further (invoking the Parseval identity)

$$
E^{\omega}(t) = \sum_{k \in M} E_k^{\omega}(t) \leq \frac{1}{\pi d \omega} \sum_{k \in M} h_k^2 = \frac{1}{\pi d \omega} \int_{\Omega} h^2 \, d\Omega, \quad t > 0.
$$

So, in this case the energy of the system is bounded by a constant independent of time and this means that ω is not a resonance frequency.

Next, if $2\pi\omega = \mu_k$ for some $k \in M$ (and moreover $h_k \neq 0$) then it follows from (3.136) that the component E_k^{ω} (and *a fortiori* the total energy) diverges for $t \to \infty$.

Thus we arrive at the simple conclusion that the impressed frequency ω is resonant if and only if

$$
\omega = \frac{1}{2\pi} \lambda_k^{\frac{1}{2}}
$$

for some $k \in M$.

3.3.6 The Fourier method and stability

We will briefly show that the Fourier method is suitable also for the investigation of stability questions. Let us consider systems described by equations of the type

$$u_{tt} + 2\alpha u_t + L\, u = g(x,t), \tag{3.137}$$

with the same assumptions on L as in § 3.3.5. Eq. (3.137) differs from Eq. (3.130) by the presence of the term $2\alpha u_t$ which corresponds to the external friction.

As having been already emphasized in § 1.1.6 the way of measuring of the magnitude of a solution at time t plays an important role in the study of stability. For the time being, we shall regard the corresponding energy $E(t)$ given by (3.131) as the "measure" of this magnitude. Again,

$$E(t) = \sum_{k\in M} (v_k'^2(t) + \lambda_k\, v_k^2(t)),$$

where now, of course, the functions v_k solve the problem (3.110). Let us first assume that $\alpha > 0$ and $\alpha^2 < \lambda_1$ where λ_1 is the smallest eigenvalue of the corresponding eigenvalue problem. After having solved equations for v_k, inserting into the expression for the energy and performing some estimates to simplify we get that for any $t > 0$

$$E(t) \le \frac{12\lambda_1}{\lambda_1 - \alpha^2} \exp(-2\alpha t)\, E(0) + \frac{9\lambda_1}{\alpha^2(\lambda_1 - \alpha^2)} G, \tag{3.138}$$

where

$$G = \sup_t \left(\int_\Omega g^2(x,t)\, d\Omega \right).$$

The inequality says that the solution of Eq. (3.137) is exponentially stable with respect to initial conditions and stable with respect to persistent disturbances. This remains true even for $\alpha > 0$, $\alpha^2 \ge \lambda_1$ while for $\alpha = 0$ the solution is stable with respect to initial conditions and unstable with respect to persistent disturbances and for $\alpha < 0$ the solution is unstable in either sense.

It is of interest to examine the rate of the exponential decay of solutions (with $g \equiv 0$) for various values of $\alpha > 0$. For $0 \le \alpha^2 < \lambda_1$ one has owing to (3.138)

$$E(t) \le a \exp(-2\alpha t)\, E(0),$$

where a is a constant and it can be shown that this estimate is optimal. For $\alpha^2 > \lambda_1$ the optimal estimate reads

$$E(t) \le a \exp(-2k(\alpha)t)\, E(0),$$

where

$$k(\alpha) = \alpha - (\alpha^2 - \lambda_1)^{\frac{1}{2}}.$$

A somewhat more complicated situation arises for $\alpha^2 = \lambda_1$. In this case for any $\varepsilon > 0$,

$$E(t) \le a(\varepsilon) \exp[-2(\alpha - \varepsilon)t]\, E(0),$$

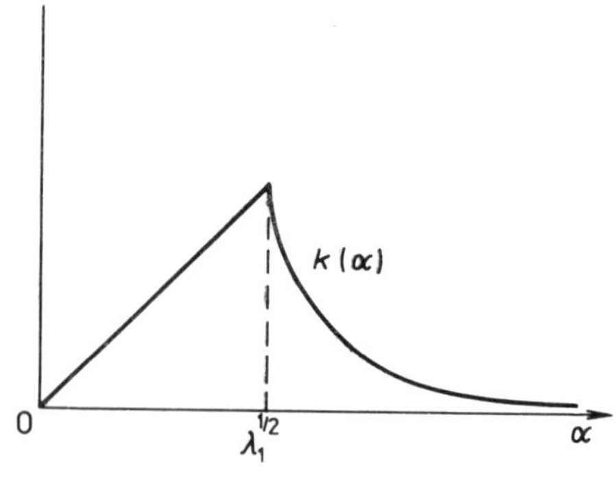

Fig. 19

However, $a(\varepsilon)$ tends to ∞ as $\varepsilon \to 0$ and thus no optimal estimate exists. The graph of the dependence of the damping exponent upon the coefficient α is drawn in Fig. 19 (with an exception for $\alpha = \lambda^{\frac{1}{2}}$).

Incidentally, it is not true that the system be damped the faster the larger be the external friction.

3.3.7 Example: the Fourier method for a general telegraph equation with constant coefficients

We look for a solution of the problem given by the equation

$$u_{tt} - c^2 u_{xx} + 2\alpha u_t + 2\beta u_x + \gamma u = g(x,t), \quad 0 < x < \ell,\, t > 0, \tag{3.139}$$

the initial conditions

$$u(x,0) = \varphi(x), \quad u_t(x,0) = \psi(x), \quad 0 < x < \ell$$

and the Dirichlet boundary conditions

$$u(0,t) = u(\ell,t) = 0, \quad t > 0.$$

The Fourier method in the form characterized in § 3.3.2 cannot be applied directly to cope with this problem (if $\beta \neq 0$). Although Eq. (3.138) might be written in the abstract form (3.107) with

$$L\,u = -c^2 u_{xx} + 2\beta u_x + \gamma u$$

there does not exist (for $\beta \neq 0$) a complete orthogonal system of eigenfunctions of the corresponding eigenvalue problem

$$\begin{cases} L\,e = \lambda\, e, \\ \text{homogeneous Dirichlet boundary conditions.} \end{cases} \tag{3.140}$$

Notwithstanding, a simple transformation (of the dependent variable) makes the problem susceptible to treatment by the Fourier method. Namely, if we put

$$u(x,t) = \exp\frac{\beta x}{c^2}\, w(x,t), \tag{3.141}$$

then the function u is a solution of the given problem if and only if the function w is a solution of the initial-boundary value problem

$$\begin{gathered} w_{tt} - c^2 w_{xx} + 2\alpha w_t + \Big(\frac{\beta^2}{c^2} + \gamma\Big)\, w = \exp\Big(-\frac{\beta x}{c^2}\Big)\, g(x,t), \\ w(x,0) = \exp\Big(-\frac{\beta x}{c^2}\Big)\, \varphi(x), \qquad w_t(x,0) = \exp\Big(-\frac{\beta x}{c^2}\Big)\, \psi(x), \quad 0 < x < \ell, \\ w(0,t) = w(\ell,t) = 0, \quad t > 0. \end{gathered}$$

By § 3.3.2 the function w is determined by the series

$$w(x,t) = \sum_{k=1}^{\infty} \Big(\frac{2}{\ell}\Big)^{\frac{1}{2}} v_k(t) \sin\frac{k\pi x}{\ell},$$

where v_k solve the ordinary differential equations

$$\begin{aligned} &v_k''(t) + 2\alpha v_k'(t) + \Big(\frac{c^2 k^2 \pi^2}{\ell^2} + \frac{\beta^2}{c^2} + \gamma\Big)\, v_k(t) \\ &= \Big(\frac{2}{\ell}\Big)^{\frac{1}{2}} \int_0^{\ell} \exp\Big(-\frac{\beta x}{c^2}\Big)\, g(x,t) \sin\frac{k\pi x}{\ell}\, dx \end{aligned}$$

with initial conditions

$$\begin{aligned} v_k(0) &= \Big(\frac{2}{\ell}\Big)^{\frac{1}{2}} \int_0^{\ell} \exp\Big(-\frac{\beta x}{c^2}\Big)\, \varphi(x) \sin\frac{k\pi x}{\ell}\, dx, \\ v_k'(0) &= \Big(\frac{2}{\ell}\Big)^{\frac{1}{2}} \int_0^{\ell} \exp\Big(-\frac{\beta x}{c^2}\Big)\, \psi(x) \sin\frac{k\pi x}{\ell}\, dx. \end{aligned}$$

For the solution u of the original problem one has an account of (3.141)

$$u(x,t) = \sum_{k=1}^{\infty} v_k(t) \Big(\frac{2}{\ell}\Big)^{\frac{1}{2}} \exp\frac{\beta x}{c^2} \sin\frac{k\pi x}{\ell}.$$

Let us remark that upon setting

$$e_k(x) = \Big(\frac{2}{\ell}\Big)^{\frac{1}{2}} \exp\frac{\beta x}{c^2} \sin\frac{k\pi x}{\ell}$$

we get a system of eigenfunctions of the problem (3.140) and hence the solution u is given formally by the same expression furnished by the Fourier method. The difference is that now the system $\{e_k\}_{k=1}^{\infty}$ is not orthogonal. Nevertheless, it may be easily shown that

$$\int_0^\ell \varrho(x)\, e_j(x)\, e_k(x)\, dx = 0, \qquad j \neq k,$$

where

$$\varrho(x) = \exp(-\frac{2\beta x}{c^2}).$$

Such a system is called *orthogonal on* $(0,\ell)$ *with the weight (function)* ϱ [more precisely: *orthogonal in the weight space* $L_{2,\varrho}(0,\ell)$ where this symbol (for any non-negative integrable function ϱ on the interval $(0,\ell)$ vanishing at at most finitely many points) denotes the Hilbert space of functions u such that $\varrho^{\frac{1}{2}} u \in L_2(0,\ell)$]. It turns out that in fact the Fourier method can be extended without great formal changes to the cases of initial-boundary value problems of the form (3.107) where the operator L together with the boundary conditions provides a system of eigenfunctions which is complete and orthogonal with a certain weight ϱ.

3.3.8 Exercises

3.3.8.1 Find a solution of the initial-boundary value problem

$$\begin{aligned} u_{tt} - u_{xx} + 2u_t &= \sin^3 x, \quad 0 < x < \pi,\ t > 0, \\ u(x,0) &= \sin x,\ u_t(x,0) = 0, \quad 0 < x < \pi, \\ u(0,t) &= u(\pi,t) = 0, \quad t > 0. \end{aligned}$$

Hint. Use the trigonometric identity

$$\sin^3 x = \frac{1}{4}(3 \sin x - \sin(3x)).$$

Answer.

$$u(x,t) = 4^{-1}\{3 + (1+t)\exp(-t)\} \sin x + 36^{-1}\{8^{-\frac{1}{2}} \sin(8^{\frac{1}{2}} t) \exp(-t) + \cos(8^{\frac{1}{2}} t) \exp(-t) - 1\} \sin 3x.$$

3.3.8.2 Determine the form that would be appropriate for a solution of the telegraph equation

$$u_{tt} - c^2 u_{xx} + 2\alpha u_t + \gamma u = g(x,t), \quad 0 < x < \ell,\ 0 < t < T,$$

satisfying the boundary conditions

$$u_x(0,t) = u_x(\ell,t) = 0.$$

Answer.

$$u(x,t) = \left(\frac{1}{\ell}\right)^{\frac{1}{2}} v_0(t) + \left(\frac{2}{\ell}\right)^{\frac{1}{2}} \sum_{k=1}^{\infty} v_k(t) \cos \frac{k\pi x}{\ell}.$$

3.4 THE WAVE AND TELEGRAPH EQUATIONS IN SEVERAL SPACE VARIABLES

The *wave equation in n space variables* is an equation of the form

$$u_{tt}(x,t) - c^2 \Delta u(x,t) = g(x,t),$$

where $x = (x_1, \dots, x_n)$ and Δ is the Laplace operator (1.11). An equation involving also lower order derivatives of the function u, for example,

$$u_{tt} + 2\alpha u_t - c^2 \Delta u = g(x,t)$$

is called, as before, the *telegraph equation*. The investigation of initial-boundary value problems for the wave and telegraph equations in higher dimensions is much more complicated compared with the case of one space variable and therefore we are compelled to restrict ourselves only to treating a few special cases. First of all, we find expressions for the solution of the initial-value problem for the wave equation in three and two space variables. Further we show that in some particular cases (regardless of the number of space variables) the Fourier method is applicable (which in turn makes it possible by the preceding subsections to examine simply among other things the stability and resonance questions).

3.4.1 On equations of the electromagnetic field

The theory of the electromagnetic field provides a wealth of examples of initial-boundary value problems for the wave and telegraph equations in three space variables. The electromagnetic field is governed by the system of Maxwell's equations

$$\begin{aligned}
\frac{\partial E}{\partial t} + \frac{\gamma}{\varepsilon} E - \frac{1}{\varepsilon\mu} \operatorname{curl} B &= -\frac{1}{\varepsilon} J, \\
\frac{\partial B}{\partial t} + \operatorname{curl} E &= 0, \\
\operatorname{div} B &= 0, \\
\operatorname{div} E &= \frac{1}{\varepsilon}\rho
\end{aligned}$$

as pointed out in § 2.1.5 where we have also already learned about the possibility to reduce this system to a system of telegraph (or wave) equations. Namely, if the vector functions E and B satisfy the above system of equations, in addition to being twice continuously differentiable, then they solve the following system of equations

$$\frac{\partial^2 E}{\partial t^2} + \frac{\gamma}{\varepsilon}\frac{\partial E}{\partial t} - \frac{1}{\varepsilon\mu}\Delta E = -\frac{1}{\varepsilon}\frac{\partial J}{\partial t} - \frac{1}{\varepsilon^2\mu}\operatorname{grad}\rho,$$

$$\frac{\partial^2 B}{\partial t^2} + \frac{\gamma}{\varepsilon}\frac{\partial B}{\partial t} - \frac{1}{\varepsilon\mu}\Delta B = -\frac{1}{\varepsilon}\operatorname{curl} J.$$

These systems split (on condition that the quantities ρ and J are known) into separate telegraph (or wave if $\gamma = 0$) equations for each individual component of the functions E and B. (Let us note that if the functions E and B are only once continuously differentiable then they represent generalized solutions of foregoing systems of telegraph equations.)

The converse problem of how solutions of these telegraph equations are connected with the solutions of the original system of Maxwell's equations will be touched upon in § 3.4.6.

3.4.2 The initial-value problem in three space dimensions

We turn now to the question of existence and uniqueness of the solution to the problem

$$u_{tt} - c^2 \Delta u = g(x_1, x_2, x_3, t) \qquad \text{in } \mathbf{R}^3 \times (0, \infty), \tag{3.142}$$

$$u = \varphi, \quad u_t = \psi \qquad \text{for } t = 0. \tag{3.143}$$

The following theorem holds.

Let $\varphi \in C^3(\mathbf{R}^3)$, $\psi \in C^2(\mathbf{R}^3)$ and g be continuous on $\mathbf{R}^3 \times \mathbf{R}^+$ together with its derivatives with respect to space variables up to the second order. Then there exists exactly one classical solution of the problem (3.142), (3.143). *This solution is given by the formula*

$$u(x,t) = \frac{1}{4\pi} \int_S \Big[\varphi(x + cty) + ct \sum_{j=1}^{3} \frac{\partial \varphi}{\partial x_j}(x + cty) y_j + t\psi(x + cty)\Big]\, d_y S$$
$$+ \frac{1}{4\pi} \int_0^t (t - \tau) \Big[\int_S g(x + c(t - \tau)y, \tau)\, d_y S\Big] d\tau, \tag{3.144}$$

where S denotes the surface of the unit three-dimensional sphere, i.e.

$$S = \{ (y_1, y_2, y_3) \mid y_1^2 + y_2^2 + y_3^2 = 1 \}.$$

A direct verification shows that the function u defined by (3.144) actually furnishes us with the solution of the problem (3.142), (3.143). Let us say a few words about why the solution must necessarily have this form (being thus determined uniquely). Assume that u is a solution of our problem, pick a definite point $x \in \mathbf{R}^3$ and define a function v for $\xi \geq 0$ and $t \in (0, T)$ by the formula

$$v(\xi, t) = \xi \int_S u(x + \xi y, t)\, d_y S.$$

The function v can be envisaged as a certain average (up to a constant factor) of the function u and accordingly this procedure is called occasionally the *averaging method*. We can check by a direct computation that the function v is a solution of the wave equation

$$v_{tt} - c^2 v_{\xi\xi} = \xi \int_S g(x + \xi y, t)\, d_y S$$

on the quarter plane $(0,\infty)\times(0,\infty)$ and satisfies the initial and boundary conditions

$$v(\xi,0)=\xi\int_S \varphi(x+\xi y)\,d_yS, \qquad \xi\geq 0,$$

$$v_t(\xi,0)=\xi\int_S \psi(x+\xi y)\,d_yS, \qquad \xi\geq 0,$$

$$v(0,t)=0, \qquad t\geq 0,$$

$$v_\xi(0,t)=4\pi u(x,t), \qquad t\geq 0.$$

The representation (3.144) can be now readily deduced by means of the relations for the solution of the wave equation we have derived in § 3.1.6.

The averaging method in the form described above yields the representation of the solution of the initial-value problem only in the case of three space dimensions and because of this fact we have not dwelt on the details. Let us remark that exact formulae for the solutions of the initial-value problem for the wave equation in arbitrary space variables are known. Since these topics are fairly complicated we omit any further discussion of methods or results. However some features of the case of two space variables are indicated in § 3.4.3.

At this point a remark is in order. Observe that the smoothness we have required of the initial functions is higher than the one the solution enjoys. It may be shown that the flaw is in the nature of things and not in the method used. The more space variables the more derivatives we "lose". (This situation does not occur when dealing with generalized solutions and this leads us to suspect that it is not very convenient to look for solutions in the class of continuously differentiable functions.)

3.4.3 The initial-value problem in two space dimensions

On the basis of the representation formula for the solution of the initial-value problem for the wave equation in three space variables we obtain easily the representation of the solution of the same problem in two space dimensions using the so-called *method of descent*. The idea behind this method is simple: if the initial functions φ and ψ do not depend upon the variable x_3 then the function u given by (3.144) does not depend on x_3 either and consequently it is (as a function of space variables x_1, x_2 and the time variable t) a solution of the wave equation in two space variables. We obtain the following theorem.

Let the region $Q=\mathbf{R}^2\times(0,\infty)$. *Then for any* $\varphi\in C^3(\mathbf{R}^3)$, $\psi\in C^2(\mathbf{R}^2)$ *and for any function* g *continuous on* $\bar{Q}$ *together with its derivatives with respect to the space variables up to the second order there exists exactly one solution of the wave equation*

$$u_{tt}-c^2\Delta u=g \qquad \textit{in } Q,$$

which satisfies the initial conditions

$$u=\varphi, \quad u_t=\psi \qquad \textit{for } t=0.$$

The solution is given by the relation

$$u(x_1,x_2,t) = \frac{1}{2\pi}\int_K \Big[\varphi(x_1+cty_1,x_2+cty_2) + ct\sum_{j=1}^{2}\frac{\partial}{\partial x_j}\varphi(x_1+cty_1,x_2+cty_2)\,y_j$$

$$+\,t\psi(x_1+cty_1,x_2+cty_2)\Big]\left(1-y_1^2-y_2^2\right)^{-\frac{1}{2}}dy_1dy_2$$

$$+\frac{1}{2\pi}\int_0^t (t-\tau)\Big[\int_K g\left(x_1+c(t-\tau)y_1,x_2+c(t-\tau)x_2,\tau\right)\left(1-y_1^2-y_2^2\right)^{-\frac{1}{2}}dy_1dy_2\Big]d\tau, \tag{3.145}$$

where K is the unit circle in $\mathbf{R}^2$,

$$K = \left\{(y_1,y_2)\mid y_1^2+y_2^2 \le 1\right\}.$$

Note that the relation (3.145) is obtained from (3.144) taking into account that for any (continuous) function f depending only on variables y_1 and y_2 one has

$$\int_S f\,d_yS = 2\int_K \frac{f(y_1,y_2)}{(1-y_1^2-y_2^2)^{\frac{1}{2}}}dy_1dy_2,$$

where S is the surface of the unit sphere in $\mathbf{R}^3$.

3.4.4 Wave propagation velocity

We can easily check that any function of the form

$$u(x_1,\dots x_n,t) = v(ct-k_1x_1-\cdots-k_nx_n)$$

is a solution of the homogeneous wave equation provided that $v\in C^2(\mathbf{R})$ and $k = (k_1,\dots,k_n)$ is a unit vector. It is natural to call such a solution the *plane wave* travelling with the velocity c in the direction of the vector k; namely, the function u maintains for any fixed t the same values on planes perpendicular to the vector k and these surfaces of constant values move, as t evolves, in the direction k with the velocity c. The constant c seems (and it is in fact justified) to mean again the *velocity of propagation* of disturbances in processes described by these equations. In the case of two and three space variables we can verify this claim directly on the basis of results of the previous two subsections. It follows from the representations (3.144) and (3.145) that the change of values of initial functions at a position y [or the value of the external force at y at time τ] can affect the value of the solution at another position x and at a later time t only if the distance between x and y is not greater than ct [or $c(t-\tau)$].

For three space variables we have actually a far stronger result. If for example the initial functions are vanishing and the external forces are different from zero only in a small neighbourhood of a position x and for small values of time then the related solution at time t is different from zero only in a small neighbourhood of the surface of the sphere centred at x and with radius ct. This means that there is no wake in the propagating wave, the wave is not diffuse. "Events" at a certain position separated in time give rise to "events" at another position (some time later) which are again separated in time. This suggests the importance enjoyed by the processes described by the wave equation in three space variables for undistorted transfer of information.

3.4.5 Initial-boundary value problems

The investigation of initial-boundary value problems for the wave (and telegraph) equation considered on regions $\Omega \times (0,T)$ where Ω is a bounded region in $\mathbf{R}^n$ can often be carried out successfully by means of the Fourier method.

To illustrate we take the problem given by the telegraph equation

$$u_{tt} + 2\alpha u_t - c^2\Delta u = g, \tag{3.146}$$

the initial conditions

$$u = \varphi, \quad u_t = \psi \qquad \text{on } \Omega \tag{3.147}$$

and the homogeneous Dirichlet condition

$$u = 0 \qquad \text{on } \partial\Omega \times (0,T),$$

where Ω is a bounded n-dimensional interval, viz.

$$\Omega = (0,\ell_1) \times (0,\ell_2) \times \cdots \times (0,\ell_n).$$

By § 3.3.2 and § 3.3.3 we shall look for a solution in the form of a series

$$u(x_1,\ldots,x_n,t) = \left(\frac{2^n}{\ell_1\cdots\ell_n}\right)^{\frac{1}{2}} \sum_{k_1,\ldots,k_n=1}^{\infty} u_{k_1,\ldots,k_n}(t)\sin\left(\frac{k_1\pi x_1}{\ell_1}\right)\cdots\sin\left(\frac{k_n\pi x_n}{\ell_n}\right), \tag{3.149}$$

where the functions $u_{k_1,\ldots,k_n}$ are solutions of the initial-value problem

$$\begin{aligned} v''(t) + 2\alpha v'(t) + \lambda_{k_1,\ldots,k_n} v(t) &= g_{k_1,\ldots,k_n}(t), \quad t > 0, \\ v(0) = \varphi_{k_1,\ldots,k_n}, \qquad v'(0) &= \psi_{k_1,\ldots,k_n}. \end{aligned} \tag{3.150}$$

Here the right-hand side and initial conditions are given as usual by the Fourier coefficients of the function g and the initial functions and

$$\lambda_{k_1,\ldots,k_n} = \frac{k_1^2\pi^2}{\ell_1^2} + \cdots + \frac{k_n^2\pi^2}{\ell_n^2}.$$

The solution of the problem (3.150) can be written out explicitly and this permits us to study in a direct fashion under what conditions on the initial functions φ, ψ and the right-hand side g the series in (3.149) is uniformly convergent on any bounded subset of $\Omega \times (0,T)$ together with term-by-term differentiated series up to the second order determining thus a twice continuously differentiable function u which is a solution of our problem. Proceeding in a very similar way as in § 3.3.4 we can check that it suffices to assume

- $\varphi \in C^{[\frac{n}{2}]+3}(\bar{\Omega}), \quad \dfrac{\partial^{2j}\varphi}{\partial x_k^{2j}} = 0 \quad \text{for } x_k = 0, \ell_k,$

  $$k = 1,\ldots,n, \quad 0 \le 2j < \left[\frac{n}{2}\right] + 3$$

- $\psi \in C^{[\frac{n}{2}]+2}(\bar{\Omega}), \quad \dfrac{\partial^{2j}\psi}{\partial x_k^{2j}} = 0 \quad \text{for } x_k = 0, \ell_k,$

  $$k = 1,\ldots,n, \quad 0 \le 2j < \left[\frac{n}{2}\right] + 2$$

- the function g is continuous on $\bar{\Omega} \times [0,T]$
  together with its partial derivatives with respect to
  the space variables $x_1, \ldots, x_n$ up to the order $\left[\frac{n}{2}\right] + 2$,
  $\dfrac{\partial^{2j} g}{\partial x_k^{2j}} = 0$ for $t \in [0,T]$, $x_k = 0, \ell_k$, $k = 1, \ldots, n$, $0 \le 2j < \left[\frac{n}{2}\right] + 2$

where $\left[\frac{n}{2}\right]$ denotes the smallest integer equal to or exceeding $\frac{n}{2}$.

The above conditions reflect the fact that with an increasing number of space variables a still higher degree of smoothness of the initial functions and the right-hand side is needed in order to ensure the existence of the classical solution. For instance, for $n = 3$ the conditions to be imposed are $\varphi \in C^4(\bar{\Omega})$, $\psi \in C^3(\bar{\Omega})$. This "loss" of smoothness is here, however, greater than indispensable (on grounds indicated in § 3.3.4). In this respect consult § 3.4.2 where the smoothness $\varphi \in C^3(\mathbf{R}^3)$ and $\psi \in C^2(\mathbf{R}^3)$ of the initial functions has been declared sufficient for the existence of the classical solution of the initial-value problem for the wave equation in three space dimensions.

3.4.6 Example: distribution of the electromagnetic field in an endless rectangular waveguide

In this subsection we show how to proceed at least in some particular cases when dealing with initial-boundary value problems for the wave equation on regions of the form $Q = \Omega \times (0,T)$ where Ω is neither the whole space $\mathbf{R}^3$ nor a bounded set. As an example, consider the problem of the distribution of the electromagnetic field in an endless waveguide (more precisely, in a uniform hollow tube of a perfectly conducting material filled with dielectric) whose cross-section is rectangular. We shall assume that inside there is a medium of zero conductivity (i.e. $\gamma = 0$) with no presence of free charges (i.e. $\rho = 0$, $J = 0$). Mathematically, the problem is to find a solution of the system of Maxwell's equations

$$\frac{\partial E}{\partial t} - \frac{1}{\varepsilon\mu}\operatorname{curl} B = 0, \tag{3.151}$$

$$\frac{\partial B}{\partial t} + \operatorname{curl} E = 0, \tag{3.152}$$

$$\operatorname{div} B = 0, \tag{3.153}$$

$$\operatorname{div} E = 0 \tag{3.154}$$

on the region $\Omega \times (0,T)$ where $\Omega = (0,a) \times (0,b) \times \mathbf{R}$. Further we shall assume the values of the functions E and B to be known for $t = 0$, that is

$$E(x,0) = \varphi(x), \qquad x \in \Omega, \tag{3.155}$$

$$B(x,0) = \sigma(x), \qquad x \in \Omega. \tag{3.156}$$

The boundary conditions will be deduced by appealing to the physics of the problem: the tangent components of the vector E and the normal components of the vector B

are known to vanish on the interface between a perfect conductor and a dielectric. The components of the vector function E we denote by E_j, $j = 1, 2, 3$ (and a similar notation is used for other vector functions). The function E_3 forms the tangent component on the whole boundary $\partial\Omega \times (0, T)$ and accordingly we get the boundary condition

$$E_3 = 0 \qquad \text{on } \partial\Omega \times (0, T). \tag{3.157}$$

Let us denote by Γ_1 the portion of the boundary $\partial\Omega$ corresponding to the front and rear walls of the waveguide and by T_2 the portion corresponding to lateral walls (see Fig. 20).

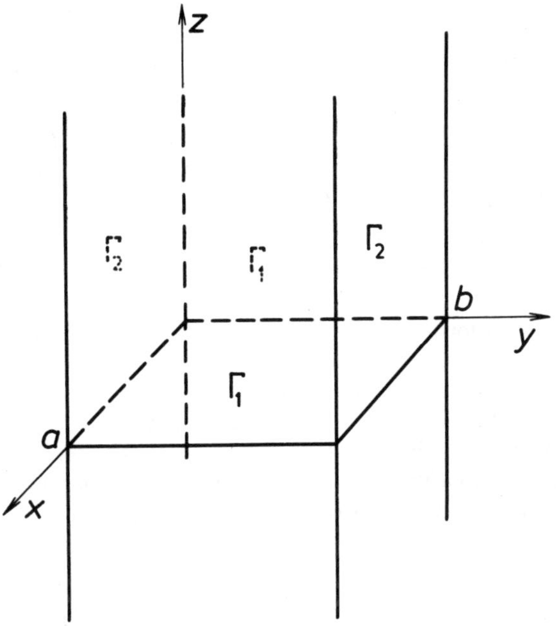

Fig. 20

The function E_1 is the tangent component of E on $\Gamma_2 \times (0, T)$ and the function E_2 on $\Gamma_1 \times (0, T)$ which leads to the boundary conditions

$$E_1 = 0 \qquad \text{on } \Gamma_2 \times (0, T), \tag{3.158}$$
$$E_2 = 0 \qquad \text{on } \Gamma_1 \times (0, T). \tag{3.159}$$

In an analogous way we obtain the boundary conditions for the components B_j of the vector function B:

$$B_1 = 0 \qquad \text{on } \Gamma_1 \times (0, T), \tag{3.160}$$
$$B_2 = 0 \qquad \text{on } \Gamma_2 \times (0, T). \tag{3.161}$$

The setting of our problem is complete.

In order to find its solution we start with the determination of the function E. We employ the fact that E must be a solution (at least generalized) of the equation

$$\frac{\partial^2 E}{\partial t^2} - \frac{1}{\varepsilon\mu}\Delta E = 0, \tag{3.162}$$

in other words, each component E_j must solve the homogeneous wave equation. One initial condition is given by relation (3.155), the other one is obtained from Eq. (3.151) which yields

$$\frac{\partial E}{\partial t}(x,0) = \frac{1}{\varepsilon\mu}\operatorname{curl}\sigma(x) \equiv \psi(x), \qquad x \in \Omega. \tag{3.163}$$

The boundary condition for E_3 is given by (3.157). Relations (3.158) and (3.159) define boundary conditions for E_1 and E_2, respectively, but only on $\Gamma_2 \times (0,T)$ and $\Gamma_1 \times (0,T)$ respectively. The conditions on the remaining parts of the boundary are deduced from Eq. (3.154):

$$\frac{\partial E_1}{\partial x_1} = \operatorname{div} E - \frac{\partial E_2}{\partial x_2} - \frac{\partial E_3}{\partial x_3} = 0 \qquad \text{on } \Gamma_1 \times (0,T), \tag{3.164}$$

$$\frac{\partial E_2}{\partial x_2} = \operatorname{div} E - \frac{\partial E_1}{\partial x_1} - \frac{\partial E_3}{\partial x_3} = 0 \qquad \text{on } \Gamma_2 \times (0,T), \tag{3.165}$$

In this way we have obtained usual initial-boundary value problems for the functions E_j.

Because of the special form of the region Ω we can take advantage of the method of extension. If for example we extend the initial functions φ_1, ψ_1 from Ω to the whole of $\mathbf{R}^3$ in the following way: with respect to x_1 as an even function to $(-a,0)$ and elsewhere periodically, with respect to x_2 as an odd function to $(-b,0)$ and elsewhere periodically, then the corresponding solution of the initial-value problem will have the same properties [as may be easily checked by a simple inspection of relation (3.144)] from which it follows easily that it must satisfy the conditions (3.158) and (3.164). The restriction of this solution to the region $\Omega \times (0,T)$ obeys all conditions imposed on the function E_1.

Similarly we can establish the functions E_2 and E_3. The only difference is in the way of extension of the initial functions. For E_2 they are to be extended as odd $2a$-periodic functions with respect to x_1 and even $2b$-periodic functions with respect to x_2 [this is a proper extension ensuring that the solution of the corresponding initial-value problem should satisfy the conditions (3.159) and (3.165)]. For E_3 we need [in order to fulfil the condition (3.157)] the initial functions to be extended as odd $2a$-periodic functions with respect to x_1 and to odd $2b$-periodic functions with respect to x_2.

The function B satisfying Eq. (3.152) together with the initial condition (3.156) is already uniquely determined by the function E defined above. This is a consequence of the evident relation

$$B(x,t) = \sigma(x) + \int_0^t \operatorname{curl} E(x,\tau)\,d\tau, \qquad x \in \Omega,\ t \in (0,T).$$

If the initial functions satisfy certain additional assumptions [demanding that the relations (3.159) through (3.161) be satisfied for $t = 0$] the functions E and B may be shown to form an actual solution of our problem. We omit the proof which is rather lengthy; let us remark only that the most peculiar part is the verification of Eq. (3.154). Neither shall we lay down all conditions on initial functions making our preceding procedure legitimate. Instead we show the uniqueness of the solution (this should be done, otherwise there is no guarantee that the solution found describes the process we had in mind). It is immediate that it suffices to prove the uniqueness of the function E (since the function B is already uniquely determined in terms of E). Consider, for example, the component E_3 and denote by u_{k_1,k_2} a function defined for $k_1, k_2 \in \mathbf{N}$, $x_3 \in \mathbf{R}$ and $t \in (0,T)$ by the formula

$$u_{k_1,k_2}(x,t) = \left(\frac{4}{ab}\right)^{\frac{1}{2}} \int_0^a \int_0^b E_3(x_1,x_2,x_2,t) \sin\frac{k_1\pi x_1}{a} \sin\frac{k_2\pi x_2}{b} dx_1 dx_2.$$

It is easily seen that the functions u_{k_1,k_2} are necessarily solutions of the telegraph equation in one space variable on the region $\mathbf{R}\times(0,T)$ with given initial functions and thus they are determined in a unique manner. Since u_{k_1,k_2} are just the Fourier coefficients of the function E_3 (for fixed x_3, t) with respect to a certain complete orthonormal system of functions on the rectangle $(0,a)\times(0,b)$, the values $E_3(x_1,x_2,x_3,t)$ are uniquely determined as well. Analogously (using other proper complete orthonormal systems) we can check the uniqueness of functions E_1 and E_2.

3.4.7 Exercises

3.4.7.1 Find the value $u(0,0,0,t) = v(t)$ of the solution u of the problem (3.142), (3.143) if
a) $\varphi(x_1,x_2,x_2) = f(x_1^2+x_2^2+x_3^2)$, $\psi \equiv 0$, $g \equiv 0$;
b) $\varphi \equiv 0$, $\psi(x_1,x_2,x_3) = f(x_1^2+x_2^2+x_3^2)$, $g \equiv 0$;
c) $\varphi \equiv 0$, $\psi \equiv 0$, $g(x_1,x_2,x_3,t) = f(x_1^2+x_2^2+x_3^2)$.
Here f is an arbitrary sufficiently smooth function.
Answer.
a) $v(t) = f(c^2t^2) + 2c^2t^2 f'(c^2t^2)$;
b) $v(t) = tf(c^2t^2)$;
c) $v(t) = \int_0^t (t-\tau) f(c^2(t-\tau)^2) d\tau$.

3.4.7.2 Find the form that would be appropriate for solving the following problem by means of the Fourier method:

$$u_{tt} - u_{xx} - u_{yy} = g(x,y,t), \qquad (x,y) \in (0,\pi)\times(0,\pi),\ t>0,$$

$$u(0,y,t) = u(\pi,y,t) = u_y(x,0,t) = u_y(x,\pi,t), \qquad t>0.$$

Answer.

$$u(x,y,t) = \frac{2}{\pi}\sum_{j,k=1}^{\infty} u_{jk}(t)\sin(jx)\cos(ky) + \frac{2^{\frac{1}{2}}}{\pi}\sum_{j=1}^{\infty} v_j(t)\sin(jx).$$

3.5 GENERAL SECOND ORDER HYPERBOLIC EQUATIONS

The wave and telegraph equations fall into the category of the so-called hyperbolic equations of the second order. A detailed investigation of these general equations would take us too far afield and therefore we content ourselves here with only giving a brief account of some of their basic properties. Among the most important and typical is the finite wave propagation velocity.

3.5.1 Hyperbolic equations

In § 1.1.3 the notion of hyperbolicity was introduced for the linear second order equation (1.5) in two independent variables. We proceed to the extension of this definition to the case of equations in $n+1$ independent variables:

$$\sum_{j,k=1}^{n+1} a_{jk}(x)\frac{\partial^2 u}{\partial x_j \partial x_k} + \sum_{j,k=1}^{n+1} b_j(x)\frac{\partial u}{\partial x_j} + c(x)\, u = g(x), \qquad x \in Q, \tag{3.166}$$

where $a_{jk}(x) = a_{jk}(x)$, $j,\ k = 1,\dots,n+1$, $x \in Q$, Q is a region in $\mathbf{R}^{n+1}$. Eq. (3.166) is called *hyperbolic at the point* $x \in Q$ if all eigenvalues of the matrix $\mathcal{A}(x) = (a_{jk}(x))_{j,k=1}^{n+1}$ are non-zero and real and all but one are of the same sign. Eq. (3.166) is called *hyperbolic in* Q if it is hyperbolic in every $x \in Q$. The condition on eigenvalues suggests that at every point $x \in Q$ there exists a certain particular direction (defined by the eigenvector of the matrix $\mathcal{A}(x)$ corresponding to that eigenvalue having the opposite sign of the others) along which the properties of hyperbolic equations differ from those in other directions. Hyperbolic equations are frequently given directly in the form (or can be turned into such a form by a change of variables) in which the variable corresponding to this privileged direction is labelled distinctly by t, viz.

$$\frac{\partial^2 u}{\partial t^2} - \sum_{j,k=1}^{n} a_{jk}\frac{\partial^2 u}{\partial x_j \partial x_k} + \alpha\frac{\partial u}{\partial t} + \sum_{j=1}^{n} \beta_j \frac{\partial u}{\partial x_j} + \gamma u = g(x,t). \tag{3.167}$$

As has already been observed the variables t and x differ also in so far as their physical meaning is concerned. The hyperbolicity of Eq. (3.167) now means that at any $(x,t) \in Q$ the matrix $\mathcal{A} = (a_{jk}(x,t))_{j,k=1}^{n}$ is symmetric and all the eigenvalues are positive.

In the sequel we shall be concerned only with equations of the form (3.167). For convenience we assume that the coefficients of the equation as well as the right-hand side are sufficiently smooth and bounded functions and that the inequality

$$\sum_{j,k=1}^{n} a_{jk}(x,t)\, \xi_j \xi_k \geq \delta \sum_{j=1}^{n} \xi_j^2$$

holds for any $(x,t) \in Q$ and all $\xi = (\xi_1,\dots,\xi_n) \in \mathbf{R}^n$ where $\delta > 0$ is a suitable constant.

For Eq. (3.167) analogous initial- and initial-boundary value problems as for the wave equation can be set up and also analogous assertions on existence and uniqueness of solutions can be proved, classical solutions (in more space variables) have in common the loss of smoothness of initial functions, and so forth. Now, however, analytic expressions for solutions are seldom available and thus the proof of the above mentioned results must be accomplished with the aid of other methods the study of which is considered to be beyond the scope of this book. An exception is paid to the Fourier method by means of which we can often successfully tackle the problems on regions of the type $\Omega \times (0,T)$ where Ω is a bounded region in $\mathbf{R}^n$ (provided, for instance, that the functions a_{jk} and γ are independent of t, α is a constant and $\beta_j \equiv 0$, as we have already seen in Section 3.3).

Next, we focus our attention on the question of the propagation velocity of wave processes described by Eq. (3.167). We start with the notion of energy inequalities (which provide also an important tool in proofs of existence and uniqueness of solutions).

3.5.2 Energy inequalities

Let us consider Eq. (3.167) on a region $Q \subset \mathbf{R}^{n+1}$. We associate with each region $D \subset Q$ and $t \in \mathbf{R}$ the set

$$D_t = \{\, x \in \mathbf{R}^n \mid (x,t) \in D\} .$$

Further for $t \in \mathbf{R}$ and any solution u of Eq. (3.167) we denote

$$E(t) = \int_{D_t} [(\frac{\partial u}{\partial t})^2 + \sum_{j,k=1}^{n} a_{jk} \frac{\partial u}{\partial x_j}\frac{\partial u}{\partial x_k}](x,t)\, dx.$$

The value $E(t)$ represents (up to a constant factor) the energy (or the leading portion of the energy) of the system described by the solution of Eq. (3.167) that is possessed by the set D_t at time t. The latter fact accounts for the term *energy inequalities* for the inequalities involving E which we are about to derive.

We have already met examples of energy inequalities (energy equalities) in § 3.3.5 and § 3.3.6. The set D_t therein has been independent of time and in their derivation the boundary conditions have been substantially exploited. Now we shall establish energy inequalities which are valid for any solution of Eq. (3.167). We obtain convenient energy inequalities under the main assumption that the region D is still narrower (or broader, respectively) as time evolves (that is, the sets D_t become either smaller or larger with increasing t).

For the sake of simplicity let us consider Eq. (3.167) on the region $Q = \mathbf{R}^n \times \mathbf{R}$. Let us pick a definite point $(x^0, t_0) \in D$ with $t_0 > 0$ and take the region D in the form

$$D = \{\,(x,t) \in Q \mid t < f(x)\}\,, \tag{3.168}$$

where f is a function defined on $\mathbf{R}^n$. Let the following assumptions be satisfied

$$f \in C(\mathbf{R}^n), \quad f \in C^1(\mathbf{R}^n \setminus \{x^0\}), \tag{3.169$_1$}$$

$$t_0 = f(x^0), \tag{3.169$_2$}$$

$$D_{t_1} \supset D_{t_2} \qquad \text{for } t_1 \le t_2. \tag{3.169$_3$}$$

The set D is nothing other than a certain "cone" with vertex at the point (x^0, t_0). We introduce further a function F defined by

$$F(t) = \int_{D_t} u^2(x,t)\,dx.$$

In view of the relation

$$u(x,t) = u(x,0) + \int_0^t \frac{\partial u(x,\tau)}{\partial t}\,d\tau$$

one easily shows that, for $0 < t < t_0$,

$$F(t) \leq k_1 F(0) + k_2 \int_0^t E(\tau)\,d\tau, \tag{3.170}$$

where the constants k_1 and k_2 are independent of the solution. A further inequality is obtained upon multiplying Eq. (3.167) by the function u_t, integrating with respect to the time variable over $(0,t)$ and with respect to the space variables. After rather lengthy manipulations which we shall not perform here and assuming moreover that

$$\sum_{j,k=1}^{n} a_{jk}(x, f(x)) \frac{\partial f}{\partial x_j} \frac{\partial f}{\partial x_k} \leq 1, \qquad x \neq x^0 \tag{3.171}$$

we arrive at the following estimate

$$E(t) \leq E(0) + k_3 \int_0^t E(\tau)\,d\tau + k_4 G + k_5 F(0), \qquad 0 < t < t_0, \tag{3.172}$$

where the constants k_3, k_4 and k_5 are independent of both u and g and

$$G = \sup\left\{\, |g(x,t)| \mid (x,t) \in D,\ 0 < t < t_0 \right\}.$$

On account of the energy inequality (3.172) we get the following result: if u is such a solution of Eq. (3.167) for which $u = u_t = 0$ in B and $g = 0$ in D^+ where

$$B = \{\,(x,0) \mid (x,0) \in D\},$$
$$D^+ = \{\,(x,0) \in D \mid 0 < t < t_0\}$$

then $u = 0$ in D^+. Namely, in this case $E(0) = F(0) = G = 0$ and the inequality (3.172) assumes the form

$$E(t) \leq k_3 \int_0^t E(\tau) d\tau, \qquad 0 < t < t_0.$$

The latter inequality together with the evident relation $E(t) \geq 0$ implies $E(t) = 0$ for $0 < t < t_0$ and by (3.170) $F(t) = 0$ for $0 < t < t_0$, whence $u = 0$ in D^+.

We can restate the previous result in another way: *the values of the solution u of Eq. (3.167) and its derivatives at the point (x^0, t_0) depend only upon the values of the* *functions u and u_t in the set B (and upon the values of the function g in D^+).* To see this, let u_1, u_2 be two solutions of Eq. (3.167) with the same initial values (functions together with their time derivatives) in B. The difference $u_1 - u_2$ is then a solution of Eq. (3.167) with $g \equiv 0$ and $u = u_t = 0$ in B and by the previous result $u_1 = u_2$ in D^+. It follows that the derivatives of the functions u_1 and u_2 coincide in the region D^+ and hence also at the boundary point (x^0, t_0).

As an illustration, if the functions a_{jk} are bounded then the conditions (3.169) and (3.171) are fulfilled for a sufficiently small $d > 0$ by the function

$$f(x) = t_0 - d\,\|x - x^0\|.$$

Thus, the value $u(x^0, t_0)$ of the solution of Eq. (3.167) depends only on values of initial functions in the sphere centred at x^0 and of radius $d^{-1}t_0$. This implies that the wave propagation velocity in processes described by Eq. (3.167) is bounded above by the number d^{-1}. We shall try to determine the propagation velocity as precisely as possible. In doing so, we need to find a set D, as small as possible, of the form (3.168) where f satisfies (3.169) and (3.171).

3.5.3 Characteristic surfaces

The set D will be the smallest one among those sets of the form (3.168) determined by a function f satisfying (3.169) and (3.171) if relation (3.171) turns into equality, that is to say,

$$\sum_{j,k=1}^{n} a_{jk}(x, f(x)) \frac{\partial f}{\partial x_j} \frac{\partial f}{\partial x_k} = 1, \qquad x \neq x^0.$$

A surface defined by a such a function [having equation $t = f(x)$] is called the backward (retrograde) characteristic conic surface [for Eq. (3.167), with vertex at the point (x^0, t_0)]. It can be shown that the corresponding set D (retrograde characteristic cone) is indeed minimal in the sense that on "smaller" sets the uniqueness theorem is no longer true. This means that the retrograde characteristic cone defines the so-called *domain of dependence* of the point (x^0, t_0); in fact, the value of the solution u at (x^0, t_0) is affected only by those values of the data (coefficients of the equation, right-hand side, initial functions) which lie in D. (Many books use the term domain of dependence for only the base of the characteristic cone which in turn characterizes the dependence only on initial functions.)

By the forward characteristic conic surface with vertex at the point (x^0, t_0) we mean a surface given by $t = f(x)$ where f is a solution of Eq. (3.173) and satisfies $(3.169_1), (3.169_2)$ and

$$D_{t_1} \subset D_{t_2} \qquad \text{for } t_1 \leq t_2. \tag{3.174}$$

It turns out that the set $\{(x, t) \mid t > f(x)\}$, the so-called forward characteristic cone, consists of the points whose domain of dependence contains the point (x^0, t_0); hence it describes just the set of points in which the solution is influenced (in general) by the values of the data at the point (x^0, t_0), in different terms the so-called *domain of influence* of the point (x^0, t_0).

Any surface given by equation $t = f(x)$ where f solves the equation

$$\sum_{j,k=1}^{n} a_{jk}(x, f(x)) \frac{\partial f}{\partial x_j} \frac{\partial f}{\partial x_k} = 1$$

is referred to as the ***characteristic surface*** of Eq. (3.167). Characteristic surfaces constitute an extremely effective tool in studying many other aspects concerning Eq. (3.167). We gave a brief account of how to find special characteristic surfaces in § 2.3.9; a more general definition was introduced in § 2.5.5 with examples in §§ 2.5.6 – 2.5.8.

3.5.4 Wave propagation velocity

In the preceding subsections we learned (roughly speaking) that the wave described by Eq. (3.167) starting to propagate at time t_0 from a point source located at the position x^0 reaches another position x^1 at a later time t_1 such that the point (x^1, t_1) lies on the forward characteristic conic surface Γ with vertex at (x^0, t_0). Assume that the surface Γ is given by the equation

$$t = f(x).$$

As explained above the wave proceeds from x^0 to a position the distance of which from x^0 in the direction of a unit vector $y \in \mathbf{R}^n$ is ξ, i.e. to $x^0 + \xi y$, after the elapse of an interval of time $f(x^0 + \xi y) - f(x^0)$, hence with the average velocity

$$\frac{\xi}{f(x^0 + \xi y) - f(x^0)}.$$

The limit of the latter expression for $\xi \to 0$ may be looked upon as the instantaneous velocity $c(x^0, t_0; y)$ of the wave in the y-direction at the position x^0 and time t_0. Thus

$$c(x^0, t_0; y) = \Big(\sum_{j,k=1}^{n} b_{jk}(x^0, t_0)\, y_j y_k \Big)^{-\frac{1}{2}}, \tag{3.175}$$

where $\big(b_{jk}(x^0, t_0)\big)_{j,k=1}^{n}$ is the inverse of the matrix $\big(a_{jk}(x^0, t_0)\big)_{j,k=1}^{n}$ which means that

$$\sum_{m=1}^{n} b_{jm}(x^0, t_0)\, a_{mk}(x^0, t_0) = \begin{cases} 0 & \text{for } j \neq k \\ 1 & \text{for } j = k. \end{cases}$$

We demonstrate this result only in a special case of constant functions a_{jk}. In this case we can easily check that the equation of the forward characteristic conic surface with vertex at (x^0, t_0) assumes the form

$$t = t_0 + \Big[\sum_{j,k=1}^{n} b_{jk}(x_j - x_j^0)(x_k - x_k^0) \Big]^{\frac{1}{2}}$$

and hence

$$\frac{\xi}{f(x^0 + \xi y) - f(x^0)} = \frac{\xi}{\big(\sum_{j,k=1}^{n} b_{jk} \xi^2 y_j y_k\big)^{\frac{1}{2}}} = \Big(\sum_{j,k=1}^{n} b_{jk}\, y_j y_k \Big)^{-\frac{1}{2}}$$

for all $\xi > 0$.

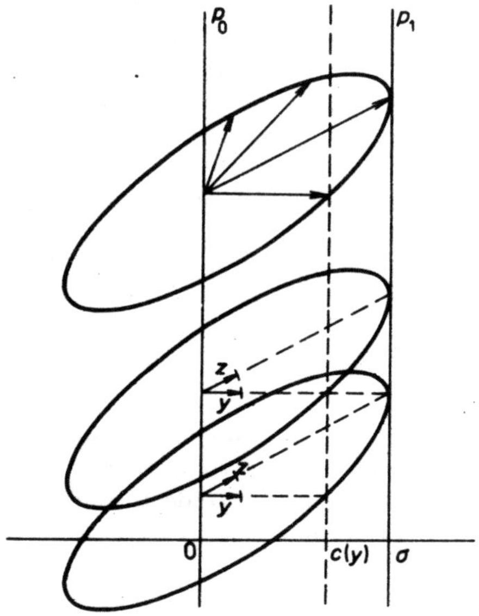

Fig. 21

It should be emphasized that the quantity $c(x^0, t_0; y)$ refers to the ***velocity of propagation*** of a wave which originates at a point source and must not be confused with other velocities which arise with wave processes. To illustrate, let us take the equation

$$\frac{\partial^2 u}{\partial t^2} - \sum_{j,k=1}^{n} a_{jk} \frac{\partial^2 u}{\partial x_j \partial x_k} = 0 \tag{3.176}$$

with constant coefficients. It is a simple matter to show that for any function $s \in C^2(\mathbf{R})$ and for any unit vector $y \in \mathbf{R}^n$ the function

$$u(x_1, \ldots, x_n, t) = s\big(\sigma t - \sum_{j=1}^{n} y_j x_j\big)$$

is a solution of Eq. (3.176) if

$$\sigma = \left(\sum_{j,k=1}^{n} a_{jk}\, y_j y_k\right)^{\frac{1}{2}}.$$

This solution is a plane wave which moves in the direction of the vector y with velocity σ which is in general greater than the velocity in the y-direction given by relation (3.175). This apparent contradiction can be explained easily by means of Fig. 21. The front of the plane wave at the time $t = 0$ is schematically drawn as the straight line p_0. At the time $t = 1$ the distance of the wave front p_1 from p_0 is greater than formula (3.175) predicts since the wave spreading out from individual points of the wave front in the z-direction outpaces the wave motion from x_0.

3.5.5 Example: system of equations of linear elasticity

As we have seen in Section 3.4 or in Chapter 2 the system of Maxwell's equations may be translated to a system of separate wave (telegraph) equations for each individual component and this makes it possible to determine in a simple manner the propagation velocity of electromagnetic waves with the aid of the above "scalar" theory. We conclude with a brief look at an example of a system for which we can draw simple conclusions concerning the wave propagation on the basis of the preceding results even though this system cannot be separated into individual equations.

We consider the system of equations of linear elasticity the form of which is (with no presence of external forces)

$$u_{tt} - c_1^2\,\Delta u - c_2^2\,\mathrm{grad}\,\mathrm{div}\,u = 0. \tag{3.177}$$

The function u is a vector function with components which are assumed to be continuously differentiable up to order three. After applying successively the operators curl and div to Eq. (3.177) we find out [owing to identities (1.26) and (1.27)] that $\mathrm{curl}\,u$ and $\mathrm{div}\,u$ are solutions of the wave equations

$$\begin{aligned}(\mathrm{curl}\,u)_{tt} - c_1^2\,\Delta(\mathrm{curl}\,u) &= 0,\\ (\mathrm{div}\,u)_{tt} - \left(c_1^2 + c_2^2\right)\,\Delta(\mathrm{div}\,u) &= 0.\end{aligned}$$

This means that the process described by Eq. (3.177) exhibits a double wave propagation – the first one (transverse waves) with velocity c_1 and the other one (longitudinal waves) with velocity $\left(c_1^2 + c_2^2\right)^{\frac{1}{2}}$.

3.5.6 Exercises

3.5.6.1 Determine when the equation

$$a_{11}u_{xx} + 2a_{12}u_{xy} + a_{22}u_{yy} = g$$

with constant coefficients is hyperbolic (in an arbitrary region $Q \subset \mathbf{R}^2$).

Answer.

The equation is hyperbolic if and only if $a_{11}a_{22} - a_{12}^2 < 0$.

3.5.6.2 Determine the time T after which a signal from the origin of coordinates arrives at a point (x, y) (the units of time and distance are settled) if the process is described by equation

$$u_{tt} - u_{xx} - 4u_{yy} = 0.$$

Answer.

$$T = \frac{1}{2}\left(4x^2 + y^2\right)^{\frac{1}{2}}.$$

CHAPTER 4

Second order parabolic equations

4.1 THE HEAT EQUATION

We shall investigate the *heat equation*

$$u_t(x,t) - a\,u_{xx}(x,t) = 0,$$

which is the simplest prototype of a parabolic equation. Its counterpart in higher space dimensions assumes the form

$$u_t(x,t) - a\sum_{j=1}^{n} u_{x_j x_j}(x,t) = 0.$$

In this section we will show that the heat equation describes the process of the heat conduction in heat conducting media. The variable u which is a function of a spatial position x and time t represents the temperature while the positive constant a is defined by the thermal properties of the material.

The heat equation is also called the *diffusion equation* for it arises in the analysis of the process of diffusion in physical chemistry.

The above equations (and parabolic equations in general) can be encountered further in electromagnetism, diffusion-reaction theory, fluid and soil mechanics, statistics, biological sciences, etc.

There is a number of monographs on the subject, for example, Cannon (1984), A. Friedman (1964), Ladyzhenskaya, Solonnikov and Ural'ceva (1968), Özisik-Necati (1968), Widder (1975).

4.1.1 The conduction of heat in solids

Let us take up the process of the conduction of heat in an isotropic solid occupying the set $\bar{\Omega}$, where Ω is a region in $\mathbf{R}^3$. Let ρ, c and λ stand for the mass density, heat capacity and (inner) heat conductivity, respectively, of the material of the body.[12]

The coefficients ρ, c and λ are in general functions of position, i.e. $\rho = \rho(x)$, $c = c(x)$, $\lambda = \lambda(x)$, $x \in \bar{\Omega}$, unless the medium is homogeneous (pure materials) in which case we consider them as constants. (These coefficients are also functions of temperature but this dependence may be neglected if the variation of temperature under consideration is not too significant. All these quantities are assumed to be time-independent as well.)

Suppose that within the body there exist sources generating heat (through electrical, chemical or radioactive processes, say) which are characterized by a function of both position and time $G = G(x,t)$ giving the amount of heat generated (or absorbed if the quantity is negative (we then speak of a negative source or a sink)) per unit time per unit volume.

To obtain a partial differential equation for the temperature $u = u(x,t)$ we set up the heat energy balance for an arbitrary control volume (region) V within Ω surrounding the point x in the time from t to $t + dt$. To this end let dQ_0 denote the amount of heat generated by sources in V during this infinitesimal time interval, dQ_1 the outflow of heat through the boundary ∂V and finally dQ_2 the increase (decrease) in heat which brings about an increase (decrease) in temperature of V. The heat energy conservation law yields

$$dQ_0 - dQ_1 = dQ_2. \tag{4.1}$$

For dQ_0 we have

$$dQ_0 = dt \int_V G(x,t)\, dV.$$

To evaluate dQ_1 and dQ_2 we apply the following two well-known experimental laws of physics.

- Fourier's law of inner heat conductivity. Let $q = q(x,t)$ denote the heat flux density vector, that is, the vector of which the direction coincides with that of the heat flow, namely with the direction of decreasing temperature, and of which the magnitude is equal to the amount of heat passing out, per unit time, through a surface of a unit area perpendicular to heat flow. The relation between q and u is given by the celebrated *Fourier law*:

$$q = -\lambda \operatorname{grad} u.$$

(The operator grad acts only on the space coordinates.)

---

[12]For common solids at ordinary temperatures the values c_p and c_v, the heat capacity at constant pressure and at constant volume, respectively, differ very little (c_p being at most of 5% greater than c_v), so we shall not distinguish between c_p and c_v writing simply c instead.

The amount of heat flowing, per unit time, across a unit area with a unit normal vector ν is equal to the projection of the vector q on the vector ν, viz.

$$q_\nu = \nu \cdot q = -\lambda \frac{\partial u}{\partial \nu}. \tag{4.2}$$

This (scalar) quantity is called the heat flux density in the direction ν. It is positive or negative according as the heat flows in the direction of ν, i.e. $\partial u/\partial \nu < 0$, or in the opposite direction, i.e. $\partial u/\partial \nu > 0$.

• The second law we need states that the change dQ of heat in a matter of mass m is proportional (through c) to m and the change du in temperature:

$$dQ = c\, m\, du.$$

By Fourier's law the amount of heat d^2Q which passes out during the time $(t, t+dt)$ through an oriented surface of area dS and of unit normal ν is given by

$$d^2Q = -\lambda \frac{\partial u}{\partial \nu}\, dt\, dS$$

and consequently

$$dQ_1 = -dt \int_S \lambda \frac{\partial u}{\partial \nu}\, dS,$$

where now $S = \partial V$ and $\partial/\partial \nu$ denotes the outward normal derivative on S. Using the divergence theorem (1.31) we can write (on condition that the function u is sufficiently smooth and the region V is regular)

$$dQ_1 = -dt \int_V \operatorname{div}(\lambda \operatorname{grad} u)\, dV.$$

By the second law the heat dQ_2 required to raise the temperature of V the quantity du is given by

$$dQ_2 = \int_V c\rho\, du\, dV.$$

Now, for the change du at the point x between the times t and $t+dt$ we have, to second order,

$$du(x,t) = dt\, \frac{\partial u(x,t)}{\partial t},$$

thus

$$dQ_2 = dt \int_V c\rho \frac{\partial u}{\partial t} dV.$$

Upon inserting into (4.1) we get

$$\int_V [c\rho \frac{\partial u}{\partial t} - \operatorname{div}(\lambda \operatorname{grad} u) - G(t,x)]\, dV = 0$$

and since $V \subset \Omega$ is arbitrary this yields (see § 1.2.4) the following partial differential equation

$$c\rho \frac{\partial u}{\partial t} - \operatorname{div}(\lambda \operatorname{grad} u) = G(x,t), \quad x \in \Omega,\, t > 0.$$

If ρ, c and λ are constants the above equation assumes the form

$$\frac{\partial u}{\partial t} - a\Delta u = g(x,t), \qquad x \in \Omega,\, t > 0, \tag{4.3}$$

where we have denoted

$$g = \frac{G}{c\rho} \quad \text{and} \quad a = \frac{\lambda}{c\rho}. \tag{4.4}$$

The coefficient a is called the temperature conductivity. Its numerical values for some materials (at room temperature) are given in Table 2.

Table 2

	a $(10^{-3}\mathrm{m}^2\,\mathrm{s}^{-1})$
Silver	0.170
Copper	0.114
Aluminium	0.086
Steel, mild	0.012
Granite	0.0013
Asbestos	0.00025
Water	0.00013

Once we know the function $u = u(x,t)$ for all $x \in \bar{\Omega}$ and $t \geq 0$ we are able to evaluate at any instant the heat flow in any direction, hence across an arbitrary surface; the problem of conduction of heat is thereby solved. Eq. (4.3) is called the *equation of conduction of heat* or simply the *heat equation.* □

Now, let us consider a solid in the form of a straight rod of uniform cross-section oriented so that the x-axis lies along the axis of the rod with the end points situated at $x = 0$ and $x = \ell$. Let us assume that

a) the lateral surface of the rod is perfectly insulated so that no heat passes across the lateral surface,

b) at any instant each cross-section of the rod is an isothermal surface, that is, the temperature is the same at all points of each cross-section.

(The latter assumption is plausible when the cross-sectional dimensions of the rod are small in comparison with its length.)

Under these assumptions the temperature u of the rod depends on merely one space coordinate x, Eq. (4.3) is reduced to

$$u_t - a\,u_{xx} = g(x,t), \qquad 0 < x < \ell,\, t > 0, \tag{4.5}$$

and one speaks then of the *one-dimensional heat equation.*

There are also other situations leading to the one-dimensional equation, namely when the heat flow in a body is permissible only in one direction. Such is the case, for example, if an infinite slab of homogeneous material is bounded by two parallel planes $x = 0$ and $x = \ell$ which are kept at constant temperatures and we may assume that the isothermal surfaces, at any instant, are planes parallel to the bounding ones. The temperature is then again a function of (in addition to t) only one space variable.

Of course, two-dimensional analogues to (4.3) are obtained in situations when there is no flow of heat in one direction, say of the z-axis.

4.1.2 The initial and boundary conditions for the heat equation

In § 4.1.1 where the equation governing conduction of heat in a three-dimensional solid has been derived the initial and boundary conditions did not have to be considered since the inner heat energy balance was unaffected by them. However, if we wish to determine uniquely the process of heat conduction, this means, if we want to know the temperature distribution within the entire body at any instant, Eq. (4.3) must be supplemented by boundary conditions defining the thermal schedule on the boundary surface of the body and by an initial condition defining the temperature distribution at some initial time, say $t = 0$. The initial condition therefore takes the form

$$u(x,0) = \varphi(x), \quad x \in \Omega, \tag{4.6}$$

where φ is a given function.

On the boundary we can prescribe, for instance, the temperature. This comes up in the case of a perfectly conducting body that is immersed into a surrounding medium of a great heat capacity, that is, the known ambient temperature remains practically unaltered by the heat flow through the boundary. The prescribed temperature is, in general, a function of both position and time so that we are led to the boundary condition

$$u(x,t) = h(x,t), \quad x \in \partial\Omega,\, t > 0. \tag{4.7}$$

Another situation arises when the heat is supplied from the exterior (with due regard to sign) through the boundary. So, let us assume we are given the heat flux density q_- on $\partial\Omega$ in the direction of the inward normal to $\partial\Omega$. Considering the heat balance on the boundary we see that this inward flux density on $\partial\Omega$ must be equal in magnitude and opposite in sign to the outward flux density q_ν on $\partial\Omega$ being due to the heat conductivity of the body which is given by (4.2), hence

$$\lambda\frac{\partial u(x,t)}{\partial \nu} = q_-(x,t), \quad x \in \partial\Omega,\, t > 0, \tag{4.8}$$

where $\partial/\partial\nu$ denotes the outward normal derivative on $\partial\Omega$.

In particular, the homogeneous boundary condition

$$\frac{\partial u(x,t)}{\partial \nu} = 0, \quad x \in \partial\Omega,\, t > 0, \tag{4.9}$$

corresponds to the case when the boundary of the solid is thermally insulated (the heat conduction in the body is an adiabatic process).

So far we have only discussed the heat transfer by conduction. It is the transfer of heat energy by means of collisions between vibrating molecules (or by means of free electrons in metals). From the microscopic point of view the medium is in motion but macroscopically viewed the medium remains at rest. This transference of heat is typical for solids.

In liquids and gases the heat transfer is due to the convection (in addition to the conduction which is, however, usually negligible). It means that the heat is transported by a stream of fluid. The third way of heat transmission is by radiation: this is transfer of heat by electromagnetic waves with or without an intervening medium.

We are commonly faced with the following situation: a solid is in contact with a fluid medium at a given temperature and the heat is dissipated by convection (and, to a lesser extent, conduction, and radiation) from the boundary to the surrounding medium. The mathematical description of this complex mechanism is rather difficult. Nonetheless, the first approximation which is borne out very well by experimental investigations (at least for moderate temperatures) is given by the *Newton law* of surface heat transfer according to which the outward heat flux density on the boundary is proportional to the difference $u - u_0$ between the temperature of the surface and that of the surroundings:

$$q_+ = \alpha(u - u_0). \tag{4.10}$$

Here α is the so-called heat transfer coefficient, the value of which varies greatly with the state, shape and temperature of the boundary surface and depends considerably on the surroundings. (The values of the coefficient α encountered in practice run from 3 to 14 000 $\mathrm{Jm^{-2}s^{-1}deg^{-1}}$.) Upon comparing (4.2) with (4.10) we get

$$\alpha(u - u_0) = -\lambda\frac{\partial u}{\partial \nu},$$

hence

$$\lambda\frac{\partial u(x,t)}{\partial \nu} + \alpha u(x,t) = \alpha u_0(x,t), \quad x \in \partial\Omega,\ t > 0, \tag{4.11}$$

which is the Newton boundary condition for the heat equation. In particular, if $\alpha = 0$ in (4.11) we get the Neumann condition (4.9) and for $\alpha \to \infty$ (4.11) becomes the Dirichlet condition (4.7).

We have mentioned the most important boundary conditions we meet in practice. Naturally, different portions of the boundary can carry different types of boundary conditions. We shall not deal with other more complicated boundary conditions, for example, with those describing heat radiation governed by the Stefan-Boltzmann radiation law which states that the heat flux density on the boundary is proportional to the difference of the fourth powers of temperatures of the surface and the surroundings. For small temperatures these boundary conditions (which make the problem nonlinear) may be approximated by (linear) Newton boundary conditions (4.11).

Nothing will be said either about the so-called *free boundary problems* in which the location of boundaries is to be determined as part of the solution [see Crank (1984)]. These problems arise in applications involving a change of phase due to melting or solidification (a typical example, the Stefan problem: water in contact with ice), in the study of a fluid flow in porous media, etc.

It appears plausible that if we investigate a "big" solid and we are interested in transient rather than steady state temperatures, the influence of boundary conditions is not appreciable in a decisive portion of the body. The abstraction of passing from a given body to another one with infinite extent in some direction(s), in particular to a body occupying the full space, may be sometimes helpful. In terms of mathematics, the latter case means that the initial-boundary value problem is replaced by the initial-value problem given by equation (4.3) and the initial condition (4.6) where now the variable x ranges over the entire space $\mathbb{R}^3$. □

In the sequel we shall be concerned primarily with the heat equation in one dimension. Namely, the heat equation does not exhibit the phenomenon we have met in the study of the wave equation where the properties of solutions were strikingly different according to dimension of the space variable. For further convenience let us write out the boundary conditions for the one-dimensional heat equation when the boundary consists of merely two points $x = 0$ and $x = \ell$. The Dirichlet boundary condition (4.7) may be written as

$$u(0,t) = h_0(t), \quad u(\ell,t) = h_1(t), \quad t > 0. \tag{4.12}$$

Since $\partial u/\partial \nu = \partial u/\partial x$ at $x = \ell$ and $-\partial u/\partial \nu = \partial u/\partial x$ at $x = 0$, the Neumann boundary condition (4.8) can be written as

$$-\lambda \frac{\partial u(0,t)}{\partial x} = q_0(t), \quad \lambda \frac{\partial u(\ell,t)}{\partial x} = q_1(t), \quad t > 0 \tag{4.13}$$

and eventually the Newton boundary condition (4.11) can be written in the form

$$\begin{aligned} -\frac{\partial u(0,t)}{\partial x} + \sigma_0 u(0,t) &= h_0(t), \\ \frac{\partial u(\ell,t)}{\partial x} + \sigma_1 u(\ell,t) &= h_1(t), \quad t > 0, \qquad (\sigma_0 > 0,\, \sigma_1 > 0). \end{aligned} \tag{4.14}$$

4.1.3 Some modifications of the heat equation

As we have shown in § 4.1.1 the one-dimensional heat equation (4.5) describes the conduction of heat in a rod on condition that its lateral surface is rendered impervious to heat. If this condition is not satisfied but there is an exchange of heat between the boundary surface of the rod and the surroundings at the temperature u_0 [governed by the Newton law (4.10)] we need not go back to the three-dimensional model as might seem at first sight. We show that the one-dimensional model is sufficient if we retain the assumption b) on p. 174 (which now, however, is not so convincing as before). Namely, the description of the thermal schedule on the lateral surface can be amalgamated into the differential equation and the boundary condition for the lateral surface drops out. We obtain a somewhat more complicated equation but again only in one space variable.

Let the cross-section of the rod have area A and perimeter o. Let us set up the heat balance in a rod segment $(x, x + dx)$ during the time $(t, t + dt)$. Using Fourier's and Newton's laws we obtain

$$dt\left[\lambda A \frac{\partial u(x,t)}{\partial x} - \lambda A \frac{\partial u(x+dx,t)}{\partial x}\right] + \alpha[u(x,t) - u_0(x,t)]o\,dx\,dt$$
$$+ c\rho\, A\, dx\, dt\, \frac{\partial u(x,t)}{\partial t} = G\, A\, dx\, dt$$

and after an easy arrangement we arrive at the following modification of the heat equation

$$u_t - a\, u_{xx} + \gamma(u - u_0) = g(x,t),$$

where $\gamma = \dfrac{\alpha o}{c\rho\, A}$ and a, g are given in (4.4).

The equation

$$u_t - a\,\Delta u + \beta_0\, s \cdot \operatorname{grad} u = g(x,t),$$

which in the case of one space variable takes the form

$$u_t - a\, u_{xx} + \beta\, u_x = g(x,t),$$

arises in situations when the heat is transferred not only by conduction but also by convection in the direction s with the velocity β_0 (for example, if a fluid flows through a tube). [This equation is obtained from the heat energy balance relation $dQ_1 + dQ_2 + dQ_3 = dQ_0$, where dQ_0, dQ_1 and dQ_2 have the same meaning as in § 4.1.1 and $dQ_3 = dt \int_S \beta_0 c \rho u\, \nu \cdot s\, dS$.]

By means of appropriate transformations (cf. § 4.2.15) the above equations can be reduced to the usual heat equation. □

Let us mention some other modifications which are more complicated, the detailed study of which lies beyond the scope of this introductory text.

In the case of an inhomogeneous material the temperature conductivity coefficient a is a function of the variable x. Actually, as we have already remarked, it also depends upon the temperature, but if the range of temperature is limited it can be taken as constant. If, however, this dependence must be considered (great changes in temperature) the equation assumes the form

$$u_t - \operatorname{div}\,[a(x,u)\operatorname{grad} u] = g(x,t)$$

and we obtain, in general, a nonlinear equation.

In isotropic solids the heat conductivity does not depend on the direction in which the heat is conducted. On the other hand, in the case of anisotropic solids (such as crystals, woods, etc.), the heat conductivity varies with direction. For such materials it is no longer true that the heat flux density vector q is a normal vector to the isothermal surface as is stated by the Fourier law. If we modify the heat conduction law writing

$$q = -\Lambda \operatorname{grad} u, \quad \Lambda = (\lambda_{jk})_{j,k=1}^3$$

(Λ is called the heat conductivity tensor), or more explicitly,

$$q_j = -\sum_{k=1}^{3} \lambda_{jk} \frac{\partial u}{\partial x_k}, \quad j = 1,2,3,$$

the pertinent equation is (cf. § 1.2.2)

$$u_t - \sum_{j,k=1}^{3} \frac{\partial}{\partial x_j}(a_{jk} \frac{\partial u}{\partial x_k}) = g(x,t),$$

written in a more compact form,

$$u_t - \operatorname{div}(\mathcal{A} \operatorname{grad} u) = g,$$

where $\mathcal{A} = (a_{jk})_{j,k=1}^3$ is a matrix the elements of which are, in general, functions of x and u.

4.1.4 Diffusion

In addition to the mathematical theory of the conduction of heat in solids there are some other situations that give rise to Eq. (4.3). As a matter of fact, we are led to this equation (with the unknown function η) whenever a certain physical quantity W (considered for $x \in \Omega$) is supposed to be expressed through its volume density η, that is the total amount of this quantity in any volume $V \subset \Omega$ at time t is given by

$$W(V,t) = \int_V \eta(x,t)\,dV,$$

and the following two postulates are valid

$$(\diamondsuit) \qquad \frac{\partial \eta}{\partial t} + \operatorname{div} w = G,$$

$$(\heartsuit) \qquad w = -a \operatorname{grad} \eta.$$

Here w stands for the flux-density vector, G is the time rate of the volume density of sources, that is the total amount of the quantity generated by sources per unit time in V at time t is known to be given by $\int_V G(x,t)\,dV$, and eventually a is a constant. The former relation is a conservation law, the latter one is furnished often on experimental grounds.

A typical example is provided by diffusion processes – spontaneous spreading of a substance (fluid) through another substance; here $\eta = u$ stands for the concentration of the diffusing substance the continuity equation of which is expressed by relation $(\diamondsuit)$. Relation $(\heartsuit)$ is known as ***Fick's law*** and represents the analogue of the Fourier law from the heat conduction theory. Other pendants are summed up in the following comparative table (in which the notation of § 4.1.1 is used).

Table 3

	Heat conduction	*Diffusion*
W	$= Q$ heat	diffusing substance
η	$= c\,\rho\,u$, u temperature	concentration
a	$= \dfrac{\lambda}{c\,\rho}$ coefficient of temperature conductivity	$= D$ coefficient of diffusion
w	$= q$ heat flux density vector	diffusion current density vector

The boundary conditions (4.7), (4.9) and (4.11) respectively correspond to the following cases

– the concentration is given at the boundary,

– the boundary is impermeable, the diffusing substance cannot penetrate the boundary,

– the boundary is semi-permeable while the diffusion across the boundary is governed by a law analogous to the Newton one, see (4.10), where α is now called the permeability of the boundary. [Let us remark that in the theory of diffusion processes we can meet boundary conditions involving (apart from u and $\partial u/\partial \nu$) also the time derivative u_t.] □

A more general equation

$$u_t - \operatorname{div}(D\,\operatorname{grad} u) + \beta_0\, s \cdot \operatorname{grad} u + \gamma\, u = 0$$

arises if some driving forces (in the direction s) are present, there is a disintegration ($\gamma > 0$) or multiplication ($\gamma < 0$) of diffusing particles at a rate proportional to the concentration and the diffusion coefficient D is a function of position and perhaps of concentration (which is quite common).

For another example of the occurrence of Eq. (4.3) see § 2.1.5.

4.1.5 Exercises

4.1.5.1 Let a diffusion process take place in an impermeable cylindrical vessel of height ℓ, cross-sectional area A and the concentration η be the same at all points of each cross-section. Then the governing equation is one-dimensional. Find the relation between η (a function of a single space variable and time) and the total amount W of the diffusing substance in the vessel.

Analogously for the heat conduction, derive an expression for the total amount of heat Q in a prismatic rod (of length ℓ and cross-sectional area A) by means of its temperature u (stipulating $Q = 0$ for $u = 0$).

Hint. Choose the coordinate system so that the x-axis lies along the axis of the vessel and the bottom (respectively, the top) corresponds to the point $x = 0$ (respectively, $x = \ell$).

Answer.

$$W(t) = A \int_0^\ell \eta(x,t)\,dx, \qquad Q(t) = c\,\rho\, A \int_0^\ell u(x,t)\,dx.$$

4.1.5.2 The homogeneous (one-dimensional) heat equation possesses infinitely many solutions. Show by direct computation that it is satisfied by the following functions

$$x^2 + 2at, \quad x^3 + 6xat, \quad x^4 + 12x^2at + 12a^2t^2$$

(the first three terms of the sequence of the so-called heat polynomials), or

$$\exp(-\mu^2 at)\cos(\mu x), \quad \exp(-\mu^2 at)\sin(\mu x),$$
$$\exp(\mu^2 at)\cosh(\mu x), \quad \exp(\mu^2 at)\sinh(\mu x),$$

where μ is an arbitrary number.

4.1.5.3 Show that the homogeneous heat equation can be reduced to an ordinary differential equation

$$v''(z) + 2z\,v'(z) = 0$$

for the independent variable z by means of the (so-called Boltzmann) transformation

$$z = \frac{x}{2(at)^{\frac{1}{2}}}.$$

Using this conclude that the function

$$u(x,t) = \operatorname{erf}\frac{x}{2(at)^{\frac{1}{2}}}$$

satisfies the homogeneous heat equation. Here erf is the error function defined in (1.49).

4.1.5.4 A particularly useful function in the theory of parabolic equations is the so-called *source function*

$$\frac{1}{2(\pi at)^{\frac{1}{2}}}\exp\left(-\frac{x^2}{4at}\right)$$

(which will be studied more thoroughly in the following section). Verify [either directly or by differentiating the function $\operatorname{erf}\frac{x}{2(at)^{\frac{1}{2}}}$ with the aid of the Leibniz rule (1.35)] that this function is a solution of the homogeneous heat equation.

Show that the heat equation in n space variables

$$u_t(x_1,\ldots,x_n,t) - a\,\Delta u(x_1,\ldots,x_n,t) = 0, \quad -\infty < x_j < \infty,\ t > 0$$

is satisfied by the function

$$\left(\frac{1}{2(\pi at)^{\frac{1}{2}}}\right)^n \exp\left(-\frac{\|x\|^2}{4at}\right).$$

4.2 THE HEAT EQUATION IN ONE SPACE VARIABLE

In this section we shall first find explicitly a solution of the initial-value problem for the one-dimensional heat equation by means of the Fourier integral transform method. Then we shall be concerned with the properties and the physical meaning of the source function. The explicit form of the solution of the initial-value problem enables us to show easily typical properties of the heat equation. The heat equation considered in other (even higher dimensional) regions than in the infinite interval $-\infty < x < \infty$ exhibits analogous properties (with technically more complicated proofs, though).

The properties of the heat equation and the wave equation we examined in the previous chapter are altogether different. The heat equation – in contrast to the wave equation – describes irreversible phenomena with "infinite velocity" of propagation. We encounter here the so-called smoothing effect (as in the theory of elliptic equations), that is, the solution is infinitely differentiable in both space and time variables for all $t > 0$ even if there is a singularity (such as a jump discontinuity) in the initial function. In other words, the initial singularities are immediately "smoothed out" – for the wave equation these singularities are "preserved" being propagated along the characteristics. In the theory of the heat equation the characteristics (the lines $t =$ const. by general definitions that we shall not take up) do not play such a decisive role as in the case of the wave equation. We show that the behaviour of a function satisfying the heat equation may in general be pathological near the characteristic $t = 0$.

In the end we give several transformations which make it possible to reduce to the heat equation some of its modifications.

4.2.1 Setting of the problem

Our task is to find the classical solution of the initial-value problem for the heat equation in one space variable. Let us recall its meaning. We look for a function $u = u(x,t)$ continuous for $-\infty < x < \infty$, $t \geq 0$ with continuous derivatives u_t and u_{xx} for $-\infty < x < \infty$, $t > 0$, satisfying the equation

$$u_t(x,t) - a\,u_{xx}(x,t) = g(x,t), \quad -\infty < x < \infty,\ t > 0 \tag{4.15}$$

and the initial condition

$$u(x,0) = \varphi(x), \quad -\infty < x < \infty. \tag{4.16}$$

Here g and φ are given functions and a is a given positive constant.

As we showed in Section 4.1 the function u defines the temperature distribution in a straight uniform rod of "infinite length" with perfectly insulated sides, if the initial temperature distribution is given by φ and for times $t > 0$ heat is generated within the rod at the rate $c\,\rho\,g(x,t)$ per unit time and unit volume, ρ is the mass density, c the heat capacity and a the temperature conductivity of the material.

4.2.2 Example

Let $g \equiv 0$ and $\varphi(x) = c_0 + c_1 x + c_2 x^2$ where c_0, c_1 and c_2 are real constants (we make no attempt to interpret the case $|c_1| + |c_2| > 0$ physically). Then a solution of the initial-value problem (4.15), (4.16) is given by

$$u(x,t) = c_0 + c_1 x + c_2 x^2 + 2c_2 a t.$$

4.2.3 Computation of the solution for the homogeneous equation

For the sake of simplicity let us first consider the homogeneous equation

$$u_t(x,t) - a\,u_{xx}(x,t) = 0, \quad -\infty < x < \infty,\ t > 0 \tag{4.15'}$$

(in physical terms: let there be no heat sources within the rod). In § 4.2.14 we return to the general case of equation (4.15) with $g \not\equiv 0$.

We cannot derive the representation of the solution so easily as we managed in § 4.2.2 for quadratic initial functions or in some cases in the preceding chapters. Perhaps the most convenient and simplest way to obtain the solution is the use of the Fourier integral transform over the space coordinate (cf. § 1.2.6). The Fourier transform $\hat{u}$ of the function $u = u(x,t)$ [defined for $-\infty < x < \infty$, $t > 0$] with respect to the variable x is defined by the relation

$$\hat{u}(y,t) = \int_{-\infty}^{\infty} \exp(-ixy)\,u(x,t)\,dx, \quad -\infty < y < \infty,\ t > 0.$$

Taking the Fourier transform to Eq. (4.15′) and the initial condition (4.16) we obtain

$$\hat{u}_t(y,t) + a\,y^2\hat{u}(y,t) = 0, \quad -\infty < y < \infty,\ t > 0,$$

$$\hat{u}(y,0) = \hat{\varphi}(y), \quad -\infty < y < \infty,$$

where

$$\hat{\varphi}(y) = \int_{-\infty}^{\infty} \exp(-ixy)\,\varphi(x)\,dx.$$

Hence, we get the initial-value problem for an ordinary differential equation of the first order with the function $\hat{u}(y,t)$ where y plays the role of a parameter. This is an essential feature of this transform technique: the partial derivatives with respect to the space variable are removed from the partial differential equation and this equation is reduced to an ordinary one with respect to the time variable. The latter problem is easily solved, viz.

$$\hat{u}(y,t) = \exp(-y^2at)\,\hat{\varphi}(y). \tag{4.17}$$

If we succeed in finding a function $e = e(x)$ (depending parametrically on t) the Fourier transform of which is $\exp(-y^2at)$ the problem will be resolved due to the following two facts:

• the convolution theorem (1.42_1) yields that the right-hand side of (4.17) is equal to the Fourier transform of the function

$$\int_{-\infty}^{\infty} e(x-\xi)\,\varphi(\xi)\,d\xi,$$

• this function and the function u having the same Fourier images are identical.

The well-known result from the Fourier transform theory says [see (4.23)] that the Fourier image of the function

$$\frac{1}{(2\pi)^{\frac{1}{2}}}\exp\Big(-\frac{x^2}{2}\Big)$$

is equal to

$$\exp\Big(-\frac{y^2}{2}\Big).$$

Using the scaling property (1.40) with $c = (2at)^{\frac{1}{2}}$ we get

$$e(x) = \frac{1}{2(\pi at)^{\frac{1}{2}}}\exp\Big(-\frac{x^2}{4at}\Big)$$

and eventually

$$u(x,t) = \frac{1}{2(\pi at)^{\frac{1}{2}}}\int_{-\infty}^{\infty} \exp\Big(-\frac{(x-\xi)^2}{4at}\Big)\,\varphi(\xi)\,d\xi. \tag{4.18}$$

In deriving the formula (4.18) we proceeded purely formally. This means: we carried out all the computations without justifying at each step that these computations were permissible [in particular, we had to impose some conditions on u, u_x and u_{xx} to be entitled to write

$$\int_{-\infty}^{\infty} u_{xx}(x,t)\exp(-ixy)\,dx = -y^2\int_{-\infty}^{\infty} u(x,t)\exp(-ixy)\,dx].$$

This is a common approach when looking for solutions of the problems of mathematical physics by means of integral transforms (e.g. the Fourier transform or the Laplace transform will be applied in the next section): we use the formalism and the end result is investigated independently of this formalism. Usually weaker conditions are sufficient than those justifying the formalism. It is worth emphasizing that this subsequent investigation should not be omitted, otherwise we obtain only a "formal solution" which can very well be meaningless.

Summing up, there are four steps to be followed in using the Fourier transform technique. In concise form:
(1) Taking the Fourier transform to the equation and to the initial condition ($x \to y$, time t a parameter).
(2) Solving the resulting initial-value problem for an ordinary differential equation (in which y is a parameter) to obtain the Fourier image $\hat{u}$.
(3) The inversion problem: finding the original function u to $\hat{u}$, often with the aid of the convolution theorem.
(4) Verifying that u is actually a solution.

In our case the detailed investigation of formula (4.18) leads to the following theorem the proof of which we defer to § 4.2.7 till we have made a few observations.

Let φ be a continuous and bounded function on the interval $-\infty < x < \infty$. Then there exists exactly one bounded classical solution of the problem (4.15′), (4.16). *This solution is given by formula* (4.18).

4.2.4 The Fourier–Poisson integral

The right-hand side of formula (4.18) is called the *Fourier–Poisson integral*. In this formula we may insert only values $t > 0$, for $t = 0$ it is not defined. At first glance, this might seem inconsistent with our statement that (4.18) gives the solution. The justification consists of the following: the function defined by (4.18) on the set $(-\infty,\infty)\times(0,\infty)$ allows a continuous extension to the set $(-\infty,\infty)\times[0,\infty)$, in short: it can be extended by continuity from the open half-plane $t > 0$ to the closed half-plane $t \geq 0$. This means exactly that there exists a (double) limit of $u(x',t')$ when (x',t') approaches the initial line $t = 0$ and, moreover, this limit is equal to φ:

$$\lim_{\substack{(x',t')\to(x,0)\\ t'>0}} \frac{1}{2(\pi a t')^{\frac{1}{2}}}\int_{-\infty}^{\infty}\exp\Big(-\frac{(x'-\xi)^2}{4at'}\Big)\,\varphi(\xi)\,d\xi = \varphi(x). \qquad (4.19)$$

The verification of (4.19) will be the crux of the proof of the theorem in § 4.2.7.

Introducing a new variable of integration η in (4.18) by means of the (Boltzmann) transformation

$$\frac{\xi - x}{2(at)^{\frac{1}{2}}} = \eta, \tag{4.20}$$

we obtain an alternate representation of the Fourier–Poisson integral in the form

$$u(x,t) = \pi^{-\frac{1}{2}} \int_{-\infty}^{\infty} \varphi(x + 2(at)^{\frac{1}{2}}\eta) \exp(-\eta^2)\, d\eta \tag{4.21}$$

(especially convenient if we are interested in the function u for small values of $t > 0$).

4.2.5 Remark on uniqueness

Note that in the theorem the boundedness of the function φ is assumed and the existence and uniqueness of a bounded solution is stated. If the boundedness of either of two functions were not required, the uniqueness would not be ensured.

Let us point out that, more generally, the uniqueness of a solution can be proved within a much broader class of functions satisfying the growth condition

$$|u(x,t)| \leq M_1 \exp(M_2 x^2), \quad -\infty < x < \infty, t > 0,$$

where M_1 and M_2 are non-negative constants (the case $M_2 = 0$ corresponds to the bounded solutions). The solution of the initial-value problem for the homogeneous heat equation exists (it is still given by the Fourier–Poisson integral) being the only one within this class of functions if the initial function φ is continuous and satisfies the growth condition

$$|\varphi(x)| \leq M_1 \exp(M_2 |x|^{1+\alpha}), \quad -\infty < x < \infty,$$

where $M_1 \geq 0$, $M_2 \geq 0$ and $\alpha < 1$.

If the function φ satisfies the latter condition with $\alpha = 1$ the existence and uniqueness of a solution (within the above-mentioned class of functions) is ensured on the interval $0 < t < (4M_2)^{-1}$. If $\alpha > 1$ there is no guarantee that the solution is unique unless it is positive [see Widder (1975)].

4.2.6 Two auxiliary results

For the sake of completeness of the exposition we demonstrate the following two formulae:

- $$\int_0^{\infty} \exp(-x^2)\, dx = \frac{1}{2} \int_{-\infty}^{\infty} \exp(-x^2)\, dx = \frac{\pi^{\frac{1}{2}}}{2} \tag{4.22}$$

and

- $$\int_{-\infty}^{\infty} \exp(-\frac{x^2}{2} - ixy)\, dx = (2\pi)^{\frac{1}{2}} \exp(-\frac{y^2}{2}). \tag{4.23}$$

To prove (4.22) we use the Fubini theorem and the transformation to polar coordinates (r, ϑ) and obtain

$$\left(\int_{-\infty}^{\infty} \exp(-x^2)\,dx\right)^2 = \int_{-\infty}^{\infty} \exp(-x^2) \int_{-\infty}^{\infty} \exp(-y^2)\,dy\,dx$$
$$= \int_{-\infty}^{\infty}\int_{-\infty}^{\infty} \exp(-x^2 - y^2)\,dx\,dy = \int_0^{2\pi}\int_{-\infty}^{\infty} r\exp(-r^2)\,dr\,d\vartheta = \pi.$$

Denoting $\hat{u}(y) = \int_{-\infty}^{\infty} \exp(-\frac{x^2}{2} - ixy)\,dx$ we will prove (4.23) by the differentiation under the integral sign technique. Applying formula (1.34) and then integrating by parts we get

$$\hat{u}'(y) = -i\int_{-\infty}^{\infty} x\exp(-\frac{x^2}{2} - ixy)\,dx$$
$$= -y\int_{-\infty}^{\infty} \exp(-\frac{x^2}{2} - ixy)\,dx = -y\,\hat{u}(y).$$

The solution of this differential equation reads

$$\hat{u}(y) = \hat{u}(0)\exp(-\frac{y^2}{2}).$$

Since $\hat{u}(0) = (2\pi)^{\frac{1}{2}}$ by (4.22), the proof is complete.

4.2.7 Proof of Theorem 4.2.3

The fact that the Fourier–Poisson integral (4.18) has continuous partial derivatives and satisfies equation (4.15′) is easily checked by means of the differentiation under the integral sign (see § 1.2.5 and Exercise 4.1.5.4) taking into account the elementary inequality

$$\exp(-y) \le \frac{p!}{y^p}, \quad y > 0, \tag{4.24}$$

which is valid for any $p \in \mathbf{N} \cup \{0\}$.

We prove that the function (4.18) can be continuously continued to the closed half-plane $t \ge 0$ and takes on the values $\varphi(x)$ on the line $t = 0$. We use the equivalent formula (4.21) for the Fourier–Poisson integral and instead of showing (4.19) we shall prove an equivalent claim that for any $x \in \mathbf{R}$

$$\lim_{\substack{(x',t')\to(x,0)\\ t'>0}} \pi^{-\frac{1}{2}} \int_{-\infty}^{\infty} \varphi(x' + 2(at')^{\frac{1}{2}}\eta)\,\exp(-\eta^2)\,d\eta = \varphi(x), \tag{4.25}$$

where φ is any continuous and bounded function on $\mathbf{R}$, say, for definiteness, $|\varphi(x)| \le M$ for all $x \in \mathbf{R}$.

Fix $x \in \mathbf{R}$. In view of (4.22) we can write

$$u(x',t') - \varphi(x) = \pi^{-\frac{1}{2}} \int_{-\infty}^{\infty} [\varphi(x' + 2(at')^{\frac{1}{2}}\eta) - \varphi(x)] \exp(-\eta^2)\, d\eta.$$

Given any $\varepsilon > 0$ we can choose $L > 0$ such that

$$\pi^{-\frac{1}{2}} \int_{|\eta| \geq L} 2M \exp(-\eta^2)\, d\eta < \frac{\varepsilon}{2}.$$

Further, there exists $\delta > 0$ such that

$$|\varphi(\xi) - \varphi(x)| < \frac{\varepsilon}{2} \quad \text{whenever} \quad |\xi - x| < \frac{\delta}{2}.$$

Hence, for

$$|x' - x| < \frac{\delta}{2}, \quad 0 < t' < \frac{1}{a}\Big(\frac{\delta}{4L}\Big)^2$$

we have

$$|u(x',t') - \varphi(x)| < \varepsilon$$

which yields the conclusion.

The boundedness of the solution is readily seen from formula (4.22). The proof of uniqueness is outlined in Exercise 4.3.16.9. □

At this point we want to call the reader's attention to a device which is often used to prove the continuity up to the boundary. If a function $u = u(x,t)$ is continuous on $\mathbf{R} \times [0, t_0)$ then clearly for any $x \in \mathbf{R}$ there exists a limit of $u(x,t')$ as $t' \to 0+$ which we temporarily denote by $\Phi(x)$, that is,

$$\lim_{t' \to 0+} u(x,t') = \Phi(x) \qquad \text{for any} \quad x \in \mathbf{R}. \tag{4.26}$$

On the other hand, if u is known to be continuous on $\mathbf{R} \times (0, t_0)$ then the condition (4.26) is not sufficient to ensure the continuity up to the boundary $t = 0$ even if Φ is supposed to be continuous on $\mathbf{R}$ (for an example see the function H in §4.2.12). However, if we strengthen (4.26) to

$$\lim_{t' \to 0+} u(x,t') = \Phi(x) \quad \text{uniformly with respect to } x \\ \text{belonging to any closed bounded interval in } \mathbf{R} \tag{4.27}$$

then not only Φ is continuous on $\mathbf{R}$ but also u is continuous on $\mathbf{R} \times [0, t_0)$. (In brief outline: let $x \in \mathbf{R}$, $|\Phi(x) - \Phi(x')| \leq |\Phi(x) - u(x,t')| + |u(x,t') - u(x',t')| + |u(x',t') - \Phi(x')|$, the first and the third terms on the right-hand side can be made small uniformly on any neighbourhood of x by requiring a $t' > 0$ sufficiently small while the second one can be made small by restricting x' sufficiently near to x. This yields the continuity of Φ. The continuity of u then follows from the inequality $|u(x',t') - \Phi(x)| \leq |u(x',t') - \Phi(x')| + |\Phi(x') - \Phi(x)|$ just as easily.)

In this setting the result applies to the above situation but, of course, various modifications are possible. Among them is the following we need in §4.3.8. Let a function $u = u(x,t)$ be continuous on $(0, x_0) \times J_0$ where J_0 is any interval in $\mathbf{R}$ (not necessarily closed nor bounded). Let there exist a limit $u(x',t)$ as $x' \to 0+$ uniformly with respect to t belonging to any closed bounded interval in J_0. Then u is continuous on $[0, x_0) \times J_0$.

4.2.8 Piecewise continuous initial functions

By inspection of the proof of relation (4.25) it is easily seen that it remains true at each point of continuity x of an arbitrary function φ which is bounded on $\mathbf{R}$ and satisfies $\int_I |\varphi(\xi)|\, d\xi$ is finite for every interval I of finite length (φ is locally integrable on $\mathbf{R}$).

In practical situations we can encounter a bounded initial function φ which is piecewise continuous on every interval of finite length. Let $\{x_j\}_{j\in D}$ be the system of its points of discontinuity. Then it is necessary to modify the requirement on continuity in the definition of the solution and demand that the solution be continuous for $-\infty < x < \infty$, $t \geq 0$ except for the points $(x_j, 0)$, bounded in neighbourhoods of these points and relation (4.16) be satisfied for every $x \neq x_j$, $j \in D$.

By the above-mentioned result the Fourier–Poisson integral defines the solution even in this case.

4.2.9 Example

Consider the conduction of heat in an infinitely long rod having a thermally insulated surface. Find an expression for the temperature $u(x,t)$ for times $t > 0$ if the initial temperature distribution in the rod is given by

$$u(x,0) = \begin{cases} 0 & \text{for } x < 0 \\ \varphi_0 = \text{const.} & \text{for } x > 0. \end{cases}$$

According to (4.21) and (4.22) we can write

$$\begin{aligned} u(x,t) &= \frac{\varphi_0}{\pi^{\frac{1}{2}}} \int_{-x/2(at)^{\frac{1}{2}}}^{\infty} \exp(-\eta^2)\, d\eta \\ &= \frac{\varphi_0}{\pi^{\frac{1}{2}}} \left(\int_0^{\infty} \exp(-\eta^2)\, d\eta + \int_{-x/2(at)^{\frac{1}{2}}}^{0} \exp(-\eta^2)\, d\eta \right) \\ &= \frac{\varphi_0}{\pi^{\frac{1}{2}}} \left(\frac{\pi^{\frac{1}{2}}}{2} + \int_0^{x/2(at)^{\frac{1}{2}}} \exp(-\eta^2)\, d\eta \right). \end{aligned}$$

Now, in view of the definition of the error function (1.49) we can write the solution in the convenient fashion

$$u(x,t) = \frac{\varphi_0}{2} \left[1 + \operatorname{erf} \frac{x}{2(at)^{\frac{1}{2}}} \right].$$

Using the Taylor expansion

$$\exp(-\eta^2) = 1 - \frac{\eta^2}{1!} + \frac{\eta^4}{2!} - \cdots$$

and integrating term-by-term we obtain an alternate representation in the form of a series,

$$u(x,t) = \frac{\varphi_0}{2} + \frac{\varphi_0}{\pi^{\frac{1}{2}}} \left[\frac{x}{2(at)^{\frac{1}{2}}} - \frac{1}{3} \left(\frac{x}{2(at)^{\frac{1}{2}}} \right)^3 + \cdots \right].$$

4.2.10 Source function and its physical meaning

The function

$$K(x,t) = \frac{1}{2(\pi a t)^{\frac{1}{2}}} \exp(-\frac{x^2}{4at}), \tag{4.28}$$

which forms the kernel of the Fourier–Poisson integral (4.18) is one of the most important functions in the theory of the heat equation. It is called the *source* or *Green's function*, the *function of an instantaneous point source* or also the *fundamental solution*. Due to its frequent occurrence it deserves special mention. Before we elucidate its physical meaning let us mention some of its analytical properties.

We start by an immediate property

$$K(x,t) > 0, \qquad x \in \mathbf{R},\, t > 0, \tag{4.29}$$

which ensures, among others, the positivity of solutions as soon as $\varphi \geq 0$ (non-vanishing identically).

Further, the function K is unbounded in any neighbourhood of the point $(0,0)$. This is immediately seen by letting a point (x,t) approach $(0,0)$ along the curve $((at)^{\frac{1}{2}}, t)$, $t > 0$ since

$$\lim_{t \to 0+} K((at)^{\frac{1}{2}}, t) = \lim_{t \to 0+} \frac{1}{2(\pi a t)^{\frac{1}{2}}} \exp(-\frac{1}{4}) = \infty.$$

The same is true if we let a point (x,t) move towards $(0,0)$ along the axis $x = 0$, since $\lim\limits_{t \to 0+} K(0,t) = \infty$. As t increases the function $K(0,t)$ decreases and vanishes as $t \to \infty$ since

$$\lim_{t \to \infty} K(0,t) = \frac{1}{2(\pi a)^{\frac{1}{2}}} \lim_{t \to \infty} \frac{1}{t^{\frac{1}{2}}} = 0.$$

For a fixed $x = x_0 \neq 0$, on the other hand, $\lim\limits_{t \to 0+} K(x_0,t) = 0$, as time evolves the function $K(x_0,t)$ increases at first, then it attains its maximum and eventually it decreases again to zero. The derivative of the function $K(x_0,t)$ is equal to

$$\frac{1}{2(\pi a t)^{\frac{1}{2}}} \left(-\frac{1}{2} + \frac{x_0^2}{4at} \right) \exp(-\frac{x_0^2}{4at}),$$

hence by Calculus the time t^* in which $K(x_0,t)$ takes on its maximum is

$$t^* = \frac{x_0^2}{2a}$$

and the corresponding maximum value is

$$K(x_0,t^*) = \frac{1}{|x_0|(2\pi e)^{\frac{1}{2}}}.$$

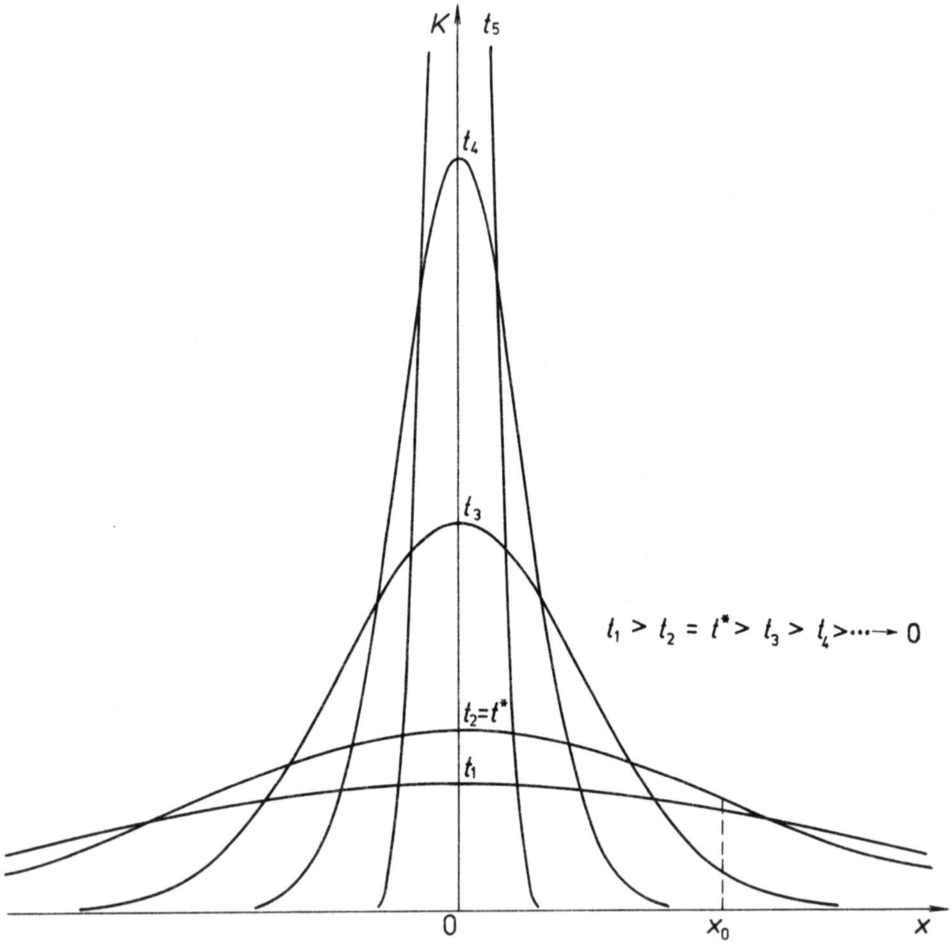

Fig. 22

Now, to visualize the function K we plot in Fig. 22 the graphs of K looked upon as a function of the variable x for a few different values of t. For $t \to 0+$ the curve becomes more and more sharply peaked. The curve corresponding to t^* is particular in the sense that for the given point x_0 the function $K(x_0, t)$ attains its maximum just on this curve.

An interesting result is obtained if we compute the area $P(t)$ of the plane surface bounded by the curve $K(x,t)$ and the x-axis for a fixed $t > 0$. Applying the substitution $\dfrac{x}{2(at)^{\frac{1}{2}}} = \eta$ and using formula (4.22) we get

$$P(t) = \int_{-\infty}^{\infty} K(x,t)\,dx$$
$$= \frac{1}{2(\pi a t)^{\frac{1}{2}}} \int_{-\infty}^{\infty} \exp\left(-\frac{x^2}{4at}\right) dx = 1, \tag{4.30}$$

that is, the area does not depend on the choice of $t > 0$.

It is natural to ask under what circumstances (if there are any) the function K determines the distribution of the temperature in our example with the infinite rod. By virtue of the second physical law introduced in § 4.1.1 (see also Exercise 4.1.5.1), the quantity

$$Q(t) = c\,\rho\,A\,P(t)$$

(A is the area of the cross-section of the rod) is equal to the total amount of heat within the rod at time t, 0 being the reference temperature. Consequently, this heat is constant for all values of $t > 0$,

$$Q(t) = Q = c\,\rho\,A.$$

Referring to Fig. 22 we see that for small $t > 0$ the heat within the rod is localized in a small neighbourhood of the point $x = 0$. We can pass to the limit to obtain

$$\lim_{t \to 0+} Q(t) = Q$$

which permits us to imagine this amount of heat as being generated at the moment $t = 0$ by an idealized source of heat concentrated at the point $x = 0$. This amount of heat spreads along the whole rod according to the law (4.28), its magnitude is constant and since $\int_{-\infty}^{\infty} K(x,t)\,dx = 1$ we speak in this context of a heat source of unit strength (simply a unit source). On the whole we speak of an *instantaneous point unit source*. Hence, the function $K(x,t)$ is the "response" at the point x to an instantaneous point unit source located at the point $x = 0$ at the time $t = 0$. This accounts for the name of $K(x,t)$ as the function of an instantaneous point source.

Analogously we can arrive at a more general conclusion that $K(x - \xi, t)$ defines the temperature of the infinite rod which is at zero temperature up to time $t = 0$ when an instantaneous point heat source of unit strength $\left(\frac{Q}{c\,\rho\,A} = 1\right)$ is introduced at the point ξ ($\in \mathbf{R}$ arbitrary). □

The reader with some basic knowledge of the theory of distributions [see, for example, Stakgold (1967)] can prove the previous results in the following way. First, the relation (4.19) yields, in particular,

$$\lim_{t \to 0+} \int_{-\infty}^{\infty} K(x - \xi, t)\,\varphi(x)\,dx = \varphi(\xi)$$

$$\text{for any } \varphi \in C_0^\infty(\mathbf{R}) \text{ and any } \xi \in \mathbf{R},$$

or written in different terms,

$$\lim_{t \to 0+} K(x - \xi, t) = \delta\,(x - \xi)$$

in the sense of distributions on $-\infty < x < \infty$, δ is Dirac's distribution. Loosely speaking, the function $K(x-\xi,t)$ is therefore "a solution of the initial-value problem for the homogeneous heat equation with the initial function $\delta(x-\xi)$":

$$\begin{aligned} u_t - a\,u_{xx} &= 0, \qquad\qquad t > 0, \\ u(x,0) &= \delta(x-\xi). \end{aligned}$$

Next, to any solution u of this problem one associates U defined, for all $t \in \mathbf{R}$, by

$$U = \chi\, u$$

where χ is the Heaviside function

$$\chi(t) = \begin{cases} 1, & t > 0 \\ 0, & t < 0. \end{cases}$$

Since (at least formally) $U_{xx} = \chi\, u_{xx}$, $U_t = \chi\, u_t + \chi_t\, u = \chi\, u_t + \delta(x-\xi)\,\delta(t-0)$, U is seen to be a solution of the inhomogeneous equation

$$\begin{aligned} u_t - a\,u_{xx} &= \delta\,(x-\xi)\,\delta\,(t-0), \quad t > 0, \\ u &\equiv 0, \qquad\qquad\qquad\qquad\quad t < 0. \end{aligned}$$

In other words, the initial impulse $\delta(x-\xi)$ for the homogeneous equation is equivalent to the source term $\delta(x-\xi)\,\delta(t-0)$ in the inhomogeneous equation and this is already in perfect agreement with the properties of the function K stated above.

4.2.11 Smoothing property of the heat equation

The solution of the initial-value problem for the homogeneous heat equation is given by the Fourier–Poisson integral (4.18), alternatively,

$$u(x,t) = \int_{-\infty}^{\infty} K(x-\xi,t)\,\varphi(\xi)\,d\xi,$$

where K is the source function (4.28) and φ is a bounded and continuous initial function on $\mathbf{R}$ (or, more generally, with bounded discontinuities, cf. § 4.2.8). In view of the rapid decay of the function K for $|x| \to \infty$ there is no difficulty in justifying differentiation under the integral sign, from which it follows that u has continuous partial derivatives of an arbitrary order for $-\infty < x < \infty$ and $t > 0$, that is, $u \in C^\infty(\mathbf{R} \times (0,\infty))$. This is the *smoothing property* (*regularizing effect*) of the heat equation: *any solution which initially is only continuous* (*and bounded*) *is arbitrarily smooth in arbitrarily short time* $t > 0$.

Compared with other types of second order equations we can say that this effect has a well-known counterpart in elliptic equations (any solution is arbitrarily smooth in the interior of its domain though it is only continuous on the boundary) but, on the other hand, it stands in marked contrast to properties of hyperbolic equations. □

This effect reveals another feature of the heat equation distinguishing it from the wave equation. Namely, in general, it is not possible to solve the heat equation for $t < 0$ with the "terminal" condition given for $t = 0$. The "terminal" function should belong to $C^\infty(\mathbf{R})$ in order that the solution should exist. Even if a solution exists (it is then unique) the problem is not well-posed due to the lack of continuous dependence of the solution on the data. In terms of physics, this phenomenon is in connection with the second law of thermodynamics which yields that certain changes of state (macroscopically viewed) are not physically realizable (heat cannot flow from a colder to a hotter part of a body). The heat conduction is an *irreversible process* in contrast to vibrations representing reversible processes where the directionality of time is of no importance.

4.2.12 Behaviour of solutions of the heat equation near the axis $t=0$

In § 4.2.11 we have shown that any solution of the initial-value problem for the homogeneous equation is of class C^∞ in the open half-plane $t > 0$. In general, the behaviour of a function satisfying the homogeneous heat equation can be singular near the axis $t = 0$ as we shall see presently. Let us consider the function

$$H(x,t) = -\frac{\partial}{\partial x} K(x,t) = \frac{x}{4\pi^{\frac{1}{2}}(at)^{\frac{3}{2}}} \exp(-\frac{x^2}{4at}), \tag{4.31}$$

which satisfies the homogeneous equation (4.15′), is of class C^∞ in the half-plane $t > 0$ and is not bounded (and much less continuous) in any neighbourhood of the point $(0,0)$.

To see this, we let (x,t) approach the origin along the curve $((at)^{\frac{1}{2}}, t)$, $t > 0$ to obtain

$$\lim_{t\to 0+} H((at)^{\frac{1}{2}}, t) = \lim_{t\to 0+} \frac{1}{4\pi^{\frac{1}{2}} at} \exp(-\frac{1}{4}) = \infty.$$

Moreover, by letting (x,t) approach $(0,0)$ along the curve $(-(at)^{\frac{1}{2}}, t)$, $t > 0$ one gets

$$\lim_{t\to 0+} H(-(at)^{\frac{1}{2}}, t) = -\infty,$$

which confirms the high degree of singularity of H at $(0,0)$.

On the other hand, the function H behaves well along the lines $x =$ const.:

$$\lim_{t\to 0+} H(x,t) = 0, \quad -\infty < x < \infty.$$

From this example it follows that the requirement of continuity up to the boundary in our definition of solution is very important. If we had replaced this requirement by a weaker one, for instance

$$\lim_{t\to 0+} u(x,t) = \varphi(x), \quad -\infty < x < \infty$$

(without having simultaneously demanded the boundedness, for example), the function identically equal to zero and the function $H(x,t)$ would have been two different solutions with the same initial data $\varphi \equiv 0$; the uniqueness would have been violated. □

The reader familiar with the elements of distribution theory can check easily that

$$\lim_{t\to 0+} H(x,t) = \delta'(x-0)$$

in the sense of distributions on $-\infty < x < \infty$, δ' being the derivative of Dirac's distribution.

4.2.13 Paradox of infinite velocity in the heat conduction

Assume we are given an initial function φ vanishing everywhere except for a certain finite interval I (however small) where it is positive. Formula (4.16) together with (4.30) makes it evident that the solution u assumes positive values for all x however small $t > 0$ may be.

This can be restated in terms introduced in the theory of hyperbolic equations (see § 2.2.2 and § 3.5.3). The domain of influence of the interval I is the whole open half-plane $t > 0$, or the domain of dependence of the solution u at (x, t) on the initial data is the whole axis $t = 0$. This stands again in sharp contrast to the properties of hyperbolic equations studied in Chapter 3 where the domain of dependence for the value $u(x,t)$ was always finite.

Physically, this phenomenon can be interpreted in such a way that the heat propagates with *infinite velocity*. This is physically impossible; the drawback is caused by the fact that upon deriving the equation of the heat conduction the phenomenological approach has been used rather than the qualitative theory taking into account the energy of vibrating molecules (or free electrons).

Anyway, for small $t > 0$ and large $|x|$ the values of $u(x,t)$ are negligibly small, which justifies the heat equation. This equation gives a good insight into the heat conductivity and has proved quite satisfactory in practice.

4.2.14 The initial-value problem for the inhomogeneous heat equation

Problems involving heat sources (or sinks) lead to the inhomogeneous heat equation. To compute the solution of the problem (4.15), (4.16) we could apply again the Fourier integral transform technique. However, we shall proceed in another way in order to illustrate the use of the Duhamel principle (cf. § 1.1.5) encountered often in applications.

First of all, since the problem under consideration is linear we can split it into two simpler ones (see again § 1.1.5). The first one ($\mathcal{P}1$) is given by (4.15) and

$$u(x,0) \equiv 0, \qquad -\infty < x < \infty,$$

the second one ($\mathcal{P}2$) by (4.15$'$) and (4.16). The sum of solutions to ($\mathcal{P}1$) and ($\mathcal{P}2$) constitutes a solution to the original problem.

It suffices to treat only ($\mathcal{P}1$) since ($\mathcal{P}2$) has already been solved (in § 4.2.3). Consider a succession of auxiliary problems

$$u_t(x,t) - a\,u_{xx}(x,t) = 0, \quad -\infty < x < \infty, \quad t > \tau,$$

$$u(x,\tau) = \varphi(x;\tau), \quad -\infty < x < \infty,$$

which differ from ($\mathcal{P}2$) only in that the process starts at $t = \tau$ instead of at $t = 0$. We associate with each positive number τ the solution $v(x,t,\tau)$ of the above problem which is, of course, available by a slight modification of the previous results:

$$v(x,t,\tau) = \int_{-\infty}^{\infty} K(x-\xi, t-\tau)\,\varphi(\xi;\tau)\,d\xi,$$

where K is the source function (4.28).

At this juncture, we set $\varphi(x;\tau) = g(x,\tau)$ and the integral

$$\int_0^t v(x,t,\tau)\,d\tau \tag{4.32}$$

may be checked to be a (formal) solution of ($\mathcal{P}1$).

Once we know the solutions of the homogeneous equation (4.15′) with the initial condition given at an "arbitrary level τ" we can construct the solution of the equation with non-vanishing right-hand side and vanishing initial function. In physical terms, the time dependent heat sources within the rod are replaced by a series of suitable initial temperature impulses at various times.

We cannot enter, here, into any detailed analysis giving the conditions on g under which (4.32) represents an actual solution. The assumptions stated in the following theorem may be somewhat weakened but the mere continuity of g (besides of boundedness) is not sufficient.

Let the function g and its partial derivative g_x be continuous for $-\infty < x < \infty$, $t > 0$ and moreover bounded on the strip $Q_T = (-\infty,\infty)\times(0,T)$ for any $T > 0$. Let φ be a continuous and bounded function on the interval $-\infty < x < \infty$. Then there exists exactly one classical solution of the problem (4.15), (4.16) *which is bounded on the strip Q_T for any $T > 0$. This solution is given by the formula*

$$\begin{aligned} u(x,t) &= \int_{-\infty}^{\infty} K(x-\xi,t)\,\varphi(\xi)\,d\xi \\ &+ \int_0^t \int_{-\infty}^{\infty} K(x-\xi,t-\tau)\,g(\xi,\tau)\,d\xi\,d\tau, \\ &\qquad -\infty < x < \infty,\ t > 0, \end{aligned} \tag{4.33}$$

where K is the source function (4.28).

4.2.15 More general equations

The equation

$$v_t(x,t) - a\,v_{xx}(x,t) + \beta\,v_x(x,t) + \gamma\,v(x,t) = 0,$$

where a, β and γ are real constants, $a > 0$, can be reduced to equation (4.15′) by means of the transformation of the dependent variable

$$u(x,t) = v(x,t)\exp\left(-\frac{\beta}{2a}x + \left(\frac{\beta^2}{4a} + \gamma\right)t\right).$$

The equation

$$v_\tau(\xi,\tau) - \alpha(\tau)\, v_{\xi\xi}(\xi,\tau) + \beta(\tau)\, v_\xi(\xi,\tau) + \gamma(\tau)\, v(\xi,\tau) = 0,$$

where α, β and γ are (continuous) functions of the variable τ and $\alpha(\tau) \geq \alpha_0 > 0$ for all $\tau \geq 0$, can be transformed to the same form (4.15′) with $a = 1$. To this end denote $\mathcal{A}(\tau) = \int_0^\tau \alpha(\eta)\, d\eta$ and $\tau = \mathcal{A}^{-1}(t)$ where $\mathcal{A}^{-1}$ is the inverse function to $t = \mathcal{A}(\tau)$; hence it follows that $\dfrac{d\,\mathcal{A}^{-1}(t)}{dt} = \dfrac{1}{\alpha(\tau)}$. The pertinent transformation is

$$u(x,t) = v(x + \int_0^t \frac{\beta(\mathcal{A}^{-1}(\sigma))}{\alpha(\mathcal{A}^{-1}(\sigma))}\, d\sigma, \mathcal{A}^{-1}(t)) \exp\left(\int_0^t \frac{\gamma(\mathcal{A}^{-1}(\sigma))}{\alpha(\mathcal{A}^{-1}(\sigma))}\, d\sigma\right).$$

4.2.16 Example: a wire carrying an electric current

Consider a wire of infinite length with cross-sectional area A and perimeter o along which, for times $t > 0$, the electric current flows at constant rate I. According to Joule's law the heat is generated within the wire at the rate

$$G = \frac{\omega I^2}{A^2}$$

per unit time per unit volume, where ω is the resistivity of the material of the wire. For simplicity we assume that the dependence of ω on the temperature may be neglected (e.g. for platiniridium). The heat is dissipated by convection into a surrounding medium at constant temperature u_0.

Using the notation as well as the results of Section 4.1 we get for the temperature u the inhomogeneous equation

$$u_t - a\, u_{xx} + \frac{\alpha o}{c\rho A}(u - u_0) = \frac{\omega I^2}{c\rho A^2}.$$

Now we employ the transformation as suggested in § 4.2.15

$$u(x,t) - u_0 = w(x,t) \exp(-\frac{\alpha o}{c\rho A} t)$$

to obtain

$$w_t - a\, w_{xx} = \frac{\omega I^2}{c\rho A^2} \exp(\frac{\alpha o}{c\rho A} t).$$

Eventually, upon applying formula (4.33) and using the result (4.30), we get

$$w(x,t) = \frac{\omega I^2}{\alpha o A}\left[\exp(\frac{\alpha o}{c\rho A} t) - 1\right]$$

and consequently

$$u(x,t) = u_0 + \frac{\omega I^2}{\alpha o A}\left[1 - \exp(-\frac{\alpha o}{c\rho A} t)\right].$$

4.2.17 Exercises

4.2.17.1 Starting with the formula

$$\int_{-\infty}^{\infty} \exp(-\alpha\, x^2)\, dx = \left(\frac{\pi}{\alpha}\right)^{\frac{1}{2}}$$

which is valid for any $\alpha > 0$ [cf. (4.22)] derive the formulae

$$\int_{-\infty}^{\infty} x^{2k} \exp(-\alpha x^2)\, dx = \frac{(2k-1)!!}{2^k} \frac{\pi^{\frac{1}{2}}}{\alpha^{k+\frac{1}{2}}}, \quad k \in \mathbf{N}.$$

Hint. Use the differentiation under the integral sign technique.

4.2.17.2 Provided that the function φ can be expanded about x into a Taylor series convergent on $\mathbf{R}$ show that the power series expansion of the solution to (4.15′), (4.16) can be written in the form (with no guarantee of convergence):

$$\sum_{k=0}^{\infty} \frac{2^k}{(2k)!!} \varphi^{(2k)}(x)(at)^k.$$

Check that the sum of the first two terms, i.e. $\varphi(x) + \varphi''(x)\, at$ gives, for any initial function of the form $\varphi(x) = c_0 + c_1 x + c_2 x^2$ (c_0, c_1 and c_2 are arbitrary real constants), the exact solution $c_0 + c_1 x + c_2 x^2 + 2c_2\, at$ (cf. § 4.2.2). Examine the connection between the above series and heat potentials from Exercise 4.1.5.2.
Hint. Use formula (4.21) and the result of the previous Exercise.

4.2.17.3 Solve the initial-value problem for the heat equation in n space variables

$$u_t(x_1, \ldots, x_n, t) - a\, \Delta u(x_1, \ldots, x_n, t) = 0, \quad -\infty < x_j < \infty,\ t > 0.$$

Answer.

$$u(x,t) = \left(\frac{1}{2(\pi a t)^{\frac{1}{2}}}\right)^n \int_{\mathbf{R}^n} \exp(-\frac{\|x-\xi\|^2}{4at})\, \varphi(\xi)\, d\xi, \quad x \in \mathbf{R}^n,\ t > 0.$$

4.2.17.4 Derive a formula for the solution of the initial-value problem for the Schrödinger equation

$$u_t(x,t) - i u_{xx}(x,t) = 0, \quad -\infty < x < \infty,\ t > 0.$$

Hint. Apply the formula (4.18), make the transformation $\tau = it$ and use the fact that the square root of the imaginary unit i is equal to $\exp(\frac{i\pi}{4})$.
Answer.

$$u(x,t) = \frac{1}{2(\pi t)^{\frac{1}{2}}} \exp(-\frac{i\pi}{4}) \int_{-\infty}^{\infty} \exp(\frac{i(x-\xi)^2}{4t})\, \varphi(\xi)\, d\xi.$$

4.3 THE INTEGRAL TRANSFORM METHOD

The solution of initial-boundary value problems for the heat equation can be obtained by a variety of techniques. Among them the methods of integral transforms are particularly effective. They make it possible to study problems with inhomogeneous time dependent boundary conditions and heat sources. Such problems can be tackled successfully also by means of various modifications of the Duhamel method the essence of which was outlined in § 4.2.14. Another important and advantageous method in solving inhomogeneous problems is the one which makes use of Green's function (see § 4.3.12). Finally, to solve problems on bounded regions with homogeneous boundary conditions the Fourier method of expanding the solution in the (complete orthonormal) system of space eigenfunctions can be employed with advantage (see Section 4.4) whenever this system of eigenfunctions is available.

In this section we shall solve inhomogeneous initial-boundary value problems by means of the Laplace transformation with respect to the time variable. We start, however, with an important and useful result in the theory of parabolic equations – the maximum principle – which provides a number of consequences (uniqueness of solutions, continuous dependence on data, positivity of Green's functions, and so forth).

4.3.1 Setting of the problem

We assume that we are treating an initial-boundary value problem for the heat equation and the variable x runs over a bounded interval of length ℓ. To be specific, let us deal with the problem given by the equation

$$u_t(x,t) - a\,u_{xx}(x,t) = g(x,t), \qquad 0 < x < \ell,\ t > 0 \tag{4.34}$$

in conjunction with the initial condition

$$u(x,0) = \varphi(x), \qquad 0 < x < \ell \tag{4.35}$$

and the inhomogeneous Dirichlet boundary conditions

$$u(0,t) = h_0(t), \quad u(\ell,t) = h_1(t), \qquad t > 0, \tag{4.36}$$

where g, φ, h_0 and h_1 are given functions.

We are to find a function $u = u(x,t)$ continuous on the semi-infinite strip $0 \le x \le \ell$, $t \ge 0$ with continuous derivatives u_t and u_{xx} for $0 < x < \ell$, $t > 0$ and satisfying (4.34) through (4.36).

Physically the problem corresponds, for example, to finding the temperature distribution within a rod of length ℓ impervious to heat on its lateral surface if initially the temperature is φ and for times $t > 0$ heat is generated in the rod at the rate $c\rho g(x,t)$ per unit time per unit volume while the ends are kept at temperatures that vary in time according to the laws h_0 and h_1, respectively.

4.3.2 The maximum principle

If the sources are not present within the rod then its temperature can neither exceed the maximum nor go below the minimum of the initial and boundary temperatures. The mathematical counterpart of this physically-evident fact is the so-called maximum (minimum) principle.

Before passing on we introduce some notation. For any $\tau > 0$ we denote

$$Q_\tau = (0, \ell) \times (0, \tau] \tag{4.37}$$

and

$$P_\tau = \bar{Q}_\tau \setminus Q_\tau = ([0, \ell] \times \{0\}) \cup (\{0, \ell\} \times [0, \tau]). \tag{4.38}$$

The set P_τ which is a part of the boundary of the rectangle $\bar{Q}_\tau$ formed by the bottom of $\bar{Q}_\tau$ and by its vertical sides is called the *parabolic boundary* of the rectangle $\bar{Q}_\tau$ (see Fig. 23).

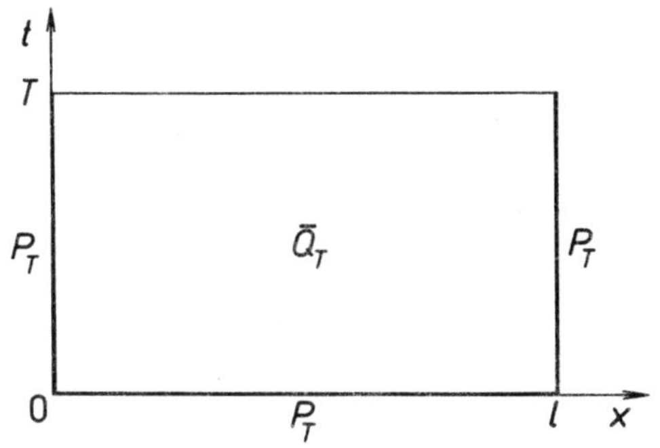

Fig. 23

The maximum principle (minimum principle). Let the function u
a) be continuous on the rectangle $\bar{Q}_T$ (for some $T > 0$)
b) have the continuous derivatives u_t and u_{xx} on Q_T satisfying the equation

$$u_t(x,t) - a\,u_{xx}(x,t) = 0 \qquad in \quad Q_T. \tag{4.39}$$

Then

$$\max_{\bar{Q}_T} u = \max_{P_T} u \qquad (\min_{\bar{Q}_T} u = \min_{P_T} u, \;\; respectively).$$

Expressed verbally: *the function u considered on $\bar{Q}_T$ assumes its largest value at least at one point of the parabolic boundary of $\bar{Q}_T$ (the same is true for the smallest value on $\bar{Q}_T$)*

Proof. For $\varepsilon > 0$ consider an auxiliary function u^ε on $\bar{Q}_T$,

$$u^\varepsilon(x,t) = u(x,t) + \varepsilon x^2,$$

satisfying

$$u_t^\varepsilon - a\, u_{xx}^\varepsilon = -2a\varepsilon < 0 \qquad \text{in} \quad Q_T. \tag{4.40}$$

If the function u^ε attains its maximum at some point (x_0, t_0) for which $0 < x_0 < \ell$ and $0 < t_0 < T$ (or $t_0 = T$) then by the elementary rules of Calculus one has

$$u_{xx}^\varepsilon(x_0,t_0) \leq 0 \quad \text{and} \quad u_t^\varepsilon(x_0,t_0) = 0 \qquad [\text{or} \quad u_t^\varepsilon(x_0,T) \geq 0].$$

In either case this is inconsistent with inequality (4.40). Thus

$$\max_{\bar{Q}_T} u^\varepsilon = \max_{P_T} u^\varepsilon.$$

Since

$$u \leq u^\varepsilon \leq u + \varepsilon\ell^2 \quad \text{in } \bar{Q}_T$$

we reach the required conclusion by letting $\varepsilon \longrightarrow 0+$.

For a minimum the result is obtained by applying the maximum principle to the function $-u$.

4.3.3 Example (the so-called comparison theorem)

Let the functions u and v be continuous on $\bar{Q}_T$ and satisfy equation (4.34) in Q_T. If $u \leq v$ on P_T then $u \leq v$ on $\bar{Q}_T$.

Upon setting $w = u - v$ one has $w_t - a\, w_{xx} = 0$ on Q_T, w is continuous on $\bar{Q}_T$ and $w \leq 0$ on P_T. By the maximum principle $w \leq 0$ on $\bar{Q}_T$ and the assertion follows.

4.3.4 Example (uniqueness theorem)

Let g be a continuous function for $0 \leq x \leq \ell$, $t \geq 0$, φ be a continuous function for $0 \leq x \leq \ell$ and h_0, h_1 be continuous functions for $t \geq 0$. Let the compatibility conditions

$$\varphi(0) = h_0(0), \qquad \varphi(\ell) = h_1(0) \tag{4.41}$$

be satisfied. Then there exists at most one classical solution of the problem (4.34) – (4.36).

Let u_1 and u_2 be two solutions. Then $w = u_1 - u_2$ is a continuous function for $0 \leq x \leq \ell$, $t \geq 0$ fulfilling the system (4.34) through (4.36) with $g = 0$, $\varphi = 0$, $h_0 = h_1 = 0$. The comparison theorem yields $w = 0$ for $0 \leq x \leq \ell$, $0 \leq t \leq T$, where $T > 0$ is arbitrary. We conclude that $w = 0$, hence $u_1 = u_2$ for $0 \leq x \leq \ell$, $t \geq 0$.

4.3.5 Remark on compatibility conditions

The compatibility conditions (4.41) must be necessarily satisfied in order that a function u fulfilling (4.34) through (4.36) should be continuous for $0 \leq x \leq \ell$, $t \geq 0$ (thus, in order that u should be a solution in the sense we have defined). If the compatibility conditions are not satisfied u cannot be continuous up to the boundary and we must abandon looking for classical solutions in favour of (more or less) generalized solutions. In any event the proof of the uniqueness of such generalized solutions cannot be performed with the aid of the maximum principle stated above. Nevertheless, in most cases another approach is feasible as we shall see in § 4.4.10.

4.3.6 Various generalizations

In § 4.3.2 we have laid down the simplest (so-called weak) maximum principle for the heat equation (4.39). By inspection of the proof it is readily seen that the result still holds if u instead of being the solution of equation (4.39) satisfies the inequality

$$u_t(x,t) - a\,u_{xx}(x,t) \leq 0 \quad \text{in } Q_T \tag{4.42}$$

(the reversed inequality for the minimum principle).

On physical grounds (and confirmed by the maximum principle) it is clear that if a function u continuous on $\bar{Q}_T$ satisfying (4.39) takes on a constant value on the parabolic boundary P_{t_1} for some $t_1 > 0$ then it is constant identically in the entire rectangle $\bar{Q}_{t_1}$. More delicate analysis shows that this is the only possible case when u is allowed to assume its maximum (minimum) at an interior point of the rectangle $\bar{Q}_T$ or on the open segment forming the top of $\bar{Q}_T$. This is just what the so-called strong maximum principle says: *no solution of equation* (4.39) [*no function satisfying inequality* (4.42)] *continuous on $\bar{Q}_T$ can assume its maximum value at a point of $\bar{Q}_T \backslash P_T$ with t-coordinate equal to t_1 unless it is constant on $\bar{Q}_{t_1}$ (similarly for the minimum).*

The maximum principle has been widely extended to more general regions, more general parabolic equations, etc. At this stage we content ourselves with stating a partial result in this context.

Let a function u be continuous on $\bar{Q}_T, u_t, u_x$ and u_{xx} be continuous on Q_T and the inequality

$$u_t(x,t) - (a(x)\,u_x(x,t))_x + \beta u_x(x,t) + \gamma u(x,t) \leq 0 \quad \text{in } Q_T$$

be satisfied, where a is a positive continuously differentiable function for $0 \leq x \leq \ell$, β and γ are constants, $\gamma \geq 0$. If $\gamma = 0$ then the above strong maximum principle holds verbatim. If $\gamma > 0$ then the result is still valid if we replace the term maximum by the term positive maximum. (For the reversed inequality the strong minimum principle holds if we replace minimum by negative minimum.)

These assertions are sufficient for the proof of theorems on uniqueness of solutions of corresponding initial-boundary value problems. A different maximum principle enables us to cope with uniqueness theorems for problems with Neumann and Newton boundary conditions. For details we refer the reader to Protter and Weinberger (1967).

Let us add that analogous results are known for second order elliptic equations [the maximum (minimum) of a solution is assumed on the boundary] but fail for hyperbolic ones.

4.3.7 Computation of the solution

After having established the uniqueness of the solution to the problem (4.34) – (4.36) we turn our attention to the existence question. In order to find a formula for the solution we apply the Laplace integral transformation over time.

The Laplace image $\tilde{u}$ of the function $u = u(x,t)$ [defined for $x \in I \subset \mathbf{R}$ and $t > 0$] with respect to the variable t is defined by

$$\tilde{u}(x,s) = \mathcal{L}\{u\}(x,s) = \int_0^\infty u(x,t)\exp(-st)\,dt$$

(where $x \in I$, $s > s_0(x)$, see § 1.2.7).

We shall follow four steps indicated in § 4.2.3 which, in fact, are common *mutatis mutandis* to any integral transform technique.

First we find the Laplace image of the problem. By means of the basic rule (1.46_1) the underlying initial-boundary value problem for the evolutionary differential equation is transformed to a boundary-value problem for a non-evolutionary equation.

This (presumably easier) boundary-value problem is solved in the second step. The formula for the Laplace image of the solution of the original problem is thereby available.

In the third step the original to this image is to be found. This will be done by consulting a table of Laplace transforms (otherwise a recourse to complex analysis techniques would be needed).

It appears to be convenient to split up the problem into two part-problems which are more easily tractable (cf. § 1.1.5). First we shall look for the solution u_1 of the problem with homogeneous differential equation, vanishing initial function and inhomogeneous boundary conditions. Second we shall seek the solution u_2 of the problem with homogeneous boundary conditions. Due to the linearity the sum $u_1 + u_2$ will be a (formal) solution to the original problem.

Since all the computations are usually carried out in a purely formal manner the fourth step consists in elucidating the conditions upon the data (differentiability requirements, the role of compatibility conditions, and so forth) which ensure that the formal solution represents an actual one (classical or generalized in some sense).

Let us remark that in § 4.4.3 we will solve the present problem (with homogeneous boundary conditions) by means of quite another approach.

Thus, to begin with, let us take the Laplace transform in (4.34) through (4.36). Formal rules of the transform Calculus yield

$$-a\,\tilde{u}_{xx}(x,s) + s\,\tilde{u}(x,s) = \varphi(x) + \tilde{g}(x,s), \quad 0 < x < \ell, \tag{4.43}$$

$$\tilde{u}(0,s) = \tilde{h}_0(s), \qquad \tilde{u}(\ell,s) = \tilde{h}_1(s), \tag{4.44}$$

where

$$\tilde{g}(x,s) = \int_0^\infty g(x,t)\exp(-st)\,dt,$$

$$\tilde{h}_j(s) = \int_0^\infty h_j(t)\exp(-st)\,dt, \qquad j = 0,1.$$

The Laplace image $\tilde{u}$ of the solution of the problem (4.34) through (4.36) satisfies the boundary-value problem (4.43), (4.44) with a second order *ordinary* differential equation in the variable x, where s plays the role of a parameter.

4.3.8 The case $\varphi \equiv 0, g \equiv 0$

Here, the solution of the problem (4.43), (4.44) is given by

$$\tilde{u}_1(x,s) = \tilde{h}_0(s)\frac{\sinh\left((\ell - x)(\frac{s}{a})^{\frac{1}{2}}\right)}{\sinh\left(\ell(\frac{s}{a})^{\frac{1}{2}}\right)} + \tilde{h}_1(s)\frac{\sinh\left(x(\frac{s}{a})^{\frac{1}{2}}\right)}{\sinh\left(\ell(\frac{s}{a})^{\frac{1}{2}}\right)}. \tag{4.45}$$

Consulting Table 1 (p. 30) and using (1.45) we find that

$$\frac{\sinh\left((\ell - x)(\frac{s}{a})^{\frac{1}{2}}\right)}{\sinh\left(\ell(\frac{s}{a})^{\frac{1}{2}}\right)} = \mathcal{L}\left\{-\frac{a}{\ell}\frac{\partial}{\partial x}\Theta\left(\frac{x}{2\ell}, \frac{at}{\ell^2}\right)\right\},$$
$$\frac{\sinh\left(x(\frac{s}{a})^{\frac{1}{2}}\right)}{\sinh\left(\ell(\frac{s}{a})^{\frac{1}{2}}\right)} = \mathcal{L}\left\{\frac{a}{\ell}\frac{\partial}{\partial x}\Theta\left(\frac{\ell - x}{2\ell}, \frac{at}{\ell^2}\right)\right\},$$

where the theta function Θ is defined in (1.50).

To obtain the original function to the product of two Laplace images we apply the convolution theorem (1.47_1) and accordingly

$$\begin{aligned} u_1(x,t) = &-\frac{a}{\ell}\int_0^t h_0(\tau)\frac{\partial}{\partial x}\Theta\left(\frac{x}{2\ell}, \frac{a(t-\tau)}{\ell^2}\right)d\tau \\ &+\frac{a}{\ell}\int_0^t h_0(\tau)\frac{\partial}{\partial x}\Theta\left(\frac{\ell - x}{2\ell}, \frac{a(t-\tau)}{\ell^2}\right)d\tau, \quad 0 < x < \ell,\ t > 0. \end{aligned} \tag{4.46}$$

Looking at (1.50) we see that the only singularity of the function $\Theta = \Theta(v,t)$ occurs at $v = t = 0$. In order to separate the "singular" and "regular" parts of Θ we will write

$$\frac{1}{2\ell}\Theta\left(\frac{x}{2\ell}, \frac{at}{\ell^2}\right) = K(x,t) + \varrho(x,t), \quad 0 < x < \ell,\ t > 0, \tag{4.47}$$

where K is the source function (4.28) and

$$\varrho(x,t) = \sum_{\substack{n=-\infty \\ n\neq 0}}^{\infty} K(x + 2n\ell, t).$$

The latter series with terms belonging to $C^\infty([0,\ell]\times[0,\infty))$ is absolutely and uniformly convergent on $[0,\ell]\times J$, where $J \subset [0,\infty)$ is an arbitrary closed bounded interval, together with the series obtained by termwise differentiation any number of times with respect to either variable. Further,

$$\frac{1}{2\ell}\frac{\partial}{\partial x}\Theta\left(\frac{x}{2\ell}, \frac{at}{\ell^2}\right) = -H(x,t) + \frac{\partial\varrho(x,t)}{\partial x}, \tag{4.48$_1$}$$
$$\frac{1}{2\ell}\frac{\partial}{\partial x}\Theta\left(\frac{\ell - x}{2\ell}, \frac{at}{\ell^2}\right) = H(\ell - x,t) - \frac{\partial\varrho(\ell - x,t)}{\partial x}, \tag{4.48$_2$}$$
$$0 < x < \ell,\ t > 0,$$

where

$$H(x,t) = \frac{x}{4\pi^{\frac{1}{2}}(at)^{\frac{3}{2}}} \exp\left(-\frac{x^2}{4at}\right)$$

and

$$\frac{\partial \varrho(x,t)}{\partial x} = - \sum_{\substack{n=-\infty \\ n\neq 0}}^{\infty} H(x+2n\ell, t).$$

(We have already met the function H in § 4.2.12.) Notice that the singularity of the derivative of Θ persists at the same point.

We prove that u_1 defined by formula (4.46) is an actual solution. To this end let us assume that h_0 and h_1 are continuous for $t \geq 0$. [More generally, near 0 the functions h_0 and h_1 may behave as $t^{-\alpha}$ for some $0 < \alpha < 1$. For details we refer to Cannon (1984), Widder (1975).]

Both integrals in (4.46) satisfy the homogeneous heat equation since the theta function Θ is a series of source functions and this series together with termwise differentiated series any number of times is absolutely and uniformly convergent on any interval $I \times J$, where $I \subset (0,\ell)$ is closed and $J \subset (0,\infty)$ is closed and bounded (so that the interchange of the order of summation and integration in both terms on the right-hand side of (4.46) is permissible).

What remains to settle is that u_1 belongs to $C([0,\ell] \times [0,\infty))$ and assumes the values on the boundary of $[0,\ell] \times [0,\infty]$ as specified.

Invoking the inequality (4.24) we get for any $\delta > 0$ the estimate

$$|H(x,t)| \leq \frac{8}{\delta^3}\left(\frac{at}{\pi}\right)^{\frac{1}{2}}$$

which is valid for any $|x| \geq \delta$, $t > 0$. Two consequences are of importance. First,

$$\lim_{t\to 0+} 2a \int_0^t H(x, t-\tau)\, h_0(\tau)\, d\tau = 0 \tag{4.49$_1$}$$

uniformly for x belonging to any closed interval $I \subset (0,\ell]$ and

$$\lim_{t\to 0+} 2a \int_0^t H(\ell - x, t-\tau)\, h_1(\tau)\, d\tau = 0 \tag{4.49$_2$}$$

uniformly for x belonging to any closed interval $I \subset [0,\ell)$. Second, the estimate together with the uniform convergence of the series for $\partial\varrho/\partial x$ on $[0,\ell] \times J$ ($J \subset [0,\infty)$ closed bounded) yields that the contributions from the regular parts $\partial\varrho/\partial x$ in (4.48) of the integrands in both integrals on the right-hand side of (4.46) disappear for $t = 0$ uniformly for $x \in (0,\ell)$. Upon recalling the concluding remark from § 4.2.7 we get: $u_1 \in C((0,\ell) \times [0,\infty))$ vanishing for $t = 0$.

Next, we turn our attention to the property $u_1 \in C([0,\ell] \times (0,\infty))$ which is most peculiar due to singularities at $(0,t)$ and (ℓ,t) of the first and second integrands respectively. The contributions stemming from the regular parts $\partial\varrho/\partial x$ in (4.48) to boundary values are zero as it is again easily seen by interchanging the order of summation and integration which is legitimate according to the uniform convergence. The same is true for the singular parts in the first integrand for $x = \ell$ and in the other one for $x = 0$. Summarizing and taking into account the remark from § 4.2.7 it suffices to prove

$$\lim_{x\to 0+} 2a\int_0^t H(x,t-\tau)\,h_0(\tau)\,d\tau = h_0(t) \tag{4.50$_1$}$$

and

$$\lim_{x\to \ell-} 2a\int_0^t H(\ell-x,t-\tau)\,h_1(\tau)\,d\tau = h_1(t) \tag{4.50$_2$}$$

in both cases uniformly for t belonging to any closed bounded interval in $(0,\infty)$.

The proof of these relations relies on the following properties of the function H the verification of which we leave to the reader as an exercise:

a) $H(x,t)\geq 0,\quad 0\leq x\leq \ell,\ t>0,\quad 2a\int_0^\infty H(x,t)\,dt = 1,\quad 0\leq x\leq \ell;$
b) $\lim_{x\to 0+} 2a\int_0^t H(x,\tau)\,d\tau = 1$ uniformly for t in any closed bounded interval in $(0,\infty)$;
c) if $\delta>0$ then $\lim_{x\to 0+}\int_\delta^t H(x,\tau)\,d\tau = 0$ uniformly for t in any closed bounded interval in (δ,∞).

In view of the property b) the relation (4.50$_1$) becomes true once we prove

$$\lim_{x\to 0+} 2a\int_0^t H(x,t-\tau)[h_0(t)-h_0(\tau)]\,d\tau = 0 \tag{4.51}$$

uniformly for t in any closed bounded interval in $(0,\infty)$. Given such an interval J and $\varepsilon>0$ then by the uniform continuity of h_0 on J there exists $\delta>0$ such that

$$|h_0(t)-h_0(\tau)| < \frac{\varepsilon}{2} \qquad \text{whenever}\quad t,\tau\in J,\quad |t-\tau|<\delta. \tag{4.52}$$

We break up the integral in (4.51) into the sum of two integrals over $(0,t-\delta)$ and $(t-\delta,t)$. The first integral then tends to zero as $x\to 0+$ uniformly for $t\in J$ by virtue of a), c) and the boundedness of h_0 on J and the second one can be estimated by $\varepsilon/2$ for sufficiently small $x>0$ uniformly for $t\in J$ by virtue of a), b) and (4.52). Hence, the first integral on the right-hand side of (4.46) belongs to $C([0,\ell]\times(0,\infty))$, assumes the values $h_0(t)$ for $x=0$ (a word of warning: the formal substitution $x=0$ leads to the zero value !) and vanishes for $x=\ell$.

An analogous reasoning proves the relation (4.50$_2$).

To recapitulate, $u_1\in C([0,\ell]\times(0,\infty))$ and the boundary conditions (4.36) are satisfied.

If, moreover, $h_0(0)=h_1(0)=0$ (these are now the compatibility conditions) then (4.49) are taken on uniformly for $x\in[0,\ell]$ and (4.50) are taken on uniformly for t belonging to any closed bounded interval of $[0,\infty)$. Consequently, $u\in C([0,\ell]\times[0,\infty))$. □

As a final remark let us state the property

$$-\frac{\partial}{\partial x}\Theta\Big(\frac{x}{2\ell},\frac{at}{\ell^2}\Big) > 0,\quad 0<x<\ell,\ t>0 \tag{4.53}$$

yielding $u_1\geq 0$ provided $h_0\geq 0$ and $h_1\geq 0$ [cf. (4.46)].

4.3.9 The case $h_0 = h_1 \equiv 0$

The solution of the problem given by Eq. (4.43) and boundary conditions

$$\tilde{u}(0,s) = \tilde{u}(\ell,s) = 0$$

can be expressed by means of its Green's function [see Roach (1982)]

$$\tilde{u}_2(x,s) = \int_0^\ell \gamma(x,\xi,s)\,\varphi(\xi)\,d\xi + \int_0^\ell \gamma(x,\xi,s)\,\tilde{g}(\xi,s)\,d\xi,$$

where

$$\gamma(x,\xi,s) = \begin{cases} \dfrac{a\sinh((\ell-\xi)(\frac{s}{a})^{\frac12})\sinh(x(\frac{s}{a})^{\frac12})}{(\frac{s}{a})^{\frac12}\sinh(\ell(\frac{s}{a})^{\frac12})}, & 0 \le x \le \xi, \\[2ex] \dfrac{a\sinh((\ell-x)(\frac{s}{a})^{\frac12})\sinh(\xi(\frac{s}{a})^{\frac12})}{(\frac{s}{a})^{\frac12}\sinh(\ell(\frac{s}{a})^{\frac12})}, & \xi \le x \le \ell. \end{cases}$$

By Table 1 (p. 30) we find that γ is the Laplace image of the function

$$\begin{aligned} \Gamma(x,\xi,t) &= \frac{1}{2\ell}\left[\Theta\Big(\frac{x-\xi}{2\ell},\frac{at}{\ell^2}\Big) - \Theta\Big(\frac{x+\xi}{2\ell},\frac{at}{\ell^2}\Big)\right] \qquad (4.54)\\ &= \frac{1}{2(\pi a t)^{\frac12}}\sum_{n=-\infty}^{\infty}\left\{\exp\Big(-\frac{(x-\xi+2n\ell)^2}{4at}\Big) - \exp\Big(-\frac{(x+\xi+2n\ell)^2}{4at}\Big)\right\} \\ &= \frac{2}{\ell}\sum_{k=1}^{\infty}\exp\Big(-\frac{k^2\pi^2 at}{\ell^2}\Big)\sin\frac{k\pi x}{\ell}\,\sin\frac{k\pi\xi}{\ell}. \end{aligned}$$

The latter two equivalent representations of the function Γ are consequences of (1.50). Using again the convolution theorem (1.47_1) one gets

$$\begin{aligned} u_2(x,t) &= \int_0^\ell \Gamma(x,\xi,t)\,\varphi(\xi)\,d\xi \\ &+ \int_0^\ell\int_0^t \Gamma(x,\xi,t-\tau)\,g(\xi,\tau)\,d\tau d\xi, \quad 0<x<\ell,\ t>0. \end{aligned} \qquad (4.55)$$

The analysis of this result is based upon breaking the function Γ into singular and regular parts

$$\Gamma(x,\xi,t) = K(x-\xi,t) + \chi(x,\xi,t), \qquad (4.56)$$

$$\chi(x,\xi,t) = -K(x+\xi,t) + \sum_{\substack{n=-\infty\\ n\ne 0}}^{\infty}[K(x-\xi+2n\ell,t) - K(x+\xi+2n\ell,t)], \qquad (4.57)$$

where K is the source function (4.28) and the series in (4.57) of functions from $C^\infty([0,\ell]\times[0,\infty))$ converges, together with any number of times term-by-term differentiated series, absolutely and uniformly on $[0,\ell]\times J$, where $J\subset[0,\infty)$ is any closed bounded interval. This permits us, in particular, to prove

$$\lim_{t\to 0+}\int_0^\ell \Gamma(x,\xi,t)\,\varphi(\xi)\,d\xi = \varphi(x)$$

provided that φ is continuous for $0 \le x \le \ell$. This limit is taken on uniformly with respect to $x \in I$, where I is any closed interval in $(0, \ell)$, or $I = [0, \ell]$ if, moreover, $\varphi(0) = \varphi(\ell) = 0$. We shall not pursue the subject further referring the interested reader to Cannon (1984). Instead, we formulate directly the corresponding assumptions ensuring that $u_1 + u_2$ is the classical solution.

4.3.10 Existence theorems

Let the function g be continuous on the semi-infinite strip $0 \le x \le \ell, t \ge 0$ and g_x be continuous for $0 < x < \ell, t > 0$. Let the functions h_0 and h_1 be continuous for $t \ge 0$. Eventually, let the compatibility conditions (4.41) be satisfied. Then there exists exactly one classical solution of the problem (4.34) through (4.36). This solution is given by the sum $u_1 + u_2$ where u_1 is given by formula (4.46) and u_2 by formula (4.55). □

We complement this theorem by an assertion which tells us, roughly speaking, that the solution possesses "better" properties if the data are "better" and further compatibility conditions are fulfilled.

Let the functions g and g_x be continuous for $0 \le x \le \ell, t \ge 0$. Let $\varphi \in C^2[0, \ell]$ and $h_j \in C^1[0, \infty)$, $j = 0, 1$. In addition to (4.41) let the following compatibility conditions be fulfilled

$$h_0'(0) - a\varphi''(0) = g(0,0), \qquad h_1'(0) - a\varphi''(\ell) = g(\ell, 0).$$

Then the classical solution u of the problem (4.34) – (4.36) (which exists and is unique by the above theorem) is continuous together with u_t and u_{xx} for $0 \le x \le \ell, t \ge 0$.

4.3.11 Lack of compatibility conditions, piecewise continuous initial functions

As mentioned earlier if the functions φ, h_0 and h_1 do not obey the compatibility conditions (4.41) the classical solution fails to exist. This might be inappropriate even in very simple physical situations. Consider the following example.

Let a metallic rod of length ℓ insulated against lateral heat conduction be heated throughout to a uniform temperature $\varphi_0 \ne 0$. Suppose that at time $t = 0$ the ends of the rod are submerged into an ice bath and subsequently they are maintained at this zero temperature. We inquire about the temperature distribution within the rod at any time $t > 0$.

In order to fit this situation we must modify the concept of a solution. In fact, a slightly more general situation involving an initial function suffering jump discontinuities can be covered at once. We replace the requirement on the continuity of the solution u in the following way (the other attributes of the solution remain unaltered): let u be continuous for all $0 \le x \le \ell, t \ge 0$ but a finite number of points $(0,0), (x_1, 0), \ldots, (x_p, 0), (\ell, 0)$, bounded in neighbourhoods of these points and the relation (4.35) be satisfied for every $0 < x < \ell, x \ne x_j, j = 1, \ldots, p$.

The sum $u_1 + u_2$ defines a unique solution in this altered sense provided that φ is a piecewise continuous function on $(0, \ell)$, the functions g, h_0 and h_1 have the same smoothness properties as above while no compatibility conditions are to be met.

Another way to cope with this situation will be discussed in § 4.4.8.

4.3.12 Green's function

The function $\Gamma = \Gamma(x, \xi, t)$ introduced in § 4.3.9 is called the *Green's function* or the *function of an instantaneous point source* [corresponding to the problem (4.34) through (4.36)]. As we shall see presently its physical interpretation is as follows: Γ defines the temperature at any point x of the rod, $0 \le x \le \ell$, at any time $t > 0$ due to a source of heat situated at the point ξ, $0 < \xi < \ell$, and releasing its heat spontaneously at time $t = 0$ in the rod which is initially at zero temperature ($x \neq \xi$) and the ends of which are thereafter kept at zero temperature. Indeed, as we know by § 4.2.10, $K(x - \xi, t)$ has the physical interpretation of being the temperature in a (virtual) infinite rod, $-\infty < x < \infty$, at time t due to an instantaneous point source located at the point ξ at time $t = 0$ while at other points the temperature is initially zero. From formula (4.56) we see that additional negative sources are placed outside the original rod at the points $-\xi$ and $-\xi + 2\ell$ to make the temperature at $x = 0$ and $x = \ell$ vanish. Now, the influence of these negative image sources at $x = 0$ and $x = \ell$ is compensated by (positive) image sources at points $\xi + 2\ell$, $\xi - 2\ell$ and so forth. Eventually, in general, to any unit (positive) image source at the point $\xi - 2n\ell$, $n = \pm 1, \ldots$ there corresponds a unit negative image source at points situated symmetrically with respect to 0 and ℓ, that is, at points $-\xi - 2n\ell$, $n = 0, \pm 1, \ldots$ (and vice versa), so that the resulting temperature at $x = 0$ and $x = \ell$ will remain zero for all $t > 0$. In other words, Γ satisfies the homogeneous boundary conditions as predicted. Moreover, its only singular point within $0 \le x \le \ell$ is $(\xi, \xi, 0)$.

We wish to emphasize the utility of the concept of a Green's function which is also introduced for other (more-dimensional) problems and other types of boundary conditions. The characteristic feature of a Green's function is that it solves the problem with homogeneous boundary conditions and with an instantaneous point source. An important fact to remember is that the Green's function gathers intrinsic properties of the problem in question, characterizes it completely, and, once it is known, solutions of corresponding inhomogeneous problems can be simply written in the form of "convolution-type integrals" – like formulae (4.55) or (4.46) [it is not incidental that the outward normal derivative (with respect to ξ-variable) of the Green's function might be recognized in (4.46), cf. Exercise 4.3.16.13]. There exists an extensive theory of Green's functions and they form an important tool both in theoretical considerations and practical applications. To find them the Laplace transform method is used most frequently. However, in some special and simple cases more straightforward methods are available.

For instance, the above physically motivated procedure can be used to *find* the Green's function (4.54). The approach consists in considering a virtual infinite medium and making "repeated amendments" of the source function $K(x - \xi, t)$ in order to meet the corresponding homogeneous boundary conditions. This is achieved by means of repeated reflection of image sources across the boundaries and we speak in this respect of the *method of reflection* (or *images*). This method has its well-known counterpart in the theory of elliptic equations. The success of this method, when generalized to higher dimensional problems, rests heavily upon a rather restrictive simple geometry of regions.

To conclude this subsection let us mention one other way to obtain the function Γ without having recourse to the Laplace transform technique. (Yet another approach will be met in § 4.4.14.) This method relies on a suitable – conformable to boundary conditions – extension of an arbitrary initial function φ outside the interval $(0, \ell)$ and on the subsequent application of the known solution of the initial-value problem for the homogeneous equation. Let us extend φ as an odd function into $-\ell < x < 0$ and then elsewhere as a periodic function of period 2ℓ. Upon using the Fourier–Poisson integral (4.18) and integrating by parts we obtain

$$u(x,t) = \int_0^\ell \sum_{n=-\infty}^{\infty} [K(x-\xi+2n\ell,t) - K(x+\xi+2n\ell,t)]\,\varphi(\xi)\,d\xi.$$

The kernel of this integral is just the Green's function Γ as formula (4.56) bears out. We have already encountered this so-called *method of extension* in § 3.1.9.

4.3.13 Initial-boundary value problems for the quarter plane; example

Initial-boundary value problems involving the variables (x,t) from the quarter plane $(0,\infty)\times(0,\infty)$ arise, for instance, if the temperature is to be found near one end of a rod and far from the other. We make the assumption that the temperature is essentially determined by the thermal schedule at the nearer end (where, for example, the Dirichlet boundary condition can be imposed) and by the initial condition and we look at the rod as being semi-infinite – its axis coincides with $0 < x < \infty$. In this way we are led, for example, to the following set-up:

$$u_t(x,t) - a\,u_{xx}(x,t) = g(x,t), \qquad x>0,\ t>0, \tag{4.59}$$

$$u(x,0) = \varphi(x), \qquad x>0, \tag{4.60}$$

$$u(0,t) = h_0(t), \qquad t>0. \tag{4.61}$$

The boundary condition at the farther end is replaced by qualifying the behaviour of a solution at ∞, for example, by requiring its boundedness.

As another illustration of the Laplace transform method we shall look for a solution of the above problem with rather special data:

$$g \equiv 0, \quad h_0 \equiv 0, \quad \varphi(x) = \varphi_0 = \text{const.}$$

(For general data we refer the reader to the next subsection and Exercise 4.3.16.2.)

Upon taking the Laplace transform over time in (4.59) through (4.61) we get a boundary-value problem for an ordinary differential equation

$$-a\tilde{u}_{xx}(x,s) + s\tilde{u}(x,s) = \varphi_0, \quad x>0 \tag{4.62}$$

in conjunction with two conditions

$$\tilde{u}(0,s) = 0 \tag{4.63}$$

and

$$\tilde{u}(x,s) \quad \text{is bounded as} \quad x \to \infty \tag{4.64}$$

($s > 0$ is a parameter). The latter condition will be seen to assure the boundedness of the original $u(x,t)$ as $x \to \infty$.

We check easily that any solution of Eq. (4.62) satisfying the boundary condition (4.63) is given by

$$\tilde{u}(x,s) = C(s)\sinh((\frac{s}{a})^{\frac{1}{2}}x) + \frac{\varphi_0}{s}\left[1 - \cosh((\frac{s}{a})^{\frac{1}{2}}x)\right],$$

where $C(s)$ is an arbitrary constant. In order to fulfil (4.64) we must put $C(s) = \frac{\varphi_0}{s}$ so that

$$\tilde{u}(x,s) = \frac{\varphi_0}{s}\left[1 - \exp(-(\frac{s}{a})^{\frac{1}{2}}x)\right].$$

By Table 1 (p. 30),

$$\frac{1}{s}\left[1 - \exp(-(\frac{s}{a})^{\frac{1}{2}}x)\right] = \mathcal{L}\left\{\operatorname{erf}\frac{x}{2(at)^{\frac{1}{2}}}\right\}.$$

Hence we conclude (with the benefit of § 4.2.9)

$$u(x,t) = \varphi_0 \operatorname{erf}\frac{x}{2(at)^{\frac{1}{2}}} = \frac{2\varphi_0}{\pi^{\frac{1}{2}}}\left[\frac{x}{2(at)^{\frac{1}{2}}} - \frac{1}{3}\left(\frac{x}{2(at)^{\frac{1}{2}}}\right)^3 + \cdots\right].$$

4.3.14 Example

We find the Green's function $D = D(x,\xi,t)$ corresponding to the problem (4.59) through (4.61).

By definition, the function D determines the temperature at any point $x > 0$ of a semi-infinite rod at any instant $t > 0$ initially at temperature 0 into which at the instant $t = 0$ a unit heat source has been placed at the point $\xi > 0$ and the end $x = 0$ is maintained at temperature 0.

We apply the method of images. Consider an infinite rod, $-\infty < x < \infty$. Taking into account once again the physical meaning of the source function $K(x-\xi,t)$ we see that in order to obtain the function D with the desired properties, it suffices to take K and add a function χ defining the distribution of temperature caused by an instantaneous point unit negative source at $t = 0$ at a point located symmetrically with respect to $x = 0$, that is, at $-\xi$. This image source and the actual one amount to zero temperature at $x = 0$. Now, $\chi(x,\xi,t) = -K(x+\xi,t)$, so that

$$D(x,\xi,t) = K(x-\xi,t) - K(x+\xi,t). \tag{4.65}$$

The solution of the problem (4.59) through (4.61) with $h_0 \equiv 0$ may be written in the form [cf. (4.55)]

$$u(x,t) = \int_0^\infty D(x,\xi,t)\varphi(\xi)\,d\xi + \int_0^t\int_0^\infty D(x,\xi,t-\tau)\,g(\xi,\tau)\,d\xi\,d\tau.$$

4.3.15 Concluding remarks on integral transform methods

We have applied two (Fourier and Laplace) integral transform methods. The utility of these transforms consists of the fact that in either case the operation of differentiation is converted into a simple *algebraic* operation performed on the image function. Both transforms can be also used for treating higher-dimensional problems.

The Fourier transform method is applicable to problems defined on the whole x-space. The Fourier method which will be discussed at some length in the next section can be considered a counterpart of the Fourier integral transform method for problems where the range of the space variables is a bounded region. The application of the Fourier transform removes the space variable thus transforming the problem into an initial-value problem for an ordinary evolutionary equation.

On the other hand, the Laplace transform method can be applied to problems involving various types of regions in $\mathbf{R}^n$. Since it removes the time variable the problem is reduced to a boundary-value problem for a stationary partial (ordinary if $n = 1$) differential equation.

The resulting problems are presumably easier to solve. The difficulty in inverting can be circumvented in practice by looking up original functions in tables or by using well-developed techniques of asymptotic expansions in case we may be interested in inverse Laplace transforms for only large values of t. We omit any further discussion of these and many related topics which lie beyond the scope and intent of this text.

Nothing will be said either about other useful integral transformations which are appropriate for special types of problems (such as the so-called Hankel transform that fits the problems involving circular geometry, and so forth).

4.3.16 Exercises

4.3.16.1 Solve the problem (4.59) through (4.61) with $g \equiv 0$, $h_0 \equiv 0$ by the method of extension of φ.

Hint. Make an odd extension of φ to $-\infty < x < 0$ by imposing the condition $\varphi(-x) = -\varphi(x)$, $x > 0$ and use the Fourier–Poisson integral (4.18).

Answer.

$$u(x,t) = \int_0^\infty [K(x-\xi,t) - K(x+\xi,t)]\,\varphi(\xi)\,d\xi.$$

4.3.16.2 Solve the problem (4.59) through (4.61) with $g \equiv 0$, $\varphi \equiv 0$ by the Laplace transform method.

Answer.

$$u(x,t) = 2a\int_0^t H(x,t-\tau)\,h_0(\tau)\,d\tau = \frac{2}{\pi^{\frac{1}{2}}}\int_{x/2(at)^{\frac{1}{2}}}^\infty h_0\Big(t - \frac{x^2}{4a\eta^2}\Big)\exp(-\eta^2)\,d\eta.$$

4.3.16.3 By means of the Laplace transform method solve the problem

$$\begin{aligned} u_t(x,t) - a u_{xx}(x,t) &= 0, \quad x > 0,\, t > 0, \\ u(x,0) &= \varphi_0, \quad x > 0, \\ u_x(0,t) - \sigma_0\, u(0,t) &= 0, \quad t > 0 \end{aligned}$$

(where φ_0 and σ_0 are constants). Interpret the boundary condition physically.
Answer.
A semi-infinite rod, $0 < x < \infty$, with the dissipation of heat by convection (radiation) at the left end into a surrounding atmosphere at zero temperature.

$$u(x,t) = \varphi_0 \operatorname{erf}\frac{x}{2(at)^{\frac{1}{2}}} + \frac{\varphi_0}{(\pi a t)^{\frac{1}{2}}} \int_x^\infty \exp(-\sigma_0(\tau - x) - \frac{\tau^2}{4at})\, d\tau.$$

4.3.16.4 Find the Green's function N corresponding to the initial-boundary value problem for the quarter plane with the Neumann boundary condition.
Hint. Use either the method of images or the method of extension (making an even extension of φ to $-\infty < x < 0$).
Answer.

$$N(x,\xi,t) = K(x-\xi,t) + K(x+\xi,t).$$

4.3.16.5 Set up the problem of heat conduction within a rod of length ℓ without sources insulated against lateral heat conduction if heat is supplied (or extracted) at its ends with the inward flux densities that vary in time according to laws $q_0(t)$ and $q_1(t)$ respectively. If initially the temperature is zero find the temperature at any subsequent time.
Hint. The boundary conditions are given by (4.13).
Answer.

$$u(x,t) = \frac{a}{\lambda \ell}\left[\int_0^t q_0(\tau)\,\Theta(\frac{x}{2\ell}, \frac{a(t-\tau)}{\ell^2})\, d\tau + \int_0^t q_1(\tau)\,\Theta(\frac{\ell - x}{2\ell}, \frac{a(t-\tau)}{\ell^2})\, d\tau\right].$$

4.3.16.6 Study the example of the previous Exercise if the heat flux boundary specification at the left end is replaced by the temperature boundary specification $u(0,t) = h_0(t)$, $t > 0$.
Answer.

$$\begin{aligned} u(x,t) &= -\frac{a}{\ell}\int_0^t h_0(\tau)\frac{\partial}{\partial x}\Theta_2(\frac{x}{2\ell}, \frac{a(t-\tau)}{\ell^2})\, d\tau \\ &+ \frac{a}{\lambda\ell}\int_0^t q_1(\tau)\,\Theta_2(\frac{\ell - x}{2\ell}, \frac{a(t-\tau)}{\ell^2})\, d\tau, \\ \Theta_2(x,t) &= \frac{1}{(\pi t)^{\frac{1}{2}}} \sum_{n=-\infty}^{\infty} (-1)^n \exp(-\frac{(v+n)^2}{t}), \quad 0 \le v \le 1. \end{aligned}$$

4.3.16.7 Study the convenience of solving initial-boundary value problems via the reduction of the heat equation to the ordinary differential equation by means of the Boltzmann transformation. (The method of reducing the equation to an equation in less independent variables is often use in solving nonlinear equations. The so-called *similarity solutions* are gained.)
Hint. See Exercise 4.1.5.3.
Answer.
Very special nature of the initial and boundary conditions is necessary.

4.3.16.8 Prove that the initial-boundary value problem (4.59) through (4.61) has at most one solution which is bounded on $(0,\infty)\times(0,T)$ for any $T>0$ [provided φ is bounded and g is bounded with respect to x, $\varphi(0)=h_0(0)$].
Hint. Let u be a solution with $g=\varphi=h_0\equiv 0$ and $|u(x,t)|\le M$ on $(0,\infty)\times(0,T)$. For any $\ell>0$ define

$$u^\ell(x,t)=u(x,t)-M\frac{x}{\ell}.$$

Apply the maximum (minimum) principle 4.3.2 on $\bar{Q}_T$ [see (4.37)] to obtain $u(x,t)\le 0$ (≥ 0) throughout $\bar{Q}_T$. Then fix (x,t) and pass to the limit $\ell\to\infty$.

4.3.16.9 Prove the results on uniqueness in theorems from § 4.2.2 and § 4.2.14.
Hint. Proceed in the same way as in Exercise 4.3.16.8 with u^ℓ replaced by

$$u^\ell(x,t)=u(x,t)-\frac{M}{\ell^2}(x^2+2at).$$

4.3.16.10 Study the continuous dependence on data of solutions to problems described in the last two Exercises.

4.3.16.11 A function continuous on the rectangle $\bar{Q}_T$ [see (4.37)] satisfying the equation

$$u_t-a\,u_{xx}+\gamma\,u=0,\qquad (x,t)\in Q_T,\quad (\gamma\in \mathbf{R})$$

can attain a value greater than its maximum taken on the parabolic boundary P_T [see (4.38)] only in either of the following two cases:
a) $\gamma\ne 0$ and u is negative throughout $\bar{Q}_T$,
b) $\gamma<0$.
Show in examples.
Answer.
a) $u(x,t)=-\cosh(x-\frac{\ell}{2})$,
b) $u(x,t)=\sin\frac{\pi x}{\ell}$.

4.3.16.12 Prove that the Green's function $\varGamma$ given by (4.54) is non-negative:

$$\varGamma(x,\xi,t)\ge 0,\qquad 0\le x,\ \xi\le\ell,\quad t>0.$$

[This result is clear on physical grounds, see § 4.3.12, but a rigorous proof is not so immediate as its analogue for the source function K was, see (4.29).]

Hint. First, let us recall a basic result concerning the approximation of functions. Let ψ be a non-negative function from $C^\infty(\mathbf{R})$ vanishing outside the interval $(-1, 1)$ for which $\int_{\mathbf{R}} \psi(\xi)\, d\xi = 1$. For example,

$$\psi(\xi) = \begin{cases} k \exp\left(-(1-\xi^2)^{-1}\right) & \text{for } \xi \in (-1, 1) \\ 0 & \text{otherwise} \end{cases}$$

with a suitably chosen constant k.

For any $\epsilon > 0$ define a function ψ_ϵ ($\in C_0^\infty(\mathbf{R})$ vanishing for $|\xi| > \epsilon$) by the formula

$$\psi_\epsilon(\xi) = \frac{1}{\epsilon}\psi\left(\frac{\xi}{\epsilon}\right).$$

Now, if $v \in C(\mathbf{R})$, $v = 0$ outside a bounded interval then denoting

$$v_\epsilon(\xi) = \int_{\mathbf{R}} \psi_\epsilon(\xi - \eta)\, v(\eta)\, d\eta$$

one has

i) $v_\epsilon \in C_0^\infty(\mathbf{R})$;
ii) $\lim\limits_{\epsilon \to 0+} v_\epsilon(\xi) = v(\xi)$ uniformly for $\xi \in \mathbf{R}$.

Second, by Theorem 4.3.10 the solution of the problem (4.3.4) through (4.3.6) with $h_0 = h_1 \equiv 0$ and $\varphi(x) = \psi_\epsilon(\xi - x)$, where $0 < \xi < \ell$ and $\epsilon \in (0, \epsilon_0]$, $\epsilon_0 > 0$ sufficiently small, is given by

$$u_\epsilon(x,t) = \int_0^\ell \Gamma(x, \eta, t)\, \psi_\epsilon(\xi - \eta)\, d\eta.$$

To conclude use a) the maximum principle 4.3.2 which yields

$$u_\epsilon(x,t) \geq 0$$

and the above result:

$$\lim_{\epsilon \to 0+} u_\epsilon(x,t) = \Gamma(x, \xi, t).$$

4.3.16.13 Prove that the solution of the problem (4.3.4) through (4.3.6) can be written by means of the Green's function (4.54) in the form

$$u(x,t) = \int_0^\ell \Gamma(x,\xi,t)\, \varphi(\xi)\, d\xi$$
$$+ \int_0^t \Gamma_\xi(x, 0, t-\tau)\, h_0(\tau)\, d\tau - \int_0^t \Gamma_\xi(x, \ell, t-\tau)\, h_1(\tau)\, d\tau$$
$$+ \int_0^t \int_0^\ell \Gamma(x, \xi, t-\tau)\, g(\xi, \tau)\, d\xi d\tau.$$

4.4 THE FOURIER METHOD

The initial-boundary value problems involving bounded regions can be attacked in a number of cases by the Fourier method. Upon expanding the sought solution in terms of a system of eigenfunctions of the given problema we get a Fourier series. This series provides a formal solution which appears to be classical under proper conditions and, on the other hand, suggests a way to define various generalized solutions. One such concept will be introduced to relax the classical continuity requirement up to the initial line $t = 0$.

4.4.1 Separation of variables

We shall be concerned with the problem given by the equation

$$u_t(x,t) - a\,u_{xx}(x,t) = g(x,t), \quad 0 < x < \ell,\, t > 0, \tag{4.66}$$

by the initial condition

$$u(x,0) = \varphi(x), \quad 0 < x < \ell \tag{4.67}$$

and by one set of the following homogeneous boundary conditions

$$u(0,t) = u(\ell,t) = 0, \quad t > 0, \tag{4.68$_1$}$$

$$\begin{aligned} -u_x(0,t) + \sigma_0\, u(0,t) &= 0, \\ u_x(\ell,t) + \sigma_1\, u(\ell,t) &= 0, \quad t > 0, \end{aligned} \tag{4.68$_2$}$$

$$\begin{aligned} -u_x(0,t) + \sigma_0\, u(0,t) &= 0, \\ u(\ell,t) &= 0, \quad t > 0, \end{aligned} \tag{4.68$_3$}$$

where σ_0 and σ_1 are real constants.

For the present we restrict ourselves to the homogeneous equation

$$u_t(x,t) - a\,u_{xx}(x,t) = 0, \quad 0 < x < \ell,\, t > 0 \tag{4.66$'$}$$

and we try to solve the problem by means of the separation-of-variables method without worrying temporarily about the initial condition. Thus, a solution to (4.66′), (4.68$_\iota$) ($\iota = 1,2,3$) is looked for in the form

$$u(x,t) = e(x)\,v(t), \tag{4.69}$$

where e and v are functions non-vanishing identically (cf. § 3.3.1).

Assume first that the product (4.69) is a solution of Eq. (4.66′) and $e(x_0) \neq 0$, $v(t_0) \neq 0$ for some $0 < x_0 < \ell$ and $t_0 > 0$. Then

$$e(x)\,v'(t) - a\,e''(x)\,v(t) = 0, \quad 0 < x < \ell,\, t > 0$$

and thus the function v satisfies the differential equation

$$v'(t) + \lambda\, a\, v(t) = 0, \quad t > 0, \tag{4.70}$$

where

$$\lambda = -\frac{e''(x_0)}{e(x_0)}.$$

Further, clearly

$$e(x)\,\frac{v'(t_0)}{a\, v(t_0)} - e''(x) = 0, \quad 0 < x < \ell,$$

so that the function e satisfies the differential equation

$$-e''(x) - \lambda\, e(x) = 0, \quad 0 < x < \ell. \tag{4.71}$$

On the other hand, if v is assumed to satisfy the differential equation (4.70) and e the differential equation (4.71), where λ is an arbitrary real number, then the function (4.69) is a solution of equation (4.66′).

The necessary and sufficient condition that (4.68_ι) be fulfilled is that e satisfies respectively

$$e(0) = e(\ell) = 0 \tag{4.72$_1$}$$

$$-e'(0) + \sigma_0\, e(0) = e'(\ell) + \sigma_1\, e(\ell) = 0, \tag{4.72$_2$}$$

$$-e'(0) + \sigma_0\, e(0) = e(\ell) = 0. \tag{4.72$_3$}$$

This in turn leads to restrictions on values of λ.

The problem of finding the functions e non-vanishing identically and the numbers λ such that the differential equation (4.71) together with the boundary conditions (4.72_ι) is to be satisfied constitutes an eigenvalue problem (cf. § 3.3.1). The present cases fall into the category of the so-called regular Sturm–Liouville problems the properties of which are well established in the mathematical literature [see B.Friedmann (1956) or § 3.3.3]. The following lemma gathers some well-known facts.

There exists a countable sequence of eigenvalues $\{\lambda_k\}_{k\in\mathbb{N}}$, the eigenvalues are all simple[13] *and can be ordered according to increasing magnitude so that*

$$\lambda_1 < \lambda_2 < \cdots < \lambda_k < \cdots, \quad \lim_{k\to\infty} \lambda_k = \infty.$$

The corresponding eigenfunctions can be chosen so as to form a complete orthonormal system $\{e_k\}_{n\in\mathbb{N}}$ on the interval $(0,\ell)$ [in the space $L_2(0,\ell)$.]

For purposes of technical convenience it is preferable to work in some particular cases to be seen below with the index set $M \subset \mathbb{N} \cup \{0,-1,-2\}$. In general considerations, however, we continue to use the set $\mathbb{N}$.

[13]The problem has a non-trivial solution if and only if $\lambda = \lambda_k$ and for each λ_k these solutions, eigenfunctions, form a one-dimensional linear space after the zero function has been added.

In § 3.3.1 we have proved that for the problem with the Dirichlet boundary conditions ($\iota = 1$)

$$\lambda_k = \Big(\frac{k\pi}{\ell}\Big)^2, \quad e_k = \Big(\frac{2}{\ell}\Big)^{\frac{1}{2}} \sin\frac{k\pi x}{\ell}, \quad k \in \mathbf{N}. \tag{4.73}$$

We pause briefly to review the results concerning the other boundary conditions. The eigenvalues of the problem (4.71), (4.72_2) need not be all positive (the problem is not always "positive definite"). It depends on the relationship between the parameters σ_0 and σ_1. We check easily that $\lambda = 0$ is an eigenvalue if and only if

$$\varrho(\sigma_0, \sigma_1) \equiv \sigma_1 + \sigma_0 + \sigma_0\sigma_1\ell = 0$$

(cf. p. 144). This relation defines an equilateral hyperbola in the plane of variables σ_0 and σ_1, namely

$$\sigma_1 = -\frac{\sigma_0}{1 + \sigma_0\ell}.$$

Its branches break the plane up into three regions $\sum_0$, $\sum_1$ and $\sum_2$ (see Fig. 24) characterized by

$$\textstyle\sum_0 = \{\, (\sigma_0, \sigma_1) \mid \varrho(\sigma_0, \sigma_1) > 0,\ \sigma_0 > -\frac{1}{\ell}\},$$

$$\textstyle\sum_1 = \{\, (\sigma_0, \sigma_1) \mid \varrho(\sigma_0, \sigma_1) < 0\},$$

$$\textstyle\sum_2 = \{\, (\sigma_0, \sigma_1) \mid \varrho(\sigma_0, \sigma_1) > 0,\ \sigma_0 < -\frac{1}{\ell}\}.$$

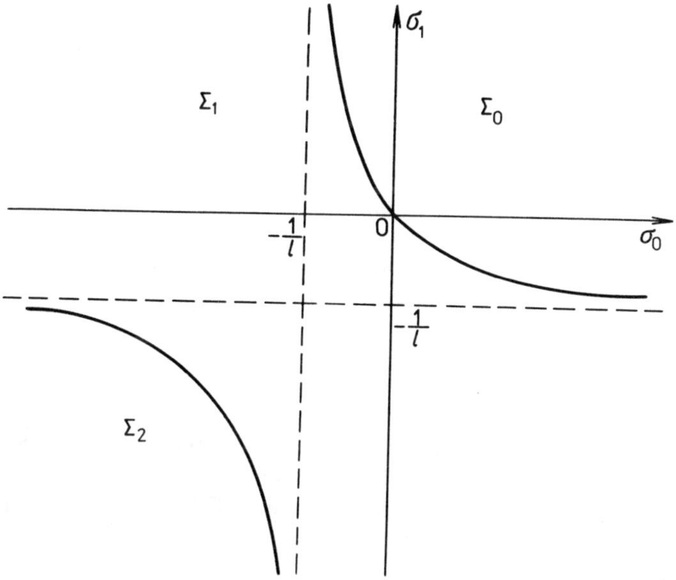

Fig. 24

If $(\sigma_0, \sigma_1) \in \sum_0$ (above the upper branch) all eigenvalues are positive. It is worthwhile to observe that this is particularly true if, simultaneously, $\sigma_0 > 0$ and $\sigma_1 > 0$.

If $(\sigma_0, \sigma_1) \in \sum_1$ (between both branches) the problem has exactly one negative eigenvalue (and all other eigenvalues are positive). Eventually, if $(\sigma_0, \sigma_1) \in \sum_2$ (under the lower branch) the problem has exactly two negative eigenvalues (and all others are positive).

Let us note that in applications we encounter mostly the (positive definite) case $(\sigma_0, \sigma_1) \in \sum_0$ (the only case giving asymptotic stability) and possibly the case $\sigma_0 = \sigma_1 = 0$. For the latter case (of Neumann boundary conditions) we set down the pertinent results:

$$\begin{aligned} \lambda_0 = 0, \qquad & \lambda_k = \left(\frac{k\pi}{\ell}\right)^2, \\ e_0(x) = \left(\frac{1}{\ell}\right)^{\frac{1}{2}}, \quad e_k(x) = & \left(\frac{2}{\ell}\right)^{\frac{1}{2}} \cos\frac{k\pi x}{\ell}, \quad k \in \mathbf{N}. \end{aligned} \tag{4.74}$$

The problem (4.71), (4.72_3) has only positive eigenvalues if and only if $\sigma_0 > -1/\ell$. If $\sigma_0 = -1/\ell$ ($\sigma_0 < -1/\ell$) then a zero eigenvalue (exactly one negative eigenvalue) emerges (the others being positive). □

To recapitulate: the existence of eigenvalues $\{\lambda_k\}_{k\in\mathbf{N}}$ and the corresponding system of eigenfunctions $\{e_k\}_{k\in\mathbf{N}}$ is ensured in any event by a general theorem. Moreover, the transcendental equations for eigenvalues and the formulae for corresponding eigenfunctions are available (see § 3.3.3). Notice, too, that all the functions e_k belong to $C^\infty[0,\ell]$.

In view of equation (4.70) any function of the form

$$\exp(-\lambda_k\, at)\, e_k(x) \tag{4.75}$$

is thereby a solution of equation (4.66′) satisfying the respective boundary conditions.

4.4.2 The initial-boundary value problem for the homogeneous equation

We have obtained solutions in a very simple form but having paid the price of leaving aside the initial condition. Since the equation and boundary conditions are linear and homogeneous any finite linear combination of functions (4.75) is also a solution of (4.66′), (4.68_ι). It is plausible to guess that the same is true for

$$\sum_{k=1}^{\infty} c_k \exp(-\lambda_k\, at)\, e_k(x), \qquad c_k \in \mathbf{R}$$

if this series converges in some appropriate sense and term-by-term differentiation is legitimate. Let us try to determine the coefficients c_k so that the sum of the latter series assumes the initial condition (4.67) where $\varphi \in C[0,\ell]$ is a given function. Upon inserting $t = 0$ we get

$$\sum_{k=1}^{\infty} c_k\, e_k(x) = \varphi(x), \quad 0 < x < \ell. \tag{4.76}$$

The same procedure as in § 3.3.2 in which the orthonormality of $\{e_k\}_{k\in\mathbb{N}}$ plays an essential role yields

$$c_k = \varphi_k = \int_0^\ell \varphi(x)\, e_k(x)\, dx, \quad k \in \mathbb{N}, \tag{4.77}$$

therefore, c_k must be taken to be the Fourier coefficient of the function φ with respect to the system $\{e_k\}_{k\in\mathbb{N}}$.

The relations (4.77) constitute a necessary condition for the validity of (4.76) [in the sense of the uniform convergence on $[0,\ell]$ or in the sense of the convergence in the space $L_2(0,\ell)$ (convergence in the mean on $(0,\ell)$)]. Since the system $\{e_k\}_{k\in\mathbb{N}}$ is complete in $L_2(0,\ell)$ this condition is also sufficient for (4.76) to be valid [in the sense of the convergence in the mean on $(0,\ell)$]. Upon superposing functions of the simple form (4.75) we are led to the presumed solution of (4.66′), (4.67), (4.68$_\iota$) in the form

$$u(x,t) = \sum_{k=1}^{\infty} u_k(t)\, e_k(x) \tag{4.78}$$

with

$$u_k(t) = \varphi_k \exp(-\lambda_k\, at). \tag{4.79}$$

One point is in order. By virtue of the completeness of the system $\{e_k\}_{k\in\mathbb{N}}$ in $L_2(0,\ell)$ one has necessarily

$$u_k(t) = \int_0^\ell u(x,t)\, e_k(x)\, dx, \quad k \in \mathbb{N}. \tag{4.80}$$

This fact serves as a starting point of the *Fourier method* which attempts (once the system $\{e_k\}_{k\in\mathbb{N}}$ is available) to find the solution in the form (4.78) with u_k given by (4.80) without appealing to the separation of variables. Only then the explicit formulae for u_k are gained by means of the equation and the initial condition (cf. § 4.4.14).

At this stage it is important to realize that up to now u is a purely formal solution. It is necessary to clarify the convergence of the series (4.78) and to elucidate what is meant by a solution. To anticipate let us say that there are difficulties, not in checking the equation nor the boundary conditions, but in how to meet the initial condition.

4.4.3 Convergence of the Fourier method

In this subsection, upon investigating the uniform convergence of the series (4.78) with respect to both variables x and t, we derive conditions ensuring the existence of the classical solution. To be specific, we shall take up the problem with the Dirichlet boundary conditions the eigenvalues and eigenfunctions of which are given in (4.73) and accordingly

$$u(x,t) = \frac{2}{\ell} \sum_{k=1}^{\infty} \int_0^\ell \varphi(\xi) \sin\frac{k\pi\xi}{\ell}\, d\xi\, \exp\Big(-\frac{k^2\pi^2 at}{\ell^2}\Big) \sin\frac{k\pi x}{\ell}. \tag{4.81}$$

The series representing formal solutions of problems with the other boundary conditions can be dealt with in essentially the same manner.

If $|\varphi|$ is integrable on $(0,\ell)$ (φ is absolutely integrable, not necessarily continuous on $[0,\ell]$) then the series (4.81) with infinitely differentiable terms on $[0,\ell]\times[0,\infty)$ is absolutely and uniformly convergent on the semi-infinite strip

$$Q^\delta = [0,\ell]\times[\delta,\infty)$$

where δ is an arbitrarily small positive number. This is a consequence of the Weierstrass M-test and the fact that on Q^δ the series (4.81) is dominated by the convergent series

$$\frac{2}{\ell}\int_0^\ell |\varphi(\xi)|\,d\xi \sum_{k=1}^{\infty} \exp\left(-\frac{k^2\pi^2 a\delta}{\ell^2}\right)$$

with terms independent of x and t. The convergence of the latter series is due to a) the elementary inequality (4.24) and b) the convergence of the series (3.127). The inequality (4.24) tells us somewhat more: the series

$$\sum_{k=1}^{\infty} k^q \exp\left(-\frac{k^2\pi^2 a\delta}{\ell^2}\right)$$

is convergent for any $q\in\mathbb{N}$, which in turn implies that the series obtained by differentiating (4.81) term-by-term any number of times both with regard to x and t converge absolutely and uniformly to the respective derivatives of u on Q^δ for any $\delta>0$. Therefore, $u\in C^\infty([0,\ell]\times[\delta,\infty))$ for any $\delta>0$ which is tantamount to saying that

$$u\in C^\infty([0,\ell]\times(0,\infty)).$$

In particular, equation (4.66′) and boundary conditions (4.68_1) are satisfied.

It is worth observing that all these conclusions have been derived under the mere assumption of the (absolute) integrability of the initial function φ. For u to meet all requirements imposed on the classical solution it remains to show that it is continuous on the semi-infinite strip

$$Q^0 = [0,\ell]\times[0,\infty)$$

and takes on the values $\varphi(x)$ for $t=0$. To this end it suffices to assume that $\varphi\in C^2[0,\ell]$ and $\varphi(0)=\varphi(\ell)=0$. Indeed, integrating by parts we get

$$\tilde{\varphi}_k = \int_0^\ell \varphi(x)\sin\frac{k\pi x}{\ell}\,dx = -\frac{\ell^2}{k^2\pi^2}\int_0^\ell \varphi''(x)\sin\frac{k\pi x}{\ell}\,dx$$

and

$$|\tilde{\varphi}_k| \le \frac{\ell^2}{k^2\pi^2}\max_{x\in[0,\ell]}|\varphi''(x)|,$$

which yields the absolute and uniform convergence of the series (4.81) on Q^0. Thus, $u\in C(Q^0)$ and, in particular, $u(x,0)=\varphi(x)$ for $0\le x\le\ell$. We have obtained the classical solution, its uniqueness follows from the maximum principle. Collecting the results we have the following theorem.

Let $\varphi\in C^2[0,\ell]$ and $\varphi(0)=\varphi(\ell)=0$. Then there exists exactly one classical solution of the problem (4.66′), (4.67), (4.68_1). *This solution is given by the series* (4.81).

4.4.4 Convergence of the Fourier method – continued

We established the result with relative ease but at the price of requiring apparently unnecessary differentiability properties of the function φ. If we use some results from advanced Calculus the conditions on φ might be weakened.

First, let us recall the Abel test of uniform convergence.

Let $\{a_k\}_{k\in\mathbb{N}}$, $\{b_k\}_{k\in\mathbb{N}}$ *be two sequences of functions defined on intervals* I *and* J *respectively. If*

1) *the series* $\sum_{k=1}^{\infty} a_k(x)$ *is uniformly convergent on* I,

2) *the sequence* $\{b_k(t)\}_{k\in\mathbb{N}}$ *is uniformly bounded on* J, *i.e. there exists* $B > 0$ *such that*

$$|b_k(t)| \le B, \qquad k \in \mathbb{N}, t \in J,$$

3) *for every* $t \in J$ *the sequence* $\{b_k(t)\}_{k\in\mathbb{N}}$ *is monotone,*
then the series

$$\sum_{k=1}^{\infty} a_k(x)\, b_k(t)$$

is uniformly convergent on $I \times J$.

Now, we will apply this result setting $I = [0,\ell]$, $J = [0,\infty)$, $a_k(x) = \varphi_k\, e_k(x)$, $b_k(t) = \exp(-\lambda_k\, at)$, where λ_k and e_k are defined in (4.73). Since the conditions 2) and 3) are automatically satisfied we arrive at the following assertion:
the series (4.81) *is uniformly convergent on* Q^0 *(and, in particular, it represents the classical solution) if and only if*

$$\sum_{k=1}^{\infty} \varphi_k\, e_k(x) \text{ is uniformly convergent on } [0,\ell]. \tag{4.82}$$

Various sets of sufficient conditions ensuring (4.82) are furnished by the theory of trigonometric Fourier series. For instance: *let* φ *be continuous on* $[0,\ell]$ *with piecewise continuous derivative and* $\varphi(0) = \varphi(\ell) = 0$. □

The assumptions on φ are weaker than in § 4.4.3, notwithstanding they are stronger than those needed in Theorem 4.3.10 derived without the help of the Fourier series technique. The point here is that no mere continuity of φ suffices to guarantee (4.82). The problem might have classical solutions (due to Theorem 4.3.10) and (4.82) fails, though. (The reader will have noticed that an analogous situation was encountered in § 3.3.4 on hyperbolic equations where the reason for this phenomenon has already been clarified.) It is interesting to observe the qualitatively different behaviour of Fourier series representing solutions for parabolic and hyperbolic equations: the monotone character of exponential terms in (4.81) stands in (far-reaching) contrast to the oscillatory character of terms in (3.121). (Loosely speaking, hyperbolic problems correspond to oscillation processes, parabolic problems to equalization ones.) □

If we had understood the classical solution in the modified sense of § 4.3.11 the existence result would have been valid under the mere assumption – φ *piecewise continuous with piecewise continuous derivative on* $(0,\ell)$. Still, the case of even more general functions φ will be encompassed by an approach we present in § 4.4.8.

4.4.5 Example

Find the classical solution of the initial-boundary value problem (4.66′), (4.67), (4.68$_1$) where $\ell = \pi$ and

a) $\varphi(x) = \sin(k_0 x)$ for some $k_0 \in \mathbb{N}$,
b) $\varphi(x) = x(\pi - x)$,

$$\text{c)}\quad \varphi(x) = \begin{cases} \dfrac{2\varphi_0}{\pi} x, & 0 \le x \le \dfrac{\pi}{2} \\ \dfrac{2\varphi_0}{\pi}(\pi - x), & \dfrac{\pi}{2} < x \le \pi \end{cases}$$

(φ_0 is a positive constant). By the result of the preceding subsections, especially by (4.81), the classical solution in all cases is given by the series

$$u(x,t) = \frac{2}{\pi} \sum_{k=1}^{\infty} \tilde{\varphi}_k \exp(k^2 at) \sin(kx),$$

where

$$\tilde{\varphi}_k = \int_0^{\pi} \varphi(\xi) \sin(k\xi)\, d\xi.$$

In case a) it is readily seen that

$$u(x,t) = \exp(-k_0^2\, at) \sin(k_0 x).$$

In case b) we have

$$\begin{aligned} \tilde{\varphi}_k &= \int_0^{\pi} \xi(\pi - \xi) \sin(k\xi)\, d\xi = \frac{2}{k^3}(1 - \cos(k\pi)) \\ &= \begin{cases} 0 & \text{for } k \text{ even} \\ \dfrac{4}{(2m-1)^3} & \text{for } k = 2m - 1,\ m \in \mathbb{N}. \end{cases} \end{aligned}$$

Hence,

$$u(x,t) = \frac{8}{\pi} \sum_{k=1}^{\infty} \frac{1}{(2k-1)^3} \exp(-(2k-1)^2 at) \sin((2k-1)x).$$

In case c) we have

$$\begin{aligned} \tilde{\varphi}_k &= \frac{2\varphi_0}{\pi} \left[\int_0^{\pi/2} \xi \sin(k\xi)\, d\xi + \int_{\pi/2}^{\pi} (\pi - \xi) \sin(k\xi)\, d\xi \right] \\ &= \frac{2\varphi_0}{\pi} \int_0^{\pi/2} \xi [\sin(k\xi) + \sin(k\pi - k\xi)]\, d\xi \end{aligned}$$

so that $\tilde{\varphi}_k = 0$ for k even and

$$\begin{aligned} \tilde{\varphi}_k &= \frac{4\varphi_0}{\pi} \int_0^{\pi/2} \xi \sin((2m-1)\xi)\, d\xi \\ &= (-1)^{m-1} \frac{4\varphi_0}{\pi(2m-1)^2} \quad \text{for} \quad k = 2m - 1,\ m \in \mathbb{N}, \end{aligned}$$

and accordingly

$$u(x,t) = \frac{8\varphi_0}{\pi^2} \sum_{k=1}^{\infty} \frac{(-1)^{k-1}}{(2k-1)^2} \exp(-(2k-1)^2 at) \sin((2k-1)x).$$

4.4.6 Example

Write down a formula for the solution of the initial-boundary value problem with the homogeneous Neumann conditions.

The associated eigenvalue problem is given by (4.71) and (4.72_2) where $\sigma_0 = \sigma_1 = 0$ and is resolved in (4.74). According to (4.78), (4.79) the (formal) solution of the problem (4.66), (4.67), (4.68_2) $[\sigma_0 = \sigma_1 = 0]$ is given by

$$u(x,t) = \frac{1}{\ell} \int_0^\ell \varphi(\xi)\, d\xi + \frac{2}{\ell} \sum_{k=1}^{\infty} \int_0^\ell \varphi(\xi) \cos\frac{k\pi\xi}{\ell}\, d\xi \, \exp(-\frac{k^2\pi^2 at}{\ell^2}) \cos\frac{k\pi x}{\ell}. \quad (4.83)$$

This formula can be fully established as the classical solution provided $\varphi \in C^1[0,\ell]$, φ'' is piecewise continuous on $(0,\ell)$ and (analogously, as in the previous example) the boundary conditions shared by eigenfunctions are captured by φ which now amounts to $\varphi'(0) = \varphi'(\ell) = 0$. [Let us point out that in accordance with our definition the solution should possess the first derivative u_x continuous up to the boundary.]

4.4.7 Asymptotic behaviour of solutions. Stabilization

By virtue of (4.81) any solution u_D of the initial-boundary value problem with homogeneous Dirichlet boundary conditions satisfies (regardless of φ)

$$\lim_{t\to\infty} u_D(x,t) = 0, \qquad 0 < x < \ell.$$

The formula (4.83) in turn yields

$$\lim_{t\to\infty} u_N(x,t) = \frac{1}{\ell} \int_0^\ell \varphi(\xi)\, d\xi, \qquad 0 < x < \ell,$$

for any solution u_N of the problem with homogeneous Neumann boundary conditions.

These results are expected from physical intuition. The zero temperature in the first case and the average of the initial temperature in the second one represent steady state temperature distributions towards which the temperatures settle down (stabilize) after the transients (due to initial temperature distributions) die out. The study of interrelations between the limiting values for $t \to \infty$ and solutions of "the corresponding elliptic boundary value problems" gave rise to an extensively developed mathematical theory (applied to much more general situations). In this context confer also § 4.4.17.

We leave it to the reader to verify that the zero solution is asymptotically stable in the case of the Dirichlet boundary conditions while it is stable (but not asymptotically) in the case of the Neumann conditions.

4.4.8 Generalized solutions – preliminaries

Our goal is to examine another type of convergence of the series (4.78) representing a formal solution of the problem (4.66′), (4.67), (4.68$_\iota$). The salient point here is the observation that (4.78) is a Fourier series with respect to the complete orthonormal system $\{e_k\}_{k\in\mathbf{N}}$ in $L_2(0,\ell)$ (with coefficients depending on t). Therefore, it is tempting to operate within the framework of the Hilbert space $L_2(0,\ell)$ where a very simple sufficient and at once necessary condition for the convergence of Fourier series is at our disposal [see, for example, Weidmann (1980)]. Namely, *the Fourier series of a function f with respect to a complete orthonormal system in a Hilbert space is convergent and its sum is equal to f if and only if the series of squares of the Fourier coefficients of f is convergent.*

Using this result together with the Dini theorem from advanced Calculus we can establish a lemma (§ 4.4.9) which enables us to clarify easily properties of the sum of series (4.78), that is, of the alleged solution. This lemma turns out to be a suitable guidance in making various definitions of generalized solutions. We introduce one such concept of a generalized solution the existence of which will be ensured under a very weak assumption on the initial function φ. Unlike the classical solution, however, this generalized solution satisfies the initial condition only in an appropriate "integral" sense.

For the reader's convenience we recall Dini's theorem.

Let $J \subset \mathbf{R}$ be a closed bounded interval, $\{f_k\}_{k\in\mathbf{N}}$ be a sequence of non-negative continuous functions on J and $\sum_{k=1}^{\infty} f_k(t)$ converge everywhere on J to a continuous function. Then the convergence of the series is uniform on J.

For the sake of definiteness and simplicity we restrict ourselves to the special case of eigenfunctions given in (4.73), that is, to the problem with the Dirichlet boundary conditions. Analogous results hold for the other cases as well. We use the notation introduced in § 1.2.9.

4.4.9 Lemma

Let $J \subset \mathbf{R}$ be a closed bounded interval. Let w be a function on $(0,\ell)\times J$ such that $w(.,t) \in L_2(0,\ell)$ for every $t \in J$. Denote

$$w_k(t) = \Big(\frac{2}{\ell}\Big)^{\frac{1}{2}} \int_0^\ell w(x,t)\,\sin\frac{k\pi x}{\ell}\,dx, \qquad k\in\mathbf{N}.$$

Then

- *$w \in C(J; L_2(0,\ell))$ if and only if $w_k \in C(J)$, $k \in \mathbf{N}$ and $\sum_{k=1}^{\infty} w_k^2(t)$ is uniformly convergent on J.*
- *w together with its (generalized) derivative $\dfrac{\partial w}{\partial t}$ belongs to $C(J; L_2(0,\ell))$ if and only if $w_k \in C^1(J)$, $k \in \mathbf{N}$ and*

$$\sum_{k=1}^{\infty} w_k^2(t),\ \sum_{k=1}^{\infty} w_k'^2(t) \ \textit{are uniformly convergent on } J.$$

• *w together with its (generalized) derivative* $\frac{\partial^r w}{\partial x^r}$, $r = 1$ *or* 2 *belongs to* $C(J; L_2(0,\ell))$ *and*

$$w(0,t) = w(\ell,t) = 0, \quad t \in J^{14}$$

if and only if $w_k \in C(J)$, $k \in \mathbb{N}$ *and* $\sum_{k=1}^{\infty} k^{2r} w_k^2(t)$ *is uniformly convergent on* J.

• *More generally: let* $r, s \in \mathbb{N} \cup \{0\}$, *then*

$$\frac{\partial^{r+s} w}{\partial x^r \partial t^s} \in C(J; L_2(0,\ell))$$

and

$$\frac{\partial^{2m+p} w(0,t)}{\partial x^{2m} \partial t^p} = \frac{\partial^{2m+p} w(\ell,t)}{\partial x^{2m} \partial t^p} = 0, \quad t \in J,\ 2m < r,\ p = 0,1,\dots,s$$

if and only if $w_k \in C^s(J)$, $k \in \mathbb{N}$ *and*

$$\sum_{k=1}^{\infty} k^{2r} \left[\frac{d^p w_k(t)}{d\,t^p}\right]^2 \quad \text{are uniformly convergent on } J$$

for any $p = 0,1,\dots,s$.

If the premise is satisfied then the series

$$\left(\frac{2}{\ell}\right)^{\frac{1}{2}} \sum_{k=1}^{\infty} \left(\frac{k\pi}{\ell}\right)^r \frac{d^s w_k(t)}{dt^s} \sin\left(\frac{k\pi x}{\ell} + \frac{r\pi}{2}\right)$$

is convergent in the mean with respect to $x \in (0,\ell)$ *uniformly with respect to* $t \in J$ *[more precisely, the series is convergent in the sense of the space* $C(J; L_2(0,\ell))$*] and its sum is equal to* $\frac{\partial^{r+s} w}{\partial x^r \partial t^s}$.

4.4.10 Generalized solutions – continued

Let us assume that $\varphi \in L_2(0,\ell)$ which is equivalent to

$$\sum_{k=1}^{\infty} \varphi_k^2 < \infty, \qquad \varphi_k = \left(\frac{2}{\ell}\right)^{\frac{1}{2}} \int_0^{\ell} \varphi(x) \sin\frac{k\pi x}{\ell}\, dx, \ k \in \mathbb{N}.$$

It is immediate that the series

$$\sum_{k=1}^{\infty} \varphi_k^2 \exp\left(-\frac{2k^2\pi^2 at}{\ell^2}\right)$$

[124]These conditions make sense since w may be shown to belong to $C([0,\ell] \times J)$, cf. § 1.2.9.

is uniformly convergent on $[0, \infty)$ and it is easily seen that the series

$$\sum_{k=1}^{\infty} k^4 \varphi_k^2 \exp(-\frac{2k^2\pi^2 at}{\ell^2})$$

is uniformly convergent on $[\delta, \infty)$ for any $\delta > 0$. [It is just because for any $\delta > 0$ there exists $C > 0$ independent of $k \in \mathbb{N}$ such that $k^4 \exp(-2k^2\pi^2\ell^{-2}a\delta) \leq C$ for all $k \in \mathbb{N}$.]

These two statements in connection with Lemma 4.4.9 say: for any $T > 0$ the series (4.78) specified by (4.73) is convergent in the mean with respect to $x \in (0, \ell)$ uniformly with respect to $t \in [0, T]$ and the termwise differentiated series twice in x and once in t are convergent in the mean with respect to $x \in (0, \ell)$ uniformly with respect to $t \in [\delta, T]$, where $0 < \delta < T$ is arbitrary. Consequently,

$$u \in \bigcap_{0<T<\infty} C([0,T];\, L_2(\Omega)) = C([0,\infty);\, L_2(0,\ell)) \tag{4.84}$$

and

$$\frac{\partial u}{\partial x}, \frac{\partial^2 u}{\partial x^2}, \frac{\partial u}{\partial t} \in \bigcap_{0<\delta<T<\infty} C([\delta,T];\, L_2(0,\ell)) = C((0,\infty);\, L_2(0,\ell)). \tag{4.85}$$

In particular, on account of (4.84),

$$\lim_{t\to 0+} \int_0^\ell [u(x,t) - \varphi(x)]^2\, dx = 0. \tag{4.86}$$

In the light of these results we are led to the following natural definition. (Other possible definitions of generalized solutions stemming from Lemma 4.4.9 are left to the reader as an exercise.) We say that a function u is a *generalized solution* of the problem (4.66), (4.67), (4.68$_\iota$) if (4.84), (4.85) take place, equation (4.66) is satisfied for every $t > 0$ and almost every $0 < x < \ell$, boundary conditions (4.68$_\iota$) are satisfied for every $t > 0$[15] and (4.86) holds.

It is a mere restatement of the above results that *the generalized solution of the problem* (4.66′), (4.67), (4.68$_1$) *exists on condition* $\varphi \in L_2(0,\ell)$ *and it is given by* (4.81). In general it is not true that the generalized solution should provide a continuous function on $[0,\ell] \times [0,\infty)$. Only its continuity on $[0,\ell] \times (0,\infty)$ can be established (as it is recalled in § 4.4.11) . However, one point worth noting is that if φ happens to satisfy the assumptions from § 4.4.3 or § 4.4.4 then there exist both, generalized and classical, solutions and they coincide.

The proof of the *uniqueness* of the generalized solution cannot be accomplished by means of the maximum principle 4.3.2 due to the above-mentioned lack of continuity up to the boundary $t = 0$. We apply another procedure which has already been encountered in § 3.5.2 in the chapter on hyperbolic equations in connection with energy inequalities. [The "energy terminology" is still used in spite of the fact that now the

---

[15] See footnote on p. 225.

energy (heat energy) is proportional to $\int_0^\ell u(x,t)\,dx$ (u absolutely scaled, say) rather than to $\int_0^\ell u^2(x,t)\,dx$.] Let us assume $\varphi \equiv 0$, $g \equiv 0$; we prove that $u(x,t) = 0$ [for every $t > 0$ and almost every $0 < x < \ell$]. Multiplying Eq. (4.66) by the function u and integrating over $(0,\ell)$ we obtain

$$\int_0^\ell u_t(x,t)\,u(x,t)\,dx - a\int_0^\ell u_{xx}(x,t)\,u(x,t)\,dx = 0, \qquad t > 0.$$

Further, using boundary conditions (4.68_1) we get

$$\frac{1}{2}\frac{d}{dt}\int_0^\ell u^2(x,t)\,dx + a\int_0^\ell u_x^2(x,t)\,dx = 0, \qquad t > 0$$

and thus

$$\frac{d}{dt}\int_0^\ell u^2(x,t)\,dx \le 0, \qquad t > 0.$$

Now, the assumed relation

$$\lim_{t\to 0+}\int_0^\ell u^2(x,t)\,dx = 0$$

implies

$$\int_0^\ell u^2(x,t)\,dx \le 0, \qquad t > 0$$

and the assertion follows readily.

4.4.11 Remark on smoothing property

The series (4.81) represents a solution of the initial-boundary value problem for the homogeneous heat equation in a certain generalized sense that has been clarified and under the only assumption $\varphi \in L_2(0,\ell)$ which admits functions fairly discontinuous. In § 4.4.3 we have shown that under the assumption of the (absolute) integrability of the function φ [which is obeyed if $\varphi \in L_2(0,\ell)$] the series can be differentiated term-by-term as often as desired yielding a function from $C^\infty([0,\ell] \times (0,\infty))$. The representation of the solution in terms of the series (4.81) shows once again the "smoothing property" enjoyed by the heat equation we have already met in § 4.2.9. One conclusion is immediate: any generalized solution of the initial-boundary problem is in fact a classical solution of the *equation*. In this respect the term generalized relates to the way in which the initial condition is taken on but the equation itself, as well as the boundary conditions, is satisfied in the classical sense. (This situation differs markedly from the analogous one for hyperbolic equations where no smoothing property is available and the concepts of generalized solutions may bring in a substantial withdrawal from the differentiability properties.)

4.4.12 Example

Find a solution of the problem (4.66′), (4.67), (4.68_1) where

$$\varphi(x) = \varphi_0 = \text{const.}$$

It can easily be checked that

$$\int_0^\ell \varphi_0 \sin\frac{k\pi x}{\ell}\,dx = \begin{cases} 0 & \text{for } k \text{ even} \\ \dfrac{2\varphi_0 \ell}{(2m-1)\pi} & \text{for } k = 2m-1,\, m \in \mathbb{N}. \end{cases}$$

Upon inserting in (4.81) we get

$$u(x,t) = \frac{4\varphi_0}{\pi}\sum_{k=1}^{\infty}\frac{1}{2k-1}\exp\Big(-\frac{(2k-1)^2\pi^2 at}{\ell^2}\Big)\sin\frac{(2k-1)\pi x}{\ell}. \tag{4.87}$$

This series represents the generalized solution in the sense of § 4.4.10 or the modified classical solution in the sense of § 4.3.11.

4.4.13 Cooling of a heated metallic rod – a few numerical results

The series (4.81) provides a means of computing an approximate solution. Each term of the series contains an exponential factor with a negative argument which induces a relatively rapid convergence of the series. Hence, only a few first terms in the series expansion are often sufficient to furnish relatively accurate values of the solution.

For illustration we shall consider the model example introduced in § 4.3.11. By (4.87) the temperature at the middle section of the rod is given by

$$u\Big(\frac{\ell}{2}, t\Big) = \frac{4\varphi_0}{\pi}\sum_{k=1}^{\infty}\frac{(-1)^{k-1}}{2k-1}\exp\Big(-\frac{(2k-1)^2\pi^2 at}{\ell^2}\Big). \tag{4.88}$$

We are interested in absolute and relative errors arising if instead of $u(\frac{\ell}{2}, t)$ we use only the first term in the expansion (4.88), viz.

$$\tilde{u}\Big(\frac{\ell}{2}, t\Big) = \frac{4\varphi_0}{\pi}\exp\Big(-\frac{\pi^2 at}{\ell^2}\Big). \tag{4.89}$$

(Let us recall that if $\tilde{u}$ is an approximation to the true value of u then the absolute error of $\tilde{u}$ is $\Delta u = u - \tilde{u}$ and the relative (percentage) error is $\varepsilon = 100\,|\frac{\Delta u}{u}|\,[\%]$.) For the sake of definiteness let us take $\ell = 0.2\,\text{m}$, $\varphi_0 = 100°\,\text{C}$ and $t = 30\,\text{s}$. Table 4 shows the results for four different materials of the rod (different values of the coefficient a).

Table 4

Material	$a(10^{-3}\mathrm{m}^2\mathrm{s}^{-1})$	$\tilde{u}(^\circ\mathrm{C})$	Absolute error ($^\circ$C)	Relative error (%)
Silver	0.170	36.1747	0.0005	0.0014
Copper	0.114	54.756	0.021	0.04
Aluminium	0.086	67.37	0.14	0.20
Steel, mild	0.012	116.50	16.54	16.5

It is seen that the approximation by $\tilde{u}$ gives satisfactory results for the first three materials. The sum of the first two terms can be shown to agree with the exact value through the displayed number of digits.

On the other hand the result for mild steel is unacceptable the percentage error being more than 16%. Incidentally, this first approximation of the temperature is paradoxically greater than the initial temperature [this is due to the alternate character of the series (4.88)]. The sum of the first two terms is $\tilde{u} = 97.42$ with the absolute error -2.54 and the relative error 2.5%. The sum of the first five terms of the series gives the first four digits correctly.

Table 5 collects other information about the approximations by the first term. For given values of ε the table displays the times t^* (in seconds) such that the relative errors resulting from the replacement of $u(\frac{\ell}{2}, t)$ by $\tilde{u}(\frac{\ell}{2}, t)$ are less than ε% for all $t \geq t^*$.

Table 5

ε(%)	Silver	Copper	Aluminium	Steel, mild
0.001	32	47	62	440
0.01	25	37	48	343
0.1	18	26	35	246
1	11	16	21	149
10	4	6	8	53

4.4.14 The initial-boundary value problem for the inhomogeneous equation

Let us now turn our attention to the existence of a solution of the problem (4.66) through (4.68_ι). As seen earlier the Fourier method consists of looking for a solution in the form

$$u(x,t) = \sum_{k=1}^{\infty} u_k(t)\, e_k(x) \tag{4.90}$$

where $\{e_k\}_{k\in\mathbb{N}}$ is the system of eigenfunctions of the eigenvalue problem (4.71), (4.72_ι) and u_k are the Fourier coefficients (4.80) of u with respect to the system $\{e_k\}_{k\in\mathbb{N}}$. The formula (4.79) shows the form of u_k as long as $g \equiv 0$. We approach the general case.

Let us use the Fourier series developments of the functions g and φ with respect to the system $\{e_k\}_{k\in\mathbb{N}}$:

$$g(x,t) = \sum_{k=1}^{\infty} g_k(t)\, e_k(x), \tag{4.91}$$

$$\varphi(x) = \sum_{k=1}^{\infty} \varphi_k \, e_k(x), \tag{4.92}$$

where

$$g_k(t) = \int_0^{\ell} g(x,t)\, e_k(x)\, dx, \qquad \varphi_k = \int_0^{\ell} \varphi(x)\, e_k(x)\, dx. \tag{4.93}$$

Inserting (4.90) through (4.92) into (4.66), (4.67), equating the coefficients at e_k we obtain for any $k \in \mathbb{N}$

$$u_k'(t) + \lambda_k a\, u_k(t) = g_k(t), \qquad t > 0, \tag{4.94}$$

$$u_k(0) = \varphi_k. \tag{4.95}$$

The solution of the problem (4.94), (4.95) is given by

$$u_k(t) = \varphi_k \exp(-\lambda_k a t) + \int_0^t \exp(-\lambda_k a(t-\tau)) g_k(\tau)\, d\tau. \tag{4.96}$$

To be specific, let us take again the problem with $\iota = 1$. Upon combining (4.73), (4.90) and (4.96) we get the resulting formula

$$\begin{aligned} u(x,t) = {} & \left(\frac{2}{\ell}\right)^{\frac{1}{2}} \sum_{k=1}^{\infty} \varphi_k \exp\left(-\frac{k^2\pi^2 at}{\ell^2}\right) \sin\frac{k\pi x}{\ell} \\ & + \left(\frac{2}{\ell}\right)^{\frac{1}{2}} \sum_{k=1}^{\infty} \int_0^t \exp\left(-\frac{k^2\pi^2 a(t-\tau)}{\ell^2}\right) g_k(\tau)\, d\tau \, \sin\frac{k\pi x}{\ell}, \end{aligned} \tag{4.97}$$

with

$$g_k(t) = \left(\frac{2}{\ell}\right)^{\frac{1}{2}} \int_0^{\ell} g(x,t) \sin\frac{k\pi x}{\ell} dx, \quad \varphi_k = \left(\frac{2}{\ell}\right)^{\frac{1}{2}} \int_0^{\ell} \varphi(x) \sin\frac{k\pi x}{\ell} dx. \tag{4.98}$$

Formula (4.97) represents solutions of different "strength" (whether classical or generalized) according as how "strong" assumptions are imposed on the functions g and φ. Before passing on to the formulation of the theorem on existence of say the generalized solution in the sense of § 4.4.10 (the modification of assumptions ensuring the existence of the classical solution is left to the reader) it is interesting to make a comparison of (4.97) with formula (4.55) expressing the solution via the Green's function. Substituting from (4.98) into (4.97) we get

$$\begin{aligned} u(x,t) = {} & \int_0^{\ell} \left\{ \frac{2}{\ell} \sum_{k=1}^{\infty} \exp\left(-\frac{k^2\pi^2 at}{\ell^2}\right) \sin\frac{k\pi x}{\ell} \sin\frac{k\pi\xi}{\ell} \right\} \varphi(\xi)\, d\xi \\ & + \int_0^t \int_0^{\ell} \left\{ \frac{2}{\ell} \sum_{k=1}^{\infty} \exp\left(-\frac{k^2\pi^2 a(t-\tau)}{\ell^2}\right) \sin\frac{k\pi x}{\ell} \sin\frac{k\pi\xi}{\ell} \right\} g(\xi,\tau)\, d\xi\, d\tau. \end{aligned} \tag{4.99}$$

Using the notation (4.54) and interchanging the order of integration we arrive exactly at formula (4.55) which has been obtained by means of a quite different procedure in Section 4.3.

Now the theorem we have promised (cf. Exercise 4.4.19.6).
Let

$$\text{a)}\quad \textit{either } g \in C([0,\ell] \times [0,\infty)),\ g(0,t) = g(\ell,t) = 0,\ t \geq 0$$
$$\textit{and } \frac{\partial g}{\partial x} \in C([0,\infty); L_2(0,\ell)),$$
$$\text{b)}\quad \textit{or}\quad g, \frac{\partial g}{\partial t} \in C([0,\infty)); L_2(0,\ell)).$$

Further, let $\varphi \in C^1[0,\ell]$, $\varphi(0) = \varphi(\ell) = 0$ *and* $\varphi'' \in L_2(0,\ell)$. *Then the problem* (4.66), (4.67), (4.68_1) *has exactly one generalized solution in the sense of* § 4.4.10. *It is given by formula* (4.97). Moreover, $\frac{\partial u}{\partial t}, \frac{\partial u}{\partial x}, \frac{\partial^2 u}{\partial x^2} \in C([0,\infty); L_2(0,\ell))$ and, in particular, $u, \frac{\partial u}{\partial x} \in C([0,\ell] \times [0,\infty))$.

4.4.15 Example

Let us consider a rod insulated against lateral heat conduction the ends of which are kept at zero temperature. Let the rate of density of heat sources distributed along the axis of the rod be $G(x,t) = c\rho g_1(x)\, g_2(t)$. Find the expression for the temperature at any time $t > 0$ if initially the temperature is zero and

a) either $g_1 \in C[0,\ell]$, $g_1(0) = g_1(\ell) = 0$, $g_1' \in L_2(0,\ell)$, $g_2 \in C[0,\infty)$,
b) or $g_1 \in L_2(0,\ell)$ and $g_2 \in C^1[0,\infty)$.

The problem is described by (4.66), (4.67) and (4.68_1), where $g(x,t) = g_1(x)\, g_2(t)$ and $\varphi = 0$. Using formula (4.97) and Theorem 4.4.14 we get the (generalized) solution in the form

$$u(x,t) = \frac{2}{\ell} \sum_{k=1}^{\infty} \Big[\int_0^{\ell} g_1(\xi) \sin \frac{k\pi\xi}{\ell}\, d\xi \Big] \gamma_k(t) \sin \frac{k\pi x}{\ell},$$

where

$$\gamma_k(t) = \int_0^t \exp\Big(-\frac{k^2\pi^2 a(t-\tau)}{\ell^2}\Big) g_2(\tau)\, d\tau.$$

Verify the following results for a particular choice of g_1. If

(1) $g_1(x) = \sin \frac{k_0 \pi x}{\ell}$ for some $k_0 \in \mathbb{N}$, then $u(x,t) = \gamma_{k_0}(t) \sin \frac{k_0 \pi x}{\ell}$;

(2) $g_1(x) = \sin^3 \frac{\pi x}{\ell}$, then (consult Hint to Exercise 3.3.8.1)

$$u(x,t) = \frac{1}{4}\big[3\gamma_1(t) \sin \frac{\pi x}{\ell} - \gamma_3(t) \sin \frac{3\pi x}{\ell}\big];$$

(3) $g_1(x) = \begin{cases} 1 & \text{for} \quad 0 < x < \frac{\ell}{2} \\ -1 & \text{for} \quad \frac{\ell}{2} < x < \ell, \end{cases}$ then

$$u(x,t) = \frac{4}{\pi} \sum_{k=1}^{\infty} \frac{1}{2k-1} \gamma_{4k-2}(t) \sin \frac{2(2k-1)\pi x}{\ell};$$

(4) $g_1(x) = \cos \frac{\pi x}{\ell}$, then

$$u(x,t) = \frac{8}{\pi} \sum_{k=1}^{\infty} \frac{k}{4k^2 - 1} \gamma_{2k}(t) \sin \frac{2k\pi x}{\ell}.$$

4.4.16 Inhomogeneous boundary conditions

The Fourier method appears to succeed only if the boundary conditions are homogeneous. If this is not the case a trick must be used in order to legitimize the Fourier method. This trick consists in transferring the inhomogeneities from the boundary conditions to the right-hand side of the equation.

We start by trying to find a function u_0 satisfying the given (inhomogeneous) boundary conditions. Once we have such a function we put $u = v + u_0$ and solve the resulting problem for the function v with already homogeneous boundary conditions. Caution: in general both the right-hand side of the equation and the initial function are changed; in particular, the homogeneous equation may be replaced by an inhomogeneous one.

In order to bring out the main idea let us consider the problem given by equation (4.66′), initial condition

$$u(x,0) = \psi(x), \qquad 0 < x < \ell$$

and boundary conditions

$$u(0,t) = h_0(t), \quad u(\ell,t) = h_1(t), \qquad t > 0.$$

One of the possible functions satisfying the boundary conditions is

$$u_0(x,t) = \frac{\ell - x}{\ell} h_0(t) + \frac{x}{\ell} h_1(t).$$

The previous results allow us to determine the function v as the solution of the problem (4.66), (4.67), (4.68$_1$) with

$$g(x,t) = \frac{x-\ell}{\ell} h_0'(t) - \frac{x}{\ell} h_1'(t)$$

and

$$\varphi(x) = \psi(x) - \frac{\ell - x}{\ell} h_0(0) - \frac{x}{\ell} h_1(0).$$

The function $u = v + u_0$ is then the solution of the original problem.

In some cases (such as in that of Exercise 4.4.19.2) we succeed in choosing the function u_0 more properly, for example, in such a way that it also satisfies the given differential equation. Then the right-hand side remains unchanged and the initial function goes over to $\psi(x) - u_0(x,0)$.

It is to be noted that the latter choice is always feasible if the inhomogeneous Dirichlet conditions are independent of t: then the function u_0 can be determined as a (unique!) solution of the "corresponding elliptic equation" (u_t is not involved in the equation). The representation of the solution as the sum $v + u_0$ has then an important physical interpretation: v defines the transient temperature distribution while u_0 the steady state temperature distribution (cf. also § 4.4.7 and § 4.4.17).

Problems with inhomogeneous boundary conditions of the other types can be dealt with in some cases in like manner.

4.4.17 Example

The lateral surface of a rod is rendered impervious to heat, the ends are maintained at constant temperatures $u(0,t) = h_0 =$ const. and $u(\ell,t) = h_1 =$ const. Find an expression for the temperature $u(x,t)$ if initially the rod is uniformly heated to $u(x,0) = \varphi_0 =$ const. and there are no heat sources.

To begin with, we remove the inhomogeneous boundary conditions by using the approach suggested in the preceding subsection. Then we use formula (4.81) and obtain

$$u(x,t) = h_0 + (h_1 - h_0)\frac{x}{\ell} + \frac{2}{\pi}\sum_{k=1}^{\infty}\frac{1}{k}\{(\varphi_0 - h_0)[1-(-1)^k]$$
$$+ (-1)^k(h_1 - h_0)\}\exp(-\frac{k^2\pi^2 at}{\ell^2})\sin\frac{k\pi x}{\ell}.$$

We see, in agreement with experience, that after the transients have passed the temperature within the rod stabilizes (no matter what the initial distribution is) at the steady state temperature

$$u_0(x) = h_0 + (h_1 - h_0)\frac{x}{\ell}.$$

4.4.18 Concluding remarks on the Fourier method

We have applied the Fourier method to the detailed study of the problem with heat equation and Dirichlet boundary conditions. Other boundary conditions and other more general equations of the form $u_t + Lu = g$, where L is "the elliptic part" of the equation, can be worked out by means of this method as well. The crucial assumption for the applicability of the Fourier method is the existence of a "complete orthonormal" system (in a sense to be specified) of eigenfunctions of the eigenvalue problem

$$\begin{cases} L\,e = \lambda e \\ \text{and homogeneous boundary conditions.} \end{cases} \tag{4.100}$$

Without attempting to enter into details we say a few words about the role of the boundedness of the underlying interval (ℓ finite). If $L = -d^2/dx^2$ and this interval extends to infinity in one or both directions the eigenvalues are seen to become closer and closer. The limiting case of unbounded intervals leads to the appearance of the so-called continuous spectrum, the eigenvalue problem no longer has any eigenvalues, no complete orthonormal system is available and the Fourier method (in the above setting) breaks down (see, however, § 4.3.15).

Various very general hypotheses on L and boundary conditions ensuring the requisite properties are well-known. For example, the statement of the lemma of § 4.4.1 holds verbatim for

$$Le(x) = -\frac{d}{dx}\left(a(x)\frac{de(x)}{dx}\right) + c(x)\,e(x), \quad 0 < x < \ell \ (< \infty),$$

where a is a continuously differentiable and positive function for $0 \le x \le \ell$ and c is a continuous function for $0 \le x \le \ell$, accompanied with the boundary conditions (4.72).

If the right-hand side of the equation in (4.100) is replaced by $\lambda \varrho e$, where ϱ is a continuous positive function for $0 \le x \le \ell$, then the statement of Lemma 4.4.1 remains true with the only exception that the complete orthonormal system the existence of which is stated is to be understood in the sense of the weight space $L_{2,\varrho}(0,\ell)$ (cf. § 3.3.7). This result does not cover, however, the (so-called singular) cases encountered very often in applications. Such is the case, for instance, of the equation

$$-\frac{d}{dx}\left(x\,\frac{de(x)}{dx}\right) = \lambda\, x\, e(x), \qquad 0 < x < \ell,$$

arising by the study of the Bessel equation of order zero, where the positivity of the functions $a(x) = \varrho(x) = x$ is violated at $x = 0$. Nonetheless, this difficulty can be overcome if we modify the boundary condition at the "singular" point in an appropriate way (cf. Exercise 4.4.19.9). We obtain a result quite analogous to that for the above "regular" case. The same remark applies to a number of other problems but we cannot dwell on details. □

The Fourier method can be applied also to equations involving several space variables. Let

$$Le(x) = -\sum_{j,k=1}^{n} \frac{\partial}{\partial x_j}\left(a_{jk}(x)\frac{\partial e(x)}{\partial x_k}\right) + c(x)\, e(x), \quad x \in \Omega,$$

where $\Omega \subset \mathbf{R}^n$ is a bounded regular region, c is a continuous function on $\bar{\Omega}$ and $\mathcal{A}(x) = (a_{jk}(x))_{j,k=1}^{n}$ is a matrix of continuously differentiable functions on $\bar{\Omega}$ which is

- symmetric, that is

$$a_{jk}(x) = a_{kj}(x), \quad x \in \bar{\Omega},\ j,k = 1,\ldots,n,$$

- positive definite uniformly with respect to $x \in \bar{\Omega}$, that is, there exists $\delta > 0$ such that

$$\sum_{j,k=1}^{n} a_{jk}(x)\, \xi_j \xi_k \ge \delta \sum_{j=1}^{n} \xi_j^2 \quad \text{for all } \xi \in \mathbf{R}^n,\ x \in \bar{\Omega}.$$

Let the boundary conditions be given, for instance, by

$$e(x) = 0, \quad x \in \partial\Omega,$$

or

$$\frac{\partial e(x)}{\partial \nu_{\mathcal{A}}} + \sigma(x)\, e(x) = 0, \quad x \in \partial\Omega,$$

where $\partial/\partial\nu_{\mathcal{A}}$ denotes the conormal derivative on $\partial\Omega$ (see § 1.2.2) and σ is a continuous function on $\partial\Omega$. Then the statement of Lemma 4.4.1 still holds except that the eigenvalues are no longer simple in general but have finite multiplicities and $L_2(0,\ell)$ is to be replaced by $L_2(\Omega)$. (The smoothness properties of eigenfunctions depend not only on how smooth are the data (as in the case $n = 1$) but also on the underlying region. For example, if the data are infinitely differentiable and Ω is "very regular" then the eigenfunctions are infinitely differentiable on $\bar{\Omega}$.) □

In the above mentioned cases appropriate systems of eigenfunctions are known to exist. So, even though their explicit knowledge may be far from obvious the Fourier method (separation into two groups of variables, time and space) can be carried out. (At this point, we refer the reader to § 5.3.3 where the conditions for the applicability of the (time-"space") Fourier method for solving initial-"boundary" value problems for equations of the types $u_t + Lu = g$ and $u_{tt} + Lu + 2\alpha u_t = g$ $(\alpha \in \mathbf{R})$ are summed up in an abstract setting.)

In turn, the problem (4.100) itself can be attempted (if $n > 1$) to be solved again by means of the Fourier method (separation into individual space variables). However, now it should be remarked that there are rather severe limitations as to the geometry of the region Ω (in as much as to the equation and the boundary conditions as well). First, Ω should be bounded (in order "to avoid the continuous spectrum"). Next, the separation of variables can be accomplished for such regions which can be transformed to "rectangular" ones by a suitable coordinate transformation (for instance, a cylinder and sphere can be transformed to a rectangular parallelepiped by means of the cylindrical and spherical coordinate transformations respectively). Even if the region is a rectangle the method of separation of variables may fail; it depends on the operator L – whether or not it separates into functions of the separate variables (cf. Section 5.3). Each case is to be treated individually. We succeed in many but by no means in all (or even in the most of) practical situations.

4.4.19 Exercises

4.4.19.1 Assume that both ends of a rod of length ℓ are insulated and there are no heat sources within the rod. Let, initially, the temperature be $\varphi_0 =$ const. throughout. Find the temperature at any $t > 0$ provided that the lateral surface
a) either is insulated as well,
b) or radiates into a surrounding medium at a constant temperature (which is normalized to be zero without loss of generality.)
Hint. In case b) to the equation $u_t - a\,u_{xx} + \gamma u = 0$ apply the transformation from § 4.2.15.
Answer.

a) $$u(x,t) = \varphi_0,$$
b) $$u(x,t) = \exp(-\gamma t)\,\varphi_0.$$

4.4.19.2 Consider the conduction of heat in an infinite slab of homogeneous material bounded by parallel planes $x = 0$ and $x = \ell$ with no heat sources. The face $x = 0$ is kept at zero temperature while the temperature of the other one is made to vary according to the law βt, β being a positive constant. Determine the distribution of temperature within the slab for times $t > 0$ assuming for $t = 0$ the temperature is zero and the planes $x =$ const. are isothermal at any later time.

Hint. Use the one-dimensional model. The function $u_0(x,t) = \dfrac{\beta x t}{\ell} + \dfrac{\beta x}{6a\ell}(x^2 - \ell^2)$ satisfies both boundary conditions and homogeneous heat equation. Further see § 4.4.15.
Answer.

$$u(x,t) = \frac{\beta x t}{\ell} + \frac{\beta x}{6a\ell}(x^2 - \ell^2) + \frac{2\beta\ell^2}{a\pi^3}\sum_{k=1}^{\infty}\frac{(-1)^{k+1}}{k^3}\exp\Big(-\frac{k\pi^2 at}{\ell^2}\Big)\sin\frac{k\pi x}{\ell}.$$

4.4.19.3. Find the temperature of a rod with insulated lateral surface, one end of which remains insulated while at the other end at $t = 0$ the supply of heat starts at a constant rate Q per unit time and previously the temperature was zero throughout. No internal sources are present.

Hint. The boundary conditions are given by (4.13) where $q_0 \equiv 0$, $q_1 = \dfrac{Q}{A}$. Remove inhomogeneous boundary conditions by choosing $u_0(x) = \dfrac{Qx^2}{2\lambda\ell A}$ and putting $u = v + u_0$. The function v is to be determined as a solution of the heat equation with the right-hand side $\dfrac{aQ}{\lambda\ell A}$, satisfying the homogeneous boundary conditions and the initial condition $v(x,0) = -u_0(x)$. Now we use formulae (4.90), (4.96) and (4.74).

Answer.

$$u(x,t) = \frac{Q}{\lambda A}\left[\frac{at}{\ell} + \frac{3x^2-\ell^2}{6\ell} + \frac{2\ell}{\pi^2}\sum_{k=1}^{\infty}\frac{(-1)^{k+1}}{k^2}\exp\Big(-\frac{k^2\pi^2at}{\ell^2}\Big)\cos\frac{k\pi x}{\ell}\right].$$

4.4.19.4 Find the temperature of a rod, sides insulated, no sources, if one end ($x = 0$) is kept at a constant temperature h_0, the other ($x = \ell$) is insulated and initially the temperature is zero.

Hint. The boundary conditions are $u(0,t) = h_0$, $u_x(0,t) = 0$, $t > 0$.

Answer.

$$u(x,t) = h_0\left[1 - \frac{4}{\pi}\sum_{k=1}^{\infty}\frac{1}{2k-1}\exp\Big(-\frac{(2k-1)\pi^2at}{\ell^2}\Big)\sin\frac{(2k-1)\pi x}{2\ell}\right].$$

4.4.19.5 Consider a closed cylindrical vessel (of height ℓ and cross-sectional area A) with impermeable sides. The pressure and temperature of the atmosphere outside and inside are the same. In the surrounding atmosphere there is present a certain gas of the concentration h_0. Use the one-dimensional model and find the amount $Q(t)$ of the gas penetrated into the vessel if at $t = 0$ one end has been opened and the other remains impermeably closed.

Hint. Mathematically the problem is the same as the foregoing one. See Exercise 4.1.5.1.

Answer.

$$Q(t) = \ell A h_0\left[1 - \frac{8}{\pi^2}\sum_{k=1}^{\infty}\frac{1}{(2k-1)^2}\exp\Big(-\frac{(2k-1)^2\pi^2at}{\ell^2}\Big)\right].$$

4.4.19.6 Prove the theorem from § 4.4.14.

Hint. Provided that $\sum\limits_{k=1}^{\infty} k^4\varphi_k^2 < \infty$ and

a) either $\sum\limits_{k=1}^{\infty} k^2 g_k^2(t)$ is uniformly convergent on $[0,T]$,

b) or $\sum\limits_{k=1}^{\infty} g^2(t)$, $\sum\limits_{k=1}^{\infty} g'^2(t)$ are uniformly convergent on $[0,T]$

show that the series

$$\sum_{k=1}^{\infty} k^4 \left\{ \exp(-2k^2\pi^2\ell^{-2}at) + \left[\int_0^t \exp(-k^2\pi^2\ell^{-2}a(t-\tau))g_k(\tau)\,d\tau \right]^2 \right\}$$

is uniformly convergent on $[0, T]$. Use Lemma 4.4.9.

4.4.19.7 *Periodic solutions.* Find the conditions under which there exists a solution of the problem

$$\begin{aligned} u_t(x,t) - a\,u_{xx}(x,t) + \gamma\,u(x,t) = g(x,t), \quad 0 < x < \ell,\ -\infty < t < \infty, \\ u(0,t) = u(\ell,t) = 0, \qquad -\infty < t < \infty, \\ u(x,t+\omega) - u(x,t) = 0, \quad 0 < x < \ell,\ -\infty < t < \infty, \end{aligned}$$

where γ is a real constant and g is a function ω-periodic with respect to t.
Hint. Write

$$g(x,t) = \sum_{j=-\infty}^{\infty} \sum_{k=1}^{\infty} g_{jk} \exp\big(ij\frac{2\pi}{\omega}t\big) \sin\frac{k\pi x}{\ell}$$

and look for a solution in the form

$$u(x,t) = \sum_{j=-\infty}^{\infty} \sum_{k=1}^{\infty} u_{jk} \exp\big(ij\frac{2\pi}{\omega}t\big) \sin\frac{k\pi x}{\ell},$$

$$g_{jk} = \frac{2}{\ell\omega} \int_0^{\omega} \int_0^{\ell} g(x,t) \exp\big(ij\frac{2\pi}{\omega}t\big) \sin\frac{k\pi x}{\ell}\,dx\,dt,$$

$$u_{jk} = \frac{2}{\ell\omega} \int_0^{\omega} u(t,x) \exp\big(ij\frac{2\pi}{\omega}t\big) \sin\frac{k\pi x}{\ell}\,dx\,dt.$$

Answer.
If $\gamma = -\dfrac{ak_0^2\pi^2}{\ell^2}$ for some $k_0 \in \mathbf{N}$ then the solution exists only if

$$\int_0^{\omega} \int_0^{\ell} g(x,t) \sin\frac{k_0\pi x}{\ell}\,dx\,dt = 0.$$

4.4.19.8 By means of the Fourier method solve the problem

$$u_t(x,y,z,t) - a\,\Delta u(x,y,z,t) = 0, \quad (x,y,z,t) \in P \times (0,\infty),$$

$$u(x,y,z,0) = \varphi(x,y,z), \qquad (x,y,z) \in P,$$

$$\begin{aligned} u(0,y,z,t) = u(b,y,z,t) = u(x,0,z,t) = u(x,c,z,t) \\ = u(x,y,0,t) = u(x,y,d,t) = 0, \qquad (x,y,z,t) \in P \times (0,\infty), \end{aligned}$$

where

$$P = \{ (x, y, z) \in \mathbb{R}^3 \mid 0 < x < b,\ 0 < y < c,\ 0 < z < d \}.$$

Hint. Look for a solution in the form

$$u(x, y, z, t) = \sum_{k=1}^{\infty} \sum_{m=1}^{\infty} \sum_{n=1}^{\infty} u_{kmn}(t) \sin \frac{k\pi x}{b} \sin \frac{m\pi y}{c} \sin \frac{n\pi z}{d}$$

with

$$u_{kmn}(t) = \frac{8}{bcd} \int_0^b \int_0^c \int_0^d u(x, y, z, t) \sin \frac{k\pi x}{b} \sin \frac{m\pi y}{c} \sin \frac{n\pi z}{d}\, dx\, dy\, dz.$$

Answer.

$$u_{kmn}(t) = \varphi_{kmn} \exp\Big[-a\pi^2 \big(\frac{k^2}{b^2} + \frac{m^2}{c^2} + \frac{n^2}{d^2} \big) t \Big],$$

$$\varphi_{kmn} = \frac{8}{bcd} \int_0^b \int_0^c \int_0^d \varphi(x, y, z) \sin \frac{k\pi x}{b} \sin \frac{m\pi y}{c} \sin \frac{n\pi z}{d}\, dx\, dy\, dz.$$

4.4.19.9 Study the heat equalization in an infinitely long right circular cylinder of radius R with axis along the z-axis whose surface is held at temperature zero and no sources are present. Assume first that the following two postulates are true:

a) the heat flows in planes perpendicular to the axis,
b) the lines of flow are radial.

Then throw away the assumption b).
Hint. Use the cylindrical coordinates, cf. § 1.2.8 and § 3.3.1. Under a) the temperature is independent of the axial direction z and we are to solve the problem for $u = u(r, \vartheta, t)$:

$$\frac{\partial u}{\partial t} - a \left[\frac{1}{r} \frac{\partial}{\partial r} \left(r \frac{\partial u}{\partial r} \right) + \frac{1}{r^2} \frac{\partial^2 u}{\partial \vartheta^2} \right] = 0, \quad 0 < r < R,\ 0 < \vartheta < 2\pi,\ t > 0,$$
$$u = 0 \quad \text{for} \quad r = R,$$
$$u = \varphi \quad \text{for} \quad t = 0.$$

The assumption b) means that φ and u are independent of ϑ and in this case the corresponding eigenvalue problem is as follows:

$$-\frac{1}{r} \frac{d}{dr} \big(r \frac{de}{dr} \big) = \lambda\, e, \quad 0 < r < R,$$
$$e(R) = 0$$

supplemented by

$$e \text{ is bounded for } r \to 0+,$$

a suitable condition discarding the singular part of the solution of the equation which reads $C\, \mathrm{J}_0(\lambda^{\frac{1}{2}} r) + D\, \mathrm{Y}_0(\lambda^{\frac{1}{2}} r)$ [C, D arbitrary constants].

The eigenvalues are $\lambda_k = \mu_k^2/R^2$ where $(\mu_k)_{k\in\mathbb{N}}$ is a sequence of positive roots of Eq. $\mathrm{J}_0(\mu) = 0$. To evaluate the normalization constants of the system of eigenfunctions [which is orthogonal and complete in $L_{2,r}(0,R)$] use the following result from the theory of Bessel functions : let $n \in \mathbb{N}\cup\{0\}$ and $(\mu_{nk})_{k\in\mathbb{N}}$ be a sequence of positive roots of Eq. $\mathrm{J}_n(\mu) = 0$ then

$$\int_0^R r\left[\mathrm{J}_n\left(\frac{\mu_{nk}}{R}r\right)\right]^2 dr = \frac{R^2}{2}\left[\mathrm{J}_{n+1}(\mu_{nk})\right]^2, \quad k\in\mathbb{N}.$$

Answer.

$$u(r,t) = \frac{2}{R^2}\sum_{k=1}^{\infty}\frac{1}{\left[\mathrm{J}_1(\mu_k)\right]^2}\int_0^R r\,\varphi(r)\,\mathrm{J}_0\left(\frac{\mu_k}{R}r\right)dr\ \mathrm{J}_0\left(\frac{\mu_k}{R}r\right)\exp\left(-\frac{a\mu_k^2 t}{R^2}\right).$$

If the assumption b) does not take place then

$$u(r,\vartheta,t) = \sum_{n=0}^{\infty}\sum_{k=1}^{\infty}\mathrm{J}_n\left(\frac{\mu_k}{R}r\right)\left[A_{nk}\cos(n\vartheta) + B_{nk}\sin(n\vartheta)\right]\exp\left(-\frac{a\mu_{nk}^2 t}{R^2}\right),$$

where

$$\left.\begin{matrix}A_{nk}\\ B_{nk}\end{matrix}\right\} = 2\pi^{-1}R^{-2}\left[\mathrm{J}_{n+1}(\mu_{nk})\right]^{-2}\int_0^R r\,\mathrm{J}_n\left(\frac{\mu_{nk}}{R}r\right)\int_0^{2\pi}\varphi(r,\vartheta)\left\{\begin{matrix}\cos(n\vartheta)\\ \sin(n\vartheta)\end{matrix}\right.d\vartheta\,dr,$$

$$A_{0k} = \pi^{-1}R^{-2}\left[\mathrm{J}_1(\mu_{0k})\right]^{-2}\int_0^R r\,\mathrm{J}_0\left(\frac{\mu_{0k}}{R}r\right)\int_0^{2\pi}\varphi(r,\vartheta)\,d\vartheta\,dr.$$

CHAPTER 5

Fourth order equations

5.1 PROBLEMS FROM THE THEORY OF ELASTICITY

So far we have been considering mostly partial differential equations of the first and second orders. In applications the equations of the fourth order can be encountered as well. The following two, occurring in the theory of elasticity, are most typical.

First of all, small transverse vibrations of an elastic beam are governed by the equation

$$u_{tt}(x,t) + a^2 u_{xxxx}(x,t) = 0.$$

Another example is the equation

$$u_{tt}(x,y,t) + a^2 \Delta^2 u(x,y,t) = 0$$

describing small transverse vibrations of a thin elastic plate.

The variable u defines the displacements of the axis of a beam or the middle surface of a plate, a is a material constant.

5.1.1 Transverse vibration of a beam

Consider an elastic prismatic beam of length ℓ, uniform cross-sectional area A, constant mass density ρ, oriented so that in its equilibrium position the x-axis lies along the axis of the beam with the end points located at $x = 0$ and $x = \ell$. When studying the vibrations of the beam the following simplifying assumptions are usually introduced: cross-sections of the beam remain plane during any deformation, each point of the axis moves solely in a plane perpendicular to the axis and all points of the axis vibrate in one common plane (the so-called plane of vibration).

Under these assumptions the displacements, u, of the points of the axis can be described by a single function of two variables, the space variable x and time t, that is $u = u(x,t)$.

Restricting ourselves to small deflections and using classical linear theory of deformation we can obtain the following equation defining the displacement u [see Timoshenko, Young and Weaver (1974), Kneschke (1960)]:

$$\rho A\, u_{tt}(x,t) + EJ\, u_{xxxx}(x,t) = f(x,t), \tag{5.1}$$

where E is the modulus of elasticity of the beam material, J is the moment of inertia of the cross-section about an axis through the centre of mass perpendicular to the plane of vibration and $f(x,t)$ is a transverse load measured in units force/length.

It is customary to write Eq. (5.1) in the form

$$u_{tt}(x,t) + a^2 u_{xxxx}(x,t) = g(x,t), \tag{5.2}$$

where

$$g(x,t) = \frac{1}{\rho A} f(x,t) \quad \text{and} \quad a = \left(\frac{EJ}{\rho A}\right)^{\frac{1}{2}}.$$

Eq. (5.2) is called the equation of transverse vibration of a beam or simply the *beam equation.* □

At this point we mention some modifications of the (homogeneous) beam equation even though we shall not be concerned with them in the sequel.

First, if the beam is nonprismatic of variable cross-section then the pertinent equation is

$$\rho A\, u_{tt} + (EJ\, u_{xx})_{xx} = 0$$

in which the flexural rigidity EJ, the area A and/or the mass density ρ are functions of x.

Another modification of the beam equation is

$$\rho A\, u_{tt} + EJ\, u_{xxxx} + \gamma u = 0$$

governing transverse vibrations of a beam on an elastic foundation. The constant γ measures the restoring force due to the elastic foundation.

If the vibrating beam is subjected to an axial force F the equation becomes

$$\rho A\, u_{tt} + EJ\, u_{xxxx} - F\, u_{xx} = 0.$$

All the above equations do not take into account damping effects. Various additional terms in Eq. (5.2) represent various types of damping, for instance, αu_t the viscous damping (external friction) and αu_{xxxxt} the so-called Kelvin–Voigt damping (internal friction). □

Eq. (5.2) is called also the *Euler–Bernoulli equation* since its derivation is based on the Euler–Bernoulli model which neglects the effects of rotary inertia and on these grounds Eq. (5.2) is reasonable only for a beam the cross-sectional dimensions of which are small in comparison with its length. On the other hand, the rotation of cross-sections is considered in the Rayleigh model which yields the equation

$$\rho A\, u_{tt} + EJ\, u_{xxxx} - \rho J\, u_{xxtt} = 0.$$

A further equation taking into account also shearing deformations is due to Timoshenko [we refer the interested reader to Timoshenko, Young and Weaver (1974)]. This equation contains also the term u_{tttt}.

5.1.2 The initial and boundary conditions for the beam equation

The description of the motion of the beam by equation (5.2) is not complete unless we add some further information. As usual we add initial and boundary conditions for the displacement $u(x,t)$.

As the equation is of second order with respect to the time variable t two initial conditions are to be prescribed, namely, the initial displacement:

$$u(x,0) = \varphi(x), \qquad 0 < x < \ell \tag{5.3}$$

and the initial velocity:

$$u_t(x,0) = \psi(x), \qquad 0 < x < \ell, \tag{5.4}$$

where φ and ψ are given functions.

The form of the boundary conditions depends on the conditions of the beam at its ends. We mention three important cases of practical interest.

a) Both ends $x = 0$ and $x = \ell$ are *clamped*. The deflection and the slope of the axis must vanish at the ends and this results in boundary conditions of the form

$$\begin{aligned} u(0,t) &= u_x(0,t) = 0, \\ u(\ell,t) &= u_x(\ell,t) = 0, \quad t > 0. \end{aligned} \tag{5.5}$$

b) Both ends are *simply supported* or *hinged*, which means that the deflection at the ends must be zero but the ends can rotate freely with respect to the axis perpendicular to the plane of vibration. In this case the bending moment $M = -EJ\,u_{xx}$ vanishes at the ends and consequently the boundary conditions take the form

$$\begin{aligned} u(x,0) &= u_{xx}(0,t) = 0, \\ u(\ell,t) &= u_{xx}(\ell,t) = 0, \quad t > 0. \end{aligned} \tag{5.6}$$

c) A cantilever beam: one end ($x = 0$) is clamped, the other ($x = \ell$) is *free*. A free end is characterized in such a way that both the bending moment M and the shear force $Q = M_x$ must be zero. Hence, the boundary conditions are given by

$$\begin{aligned} u(0,t) = u_x(0,t) \quad &= 0, \\ u_{xx}(\ell,t) = u_{xxx}(\ell,t) &= 0, \quad t > 0. \end{aligned} \tag{5.7}$$

5.1.3 Transverse vibration of a plate

The analogue of Eq. (5.1) in two space variables is (the so-called *plate equation*)

$$\rho h\, u_{tt}(x,y,t) + D\Delta^2 u(x,y,t) = f(x,y,t), \tag{5.8}$$

where

$$\Delta^2 = \frac{\partial^4}{\partial x^4} + 2\frac{\partial^4}{\partial x^2\,\partial y^2} + \frac{\partial^4}{\partial y^4},$$

governing small transverse vibrations of a thin homogeneous elastic plate subject to a transverse load [see Timoshenko and Woinowski-Krieger (1959), Timoshenko, Young and Weaver (1974)]. Here h is the uniform thickness of the plate, ρ its density, $D = Eh^3/12(1-\mu^2)$, E the modulus of elasticity, μ Poisson's ratio, f a transverse load measured in units force/area. The middle surface of the plate at rest is supposed to occupy a set $\bar{\Omega}$ where Ω is a region in the plane of variables x and y and the function u measures its small deflections in the direction perpendicular to the xy-plane.

The operator Δ^2 is called the biharmonic operator which is why Eq. (5.8) is sometimes called the ***biharmonic wave equation*** (although it does not share completely the properties of hyperbolic equations). □

In the case of a circular plate, that is a plate for which $\Omega = \{(x,y) \mid x^2+y^2 < R^2\}$, one commonly uses polar coordinates r and ϑ instead of the Cartesian ones x and y. The biharmonic operator Δ^2 is obtained by applying twice the Laplace operator the form of which in terms of polar coordinates is [see (1.52)]

$$\Delta = \frac{1}{r}\frac{\partial}{\partial r}\Big(r\frac{\partial}{\partial r}\Big) + \frac{1}{r^2}\frac{\partial^2}{\partial \vartheta^2}.$$

5.1.4 The initial and boundary conditions for the plate equation

The initial conditions read

$$u(x,y,0) = \varphi(x,y) \tag{5.9}$$

$$u_t(x,y,0) = \psi(x,y), \quad (x,y) \in \Omega. \tag{5.10}$$

The boundary conditions

a) on a clamped edge become

$$u = \frac{\partial u}{\partial \nu} = 0, \tag{5.11}$$

where $\partial/\partial\nu$ denotes the outward normal derivative. (The region Ω is supposed to have unit normal vectors at all points of the boundary except, possibly, for a finite number of them.)

b) on a simply supported edge take the form

$$u = Mu = 0, \tag{5.12}$$

where

$$Mu = -D[(u_{xx} + \mu u_{yy})\cos^2\theta + (u_{yy} + \mu u_{xx})\sin^2\theta + (1-\mu)u_{xy}\sin 2\theta] \tag{5.13}$$

(θ is here the angle between the outward normal vector and the vector of the positive x-axis) is the bending moment at the boundary of the plate. □

Two particular cases, namely rectangular and circular plates, deserve special mention [cf. McFarland, Smith and Bernhart (1972)].

• Let $\bar{\Omega}$ be a rectangle with edges parallel to the coordinate axes. The form of boundary conditions on
(i) an edge parallel to the y-axis

– clamped	$u = 0,$	$u_x = 0$
– simply supported	$u = 0,$	$u_{xx} = 0$
– free	$u_{xx} + \mu u_{yy} = 0,$	$u_{xxx} + (2-\mu)u_{xyy} = 0$

(ii) an edge parallel to the x-axis

– clamped	$u = 0,$	$u_y = 0$
– simply supported	$u = 0,$	$u_{yy} = 0$
– free	$\mu u_{xx} + u_{yy} = 0,$	$u_{yyy} + (2-\mu)u_{xxy} = 0$

• Let $\bar{\Omega}$ be a circle centred at $(0,0)$ with radius R. The form of boundary conditions for the function $u = u(r,\vartheta,t)$ on
(i) an edge at $r = R$ and any ϑ

– clamped	$u = 0,$	$u_r = 0$
– simply supported	$u = 0,$	$u_{rr} + u_r/r = 0$

(ii) an edge along a radius at $\vartheta = \text{const.}$ and any r

– clamped	$u = 0,$	$u_\vartheta = 0$
– simply supported	$u = 0,$	$u_{\vartheta\vartheta} = 0.$

5.2 THE BEAM EQUATION

The Fourier method will be applied in solving initial-boundary value problems with boundary conditions (5.5) and (5.6). The problem with the boundary conditions (5.7) as well as the initial-value problem is touched upon in Exercises.

5.2.1 Free vibration of a clamped beam

Let us apply the method of separation of variables to the problem given by the equation

$$u_{tt}(x,t) + a^2 u_{xxxx}(x,t) = 0, \quad 0 < x < \ell, \quad t > 0, \tag{5.14}$$

and by the boundary conditions (5.5).

Since the method is the same as in the preceding two chapters we shall proceed more quickly. Letting

$$u(x,t) = e(x)\,v(t), \quad 0 < x < \ell, \quad t > 0, \tag{5.15}$$

we get upon inserting into (5.14) and (5.5)

$$v''(t) + a^2\lambda\, v(t) = 0, \quad t > 0, \tag{5.16}$$
$$e''''(x) - \lambda\, e(x) = 0, \quad 0 < x < \ell, \quad (\lambda \in \mathbf{R}), \tag{5.17}$$
$$e(0) = e'(0) = 0, \tag{5.18$_1$}$$
$$e(\ell) = e'(\ell) = 0. \tag{5.18$_2$}$$

The existence of a number λ such that the relations (5.16) through (5.18) are satisfied is a necessary and sufficient condition that the function (5.15) should satisfy (5.14) and (5.5).

It is easy to check the following two properties relating to the eigenvalue problem (5.17), (5.18) [in fact, the same properties are valid for the other types of boundary conditions mentioned in § 5.1.2]:

(♣) if $e,\ f \in C^4[0,\ \ell]$ then

$$\int_0^\ell (e\, f'''' - e''''\, f)\, dx = 0$$

for all e satisfying the boundary conditions (5.18) if and only if f satisfies the same boundary conditions;

(♠) if $e \in C^4[0,\ \ell]$ satisfies the boundary conditions (5.18) then

$$\int_0^\ell e\, e''''\, dx \geq 0.$$

Indeed, integrating by parts we get

$$\int_0^\ell (e\, f'''' - e''''\, f)\, dx = \left[(e\, f''' - e'''\, f) - (e' f'' - e'' f')\right]_{x=0}^{x=\ell}$$

and

$$\int_0^\ell e\, e''''\, dx = \left[e\, e''' - e' e''\right]_{x=0}^{x=\ell} + \int_0^\ell e''^2\, dx,$$

and these relations imply (♣) and (♠) immediately.

By results from § 4.3.3 we can state that there exists a countable sequence of eigenvalues $\{\lambda_k\}_{k\in\mathbf{N}}$ and the corresponding system of eigenfunctions $\{e_k\}_{k\in\mathbf{N}}$ which is orthonormal and complete in $L_2(0,\ell)$. The eigenvalues cannot be negative: assume on the contrary that there exist a function e not identically equal to zero satisfying the boundary conditions (5.18) and $\lambda < 0$ such that $e'''' = \lambda\, e$, then in virtue of (♠) we have $\lambda \int_0^\ell e^2 dx \geq 0$, a contradiction. □

Now, let us take up the problem how to evaluate the eigenvalues and eigenfunctions. Let us begin with the case $\lambda = 0$. The general solution of Eq. (5.17) is

$$e(x) = C_0 + C_1 x + C_2 x^2 + C_3 x^3,$$

where C_0, C_1, C_2, C_3 are arbitrary real constants. Substitution into the boundary conditions (5.18) yields

$$C_0 = C_1 = C_2 = C_3 = 0,$$

hence the value of $\lambda = 0$ must be also excluded.

For $\lambda > 0$ the general solution of Eq. (5.17) takes the form

$$e(x) = C_0 \cosh(\lambda^{\frac{1}{4}} x) + C_1 \sinh(\lambda^{\frac{1}{4}} x) + C_2 \cos(\lambda^{\frac{1}{4}} x) + C_3 \sin(\lambda^{\frac{1}{4}} x), \tag{5.19}$$

where C_0, C_1, C_2, $C_3 \in \mathbf{R}$. Upon inserting into the boundary conditions (5.18) and setting

$$\mu = \lambda^{\frac{1}{4}} \ell, \tag{5.20}$$

we get the linear algebraic system for C_0, C_1, C_2, C_3:

$$\begin{aligned}
C_0 \qquad\qquad\qquad + C_2 \qquad\qquad\qquad &= 0\\
C_1 \qquad\qquad\qquad + C_3 \qquad &= 0\\
C_0 \cosh\mu + C_1 \sinh\mu + C_2 \cos\mu + C_3 \sin\mu &= 0\\
C_0 \sinh\mu + C_1 \cosh\mu - C_2 \sin\mu + C_3 \cos\mu &= 0.
\end{aligned}$$

By the theory of linear algebraic equations this system has a non-trivial solution if and only if the determinant of the system vanishes, viz.

$$\det \begin{pmatrix} 1 & 0 & 1 & 0 \\ 0 & 1 & 0 & 1 \\ \cosh\mu & \sinh\mu & \cos\mu & \sin\mu \\ \sinh\mu & \cosh\mu & -\sin\mu & \cos\mu \end{pmatrix} = 0.$$

We are thus led to a transcendental equation for μ

$$\cos\mu \cosh\mu = 1 \tag{5.21}$$

This equation has infinitely many positive roots μ_k, $k \in \mathbf{N}$ (see Fig. 25). Their approximate values are

$$\mu_1 \doteq 4.730, \quad \mu_2 \doteq 7.853, \quad \mu_3 \doteq 10.996, \quad \mu_k \doteq \frac{\pi}{2}(2k+1),\ k > 3.$$

For any $k \in \mathbf{N}$ there exist numbers C_0, C_1, C_2, C_3 (determined uniquely up to a common constant factor) satisfying the above algebraic system and substituting into (5.19) the eigenfunctions are then given by the formula

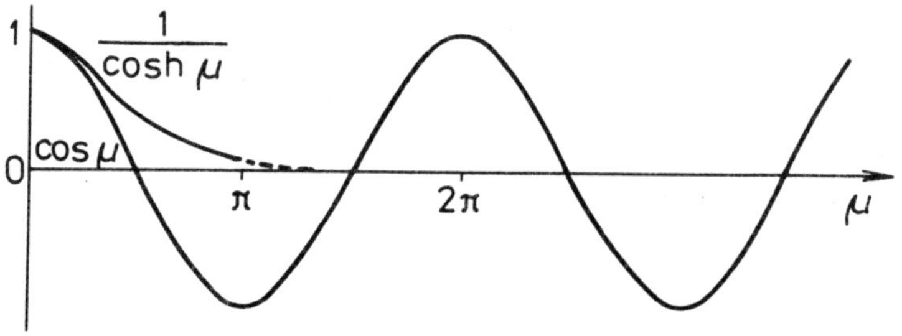

Fig. 25

$$e_k(x) = N_k\, e(x, \mu_k) \tag{5.22}$$

where

$$e(x, \mu_k) = \cosh\frac{\mu_k x}{\ell} - \cos\frac{\mu_k x}{\ell} - \frac{\cosh\mu_k - \cos\mu_k}{\sinh\mu_k - \sin\mu_k}\Big(\sinh\frac{\mu_k x}{\ell} - \sin\frac{\mu_k x}{\ell}\Big) \tag{5.23}$$

and N_k is an arbitrary non-zero real constant. It only remains to choose the constant N_k in such a way that the system $\{e_k\}_{k\in\mathbb{N}}$ [complete orthogonal system in $L_2(0,\ell)$] be orthonormal, that is $\int_0^\ell e_k^2(x)\,dx = 1$ for all $k \in \mathbb{N}$. We will show a classical approach of evaluating these constants which is more convenient to use than the direct, rather tedious, calculation of the integral of $e^2(x, \mu_k)$.

For any parameter $\mu > 0$ we denote

$$e(x, \mu) = \cosh\frac{\mu x}{\ell} - \cos\frac{\mu x}{\ell} - \frac{\cosh\mu - \cos\mu}{\sinh\mu - \sin\mu}\Big(\sinh\frac{\mu x}{\ell} - \sin\frac{\mu x}{\ell}\Big). \tag{5.24}$$

Then

$$\frac{\partial^4}{\partial x^4}\, e(x,\mu) - \Big(\frac{\mu}{\ell}\Big)^4 e(x,\mu) = 0, \qquad 0 < x < \ell, \tag{5.25}$$

$$e(0,\mu) = \frac{\partial}{\partial x}\, e(0,\mu) = e(\ell,\mu) = 0. \tag{5.26}$$

Multiplying Eq. (5.25) by the function $e(x, \mu_k)$, the equation

$$\frac{\partial^4}{\partial x^4}\, e(x,\mu_k) - \Big(\frac{\mu_k}{\ell}\Big)^4 e(x,\mu_k) = 0$$

by the function $e(x, \mu)$, subtracting the resulting relations and integrating by parts we obtain

$$\frac{\mu^4 - \mu_k^4}{\ell^4}\int_0^\ell e(x,\mu)\, e(x,\mu_k)\, dx = \Big[e(x,\mu_k)\frac{\partial^3}{\partial x^3}e(x,\mu) - \frac{\partial}{\partial x}e(x,\mu_k)\frac{\partial^2}{\partial x^2}e(x,\mu)$$
$$+\frac{\partial^2}{\partial x^2}e(x,\mu_k)\frac{\partial}{\partial x}e(x,\mu) - \frac{\partial^3}{\partial x^3}e(x,\mu_k)e(x,\mu)\Big]_{x=0}^{x=\ell} = \frac{\partial^2}{\partial x^2}e(\ell,\mu_k)\frac{\partial}{\partial x}e(\ell,\mu).$$

Upon multiplying by the fraction $\ell^4/(\mu^4 - \mu_k^4)$ and passing to the limit $\mu \to \mu_k$ with the aid of L'Hospital's rule we get

$$\int_0^\ell e^2(x, \mu_k)\, dx = \frac{\ell^5}{4\mu_k^4}\left[\frac{\partial^2}{\partial x^2} e(\ell, \mu_k)\right]^2$$

[We take into account that

$$\frac{\partial}{\partial x} e(\ell, \mu_k) = 0 \quad \text{and} \quad \frac{\partial^2}{\partial \mu \partial x} e(\ell, \mu_k) = \frac{\ell}{\mu_k} \frac{\partial^2}{\partial x^2} e(\ell, \mu_k).]$$

Hence, the proper choice of N_k in (5.22) is

$$N_k = \frac{2\mu_k^2}{\ell^{\frac{5}{2}}} \left| \frac{\partial^2}{\partial x^2} e(\ell, \mu_k) \right|^{-1}, \quad k \in \mathbb{N}. \square$$

In consequence of Eq. (5.16) and relation (5.20) we see that any function of the form

$$[A_k \cos(\omega_k t) + B_k \sin(\omega_k t)]\, e_k(x), \tag{5.27$_1$}$$

where

$$\omega_k = \frac{a\mu_k^2}{\ell^2} = \frac{\mu_k^2}{\ell^2}\left(\frac{EJ}{\rho A}\right)^{\frac{1}{2}} \tag{5.27$_2$}$$

and A_k, B_k are arbitrary real constants, is a solution of the homogeneous Eq. (15.14) and obeys the boundary conditions (5.5).

In physical terms: for any $k \in \mathbb{N}$ the function (5.27$_1$) represents the transverse vibration of the beam in one of its natural modes (free transverse vibration). The (normalized) function e_k defines the shape of the natural mode (mode shape). The numbers ω_k are the *natural angular frequencies* and because of relation (5.27$_2$) we speak of Eq. (5.21) as the *frequency equation*. In summary: the solution of the eigenvalue problem provides us with natural frequencies and natural mode shapes. A general form of the free transverse vibration of the beam is obtained by the superposition of natural modes (5.27$_1$).

5.2.2 Forced vibration of a clamped beam

The total transverse response of a beam due to the initial displacement φ, initial velocity ψ and a distributed transverse load characterized by the function g is the solution of the initial-boundary value problem given by the inhomogeneous equation

$$u_{tt}(x,t) + a^2 u_{xxxx}(x,t) = g(x,t), \qquad 0 < x < \ell,\ t > 0, \tag{5.28}$$

initial conditions (5.3), (5.4) and boundary conditions (5.5). It can be found by applying the Fourier method, that is by expressing the solution as a Fourier series with respect to the system $\{e_k\}_{k\in\mathbb{N}}$. Since the procedure is again quite analogous as in

Chapter 3 for the wave equation we state only the result. The formal solution is given by the series

$$u(x,t) = \sum_{k=1}^{\infty} u_k(t)\, e_k(x), \tag{5.29}$$

where

$$u_k(t) = \varphi_k \cos(\omega_k t) + \frac{\psi_k}{\omega_k} \sin(\omega_k t) + \frac{1}{\omega_k} \int_0^t g_k(\tau) \sin(\omega_k(t-\tau))\, d\tau, \tag{5.30}$$

and

$$\varphi_k = \int_0^\ell \varphi(x)\, e_k(x)\, dx, \quad \psi_k = \int_0^\ell \psi(x)\, e_k(x)\, dx, \quad g_k(t) = \int_0^\ell g(x,t)\, e_k(x)\, dx.$$

In conclusion we formulate the assumptions on the functions φ, ψ and g ensuring the uniform convergence of the series (5.29) for $0 \le x \le \ell$, $0 \le t \le T$ (for any $T \in \mathbf{R}^+$) together with the series termwise differentiated four times with regard to x and twice with regard to t respectively. In particular, these assumptions guarantee the existence of the classical solution. For the uniqueness assertion see Exercise 5.2.5.3.

Let

- $\varphi \in C^5[0,\ell], \quad \varphi(0) = \varphi'(0) = \varphi''''(0) = \varphi(\ell) = \varphi'(\ell) = \varphi''''(\ell) = 0$
- $\psi \in C^3[0,\ell], \quad \psi(0) = \psi'(0) = \psi(\ell) = \psi'(\ell) = 0$
- *g together with g_{xxx} be continuous for* $0 \le x \le \ell,\ t \ge 0$ *and* $g(0,t) = g_x(0,t) = g(\ell,t) = g_x(\ell,t) = 0, \quad t \ge 0.$

Then there exists exactly one classical solution of the problem (5.28), (5.3) *through* (5.5). *The solution is given by the series* (5.29).

5.2.3 Remark on generalized solutions

The Fourier series (5.29) provides a means of defining generalized solutions. We have dealt with an analogous situation in § 4.4.8. Without going into details we state directly a theorem on existence of one type of generalized solutions.

Let

- $\varphi \in C^3[0,\ell],\ \varphi'''' \in L_2(0,\ell),\ \varphi(0) = \varphi'(0) = \varphi(\ell) = \varphi'(\ell) = 0$
- $\psi \in C^1[0,\ell],\ \psi'' \in L_2(0,\ell),\ \psi(0) = \psi(\ell) = 0$
- $g, g_x \in C([0,\ell] \times [0,\infty)),\ g_{xx} \in C([0,\infty);\ L_2(0,\ell))$ and $g(0,t) = g_x(0,t) = g(\ell,t) = g_x(\ell,t) = 0, \quad t \ge 0.$

Then the series (5.29) *determines a generalized solution of the problem* (5.28), (5.3) *through* (5.5) *from the space of functions* $u \in C([0,\infty);\ L_2(0,\ell))$ *such that* u_{tt} *and* u_{xxxx} *belong also to* $C([0,\infty);\ L_2(0,\ell))$.

5.2.4 Vibration of a simply supported beam

Let us turn to the problem (5.1), (5.6). We shall proceed in the same vein as in the preceding case. The function e from the expression (5.15) should satisfy the eigenvalue problem

$$\begin{aligned} e''''(x) &= \lambda\, e(x), \qquad 0 < x < \ell, \\ e(0) &= e''(0) = 0, \\ e(\ell) &= e''(\ell) = 0. \end{aligned}$$

Inserting (5.19) into the boundary conditions we get

$$C_0 = C_1 = C_2 = 0, \quad C_3 \sin(\lambda^{\frac{1}{4}}\ell) = 0.$$

The equation for the eigenvalues (the frequency equation) takes the form

$$\sin\mu = 0 \quad \text{where} \quad \mu = \lambda^{\frac{1}{4}}\ell, \tag{5.31}$$

whence

$$\mu_k = k\pi, \quad k \in \mathbb{N} \tag{5.32}$$

and

$$e_k(x) = N_k \sin\frac{k\pi x}{\ell}, \quad k \in \mathbb{N}. \tag{5.33}$$

The value $\lambda = 0$ giving the function identically equal to zero is again excluded. (The possibility of negative eigenvalues may be ignored, as we know from the § 5.2.1.)

The choice of normalizing constants $N_k = (2/\ell)^{\frac{1}{2}}$ leads to an orthonormal system of eigenfunctions in $L_2(0,\ell)$ which is well-known to be complete.

The equation for v in the expression (5.15) is formally the same as before (the only difference is the concrete value of λ). Thus, to any $k \in \mathbb{N}$ there corresponds a function (mode shape) (5.33) which solves the problem (5.1), (5.6). Natural modes are given by (5.27_1) where now the angular frequencies ω_k are given by

$$\omega_k = \frac{k^2\pi^2}{\ell^2}\left(\frac{EJ}{\rho A}\right)^{\frac{1}{2}}, \quad k \in \mathbb{N}. \tag{5.34}$$

The formula (5.29) together with (5.30) gives the formal solution of the initial-boundary value problem (5.2), (5.3), (5.4), (5.6). This solution is a classical one under analogous assumptions on functions φ, ψ and g as those given in Theorem 5.2.2 (the difference is only in the requirement on vanishing *second* derivatives at $x = 0$ and $x = \ell$ instead of the first ones).

5.2.5 Exercises

5.2.5.1 Study the free vibration of a "tuning-bar".
Hint. By means of the separation of variables solve the problem (5.1), (5.7). The frequency equation now assumes the form

$$\cos\mu \cosh\mu = -1 \quad (\mu = \lambda^{\frac{1}{4}}\ell).$$

The roots (draw a picture) are approximately equal to

$$\mu_1 \doteq 1,875,\ \mu_2 \doteq 4,694,\ \mu_3 \doteq 7,854,\ \mu_k \doteq \frac{\pi}{2}(2k-1) \quad \text{for} \quad k > 3.$$

Answer.
The solutions are given by (5.27_1), (5.27_2) where

$$e_k(x) = (\sin\mu_k + \sinh\mu_k)(\cos\frac{\mu_k x}{\ell} - \cosh\frac{\mu_k x}{\ell}) + (\cos\mu_k + \cosh\mu_k)(\sin\frac{\mu_k x}{\ell} - \sinh\frac{\mu_k x}{\ell}).$$

5.2.5.2 Determine the transverse response of a hinged beam (initially at rest) due to a uniformly distributed force f per unit length if
a) $f(x,t) = P_0$,
b) $f(x,t) = P_0 \sin(\omega_0 t)$
(P_0 and ω_0 are constants). In particular, find an approximate formula for the solution if the pulsating load $P_0 \sin(\omega_0 t)$ is varying very slowly, that is if ω_0 is small in comparison with natural angular frequencies ω_k given in (5.34).
Answer.

a)
$$u(x,t) = \frac{4P_0\ell^4}{EJ\pi^5} \sum_{k=1,3,\cdots} \frac{1}{k^5}(1 - \cos(\omega_k t)) \sin\frac{k\pi x}{\ell},$$

b)
$$u(x,t) = \frac{4P_0\ell^4}{EJ\pi^5} \sum_{k=1,3,\cdots} \frac{1}{k^5} \frac{\sin(\omega_0 t) - \dfrac{\omega_0}{\omega_k}\sin(\omega_k t)}{1 - \left(\dfrac{\omega_0}{\omega_k}\right)^2} \sin\frac{k\pi x}{\ell}.$$

In particular, if in *b)*, $\omega_0 \ll \omega_k$, then

$$u(x,t) = \frac{4P_0\ell^4 \sin(\omega_0 t)}{EJ\pi^5} \sum_{k=1,3,\cdots} \frac{1}{k^5} \sin\frac{k\pi x}{\ell}.$$

5.2.5.3 Prove the uniqueness of solutions of initial-boundary value problems formulated in the present section.
Hint. Let u be a solution of the problem with $g \equiv 0$, $\varphi \equiv 0$, $\psi \equiv 0$. Then

$$0 = \int_0^\ell u_t(u_{tt} + a^2 u_{xxxx})\,dx = \frac{1}{2}\frac{d}{dt}\int_0^\ell (u_t^2 + a^2 u_{xx}^2)\,dx,$$

hence

$$\int_0^\ell (u_t^2(x,t) + a^2 u_{xx}^2(x,t))\,dx = 0, \quad t \geq 0,$$

thus $u \equiv 0$.

5.2.5.4 Write down a formula for the solution of the initial-value problem

$$u_{tt}(x,t) + a^2 u_{xxxx}(x,t) = g(x,t), \quad -\infty < x < \infty, \quad t > 0,$$
$$u(x,0) = \varphi(x), \quad u_t(x,0) = \psi(x), \quad -\infty < x < \infty.$$

Hint. Use the Fourier integral transformation as in Section 4.2 and the formulae

$$\frac{1}{2\pi}\int_{-\infty}^{\infty} \cos(ay^2 s)\exp(iyx)\,dy = \frac{1}{2(2\pi as)^{\frac{1}{2}}}\left(\cos\frac{x^2}{4as} + \sin\frac{x^2}{4as}\right),$$
$$\frac{1}{2\pi}\int_{-\infty}^{\infty} \sin(ay^2 s)\exp(iyx)\,dy = \frac{1}{2(2\pi as)^{\frac{1}{2}}}\left(\cos\frac{x^2}{4as} - \sin\frac{x^2}{4as}\right).$$

(where $s > 0$, $x \in \mathbf{R}$) which easily follow [see Budak, Samarskiĭ and Tikhonov (1964), Chap. II, Ex. 188] from the known values of the so-called Fresnel integrals

$$\int_0^{\infty} \cos x^2\,dx = \int_0^{\infty} \sin x^2\,dx = \frac{1}{2}\left(\frac{\pi}{2}\right)^{\frac{1}{2}}.$$

Answer.

$$u(x,t) = \frac{1}{(2\pi)^{\frac{1}{2}}}\int_{-\infty}^{\infty} \varphi(x - 2(at)^{\frac{1}{2}}\lambda)(\cos\lambda^2 + \sin\lambda^2)\,d\lambda$$
$$-\frac{1}{a(2\pi)^{\frac{1}{2}}}\int_{-\infty}^{\infty} \Psi(x - 2(at)^{\frac{1}{2}}\lambda)(\cos\lambda^2 - \sin\lambda^2)\,d\lambda$$
$$-\frac{1}{a(2\pi)^{\frac{1}{2}}}\int_0^t\int_{-\infty}^{\infty} G(x - 2[a(t-\tau)]^{\frac{1}{2}}\lambda, \tau)(\cos\lambda^2 - \sin\lambda^2)\,d\lambda,$$

where

$$\Psi(x) = \int_0^x\int_0^{\xi} \psi(\eta)\,d\eta d\xi, \quad G(x,t) = \int_0^x\int_0^{\xi} g(\eta,t)\,d\eta d\xi.$$

5.2.5.5 Show that if w is a (sufficiently smooth) solution of the initial-value problem

$$w_t(x,t) - iw_{xx}(x,t) = 0, \quad -\infty < x < \infty, \quad t > 0,$$
$$w(x,0) = \varphi(x), \quad -\infty < x < \infty,$$

where φ is a real function, then

$$u(x,t) = \operatorname{Re} w(x,t) \qquad (\operatorname{Re}\int_0^t w(x,\tau)\,d\tau) \text{ respectively)}$$

is a solution of the initial-value problem

$$u_{tt}(x,t) + u_{xxxx}(x,t) = 0, \quad -\infty < x < \infty, \quad t > 0,$$
$$u(x,0) = \varphi(x), \quad u_t(x,0) = 0, \quad -\infty < x < \infty$$
$$(u(x,0) = 0, \quad u_t(x,0) = \varphi(x), \quad -\infty < x < \infty \text{ respectively.})$$

Using this result solve Exercise 5.2.5.4.
Hint. See Exercise 4.2.17.4.

5.3 THE PLATE EQUATION

In this final section we give a brief outline of how to proceed in a somewhat more complicated problem concerning free transverse vibration of a plate. The case of a clamped circular plate can be solved by means of the separation of space variables. On the other hand, this method fails for a clamped rectangular plate since the equation $\Delta^2 u = 0$ with Dirichlet boundary conditions is not separable in rectangular coordinates. However, the separation-of-variables method can be used for a rectangular plate with all edges simply supported (Exercise 5.3.4).

5.3.1 Circular symmetric case

We shall deal with free transverse vibrations of a circular plate with the clamped edge. We restrict ourselves to the circular symmetric case which means that both the initial position φ and initial velocity ψ depend only on the distance from the centre of the plate (i.e. they have no angular space dependence) and u is looked for also as a function independent of ϑ, $u = u(r,t)$.

By Section 5.1 the problem is described by the following equations:

$$u_{tt}(r,t) + a^2\Delta_r^2(r,t) = 0, \qquad 0 < r < R,\ t > 0, \quad \left(a^2 = \frac{D}{\rho h}\right) \tag{5.35}$$

$$u(r,0) = \varphi(r), \quad 0 < r < R, \tag{5.36}$$

$$u_t(r,0) = \psi(r), \quad 0 < r < R, \tag{5.37}$$

$$u(R,t) = u_r(R,t) = 0, \qquad t > 0 \tag{5.38}$$

and

$$\Delta_r^2 = \frac{1}{r}\frac{\partial}{\partial r}\left\{r\frac{\partial}{\partial r}\left[\frac{1}{r}\frac{\partial}{\partial r}\left(r\frac{\partial}{\partial r}\right)\right]\right\}.$$

At first sight Eq. (5.35) might seem to be more difficult to treat than Eq. (5.8) but what is important is that it involves only a single independent space variable and the separation of variables can be applied.

5.3.2 Separation of variables

Setting

$$u(r,t) = e(r)\,v(t), \quad 0 < r < R,\ t > 0 \tag{5.39}$$

the following differential equations are to be satisfied

$$v''(t) + a^2\lambda\, v(t) = 0, \quad t > 0, \tag{5.40}$$

$$\Delta_r^2 e(r) - \lambda\, e(r) = 0, \quad 0 < r < R, \tag{5.41}$$

(where we consider $\lambda > 0$ without any loss of generality since analogously as in § 5.2.1 it can be shown that no eigenfunction is lost).

We shall first solve Eq. (5.41). It can be written as

$$(\Delta_r - \lambda^{\frac{1}{2}})(\Delta_r + \lambda^{\frac{1}{2}})\, e(r) = 0$$

or

$$(\Delta_r + \lambda^{\frac{1}{2}})(\Delta_r - \lambda^{\frac{1}{2}})\, e(r) = 0.$$

Hence, e is a solution of Eq. (5.41) if it solves either

$$(\Delta_r + \lambda^{\frac{1}{2}})\, e(r) = 0$$

or

$$(\Delta_r - \lambda^{\frac{1}{2}})\, e(r) = 0.$$

Introducing a new variable

$$z = \lambda^{\frac{1}{4}} r$$

and putting

$$f(z) = e\left(\frac{z}{\lambda^{\frac{1}{4}}}\right) = e(r),$$

we get, for f, the Bessel equation (1.53) and Eq. (1.54) respectively, both with $\alpha = 0$. A real fundamental set of solutions of Eq. (5.41) is formed by the functions $\mathrm{J}_0(\lambda^{\frac{1}{4}} r)$, $\mathrm{Y}_0(\lambda^{\frac{1}{4}} r)$, $\mathrm{I}_0(\lambda^{\frac{1}{4}} r)$ and $\mathrm{K}_0(\lambda^{\frac{1}{4}} r)$ [cf. § 1.2.8].

Up to now we have left aside the question of what boundary conditions are to be satisfied by the function e. First, with a view to physical relevance we take into account only solutions which remain bounded for $r \to 0+$ together with their first and second derivatives. So we discard the functions Y_0 and K_0 since any linear combination $\alpha \mathrm{Y}_0 + \beta \mathrm{K}_0$ is unbounded in the neighbourhood of the origin except for the case $2\alpha = \pi\beta$ when it is bounded together with its first derivative, however, the second one is unbounded. In this way, the solution of Eq. (5.41) will be looked for in the form

$$e(r) = C\, \mathrm{J}_0(\lambda^{\frac{1}{4}} r) + D\, \mathrm{I}_0(\lambda^{\frac{1}{4}} r).$$

The boundary conditions (5.38) require

$$e(R) = e'(R) = 0.$$

In order to determine so far arbitrary real constants C and D we must solve the system of equations

$$C\, \mathrm{J}_0(\mu) + D\, \mathrm{I}_0(\mu) = 0,$$
$$C\, \mathrm{J}_0'(\mu) + D\, \mathrm{I}_0'(\mu) = 0$$

where

$$\mu = \lambda^{\frac{1}{4}} R.$$

Since we are looking for a non-trivial solution the constants C and D must not be zero simultaneously, hence the determinant of the system must vanish. This leads to the transcendental equation

$$\mathrm{J}_0(\mu)\, \mathrm{I}_0'(\mu) - \mathrm{J}_0'(\mu)\, \mathrm{I}_0(\mu) = 0$$

to determine the values of μ. It may be proved that there exist infinitely many positive roots μ_k, $k \in \mathbb{N}$ and by Tables [for example, Jahnke and Emde (1945)] we find that approximate values of the first three roots are

$$\mu_1 \doteq 3.190, \quad \mu_2 \doteq 6.306, \quad \mu_3 \doteq 9.425.$$

It is easy to see that the eigenfunctions are given by

$$e_k(r) = N_k \left[\mathrm{J}_0\left(\mu_k \frac{r}{R}\right) - \frac{\mathrm{J}_0(\mu_k)}{\mathrm{I}_0(\mu_k)} \mathrm{I}_0\left(\mu_k \frac{r}{R}\right)\right], \quad k \in \mathbb{N}, \tag{5.42}$$

where N_k are arbitrary non-zero real constants. A simple calculation gives [see Budak, Samarskiĭ and Tikhonov (1964), Chap. VI, Ex. 71] that the system of these functions is orthogonal on $(0, R)$ with the weight r, i.e.

$$\int_0^R r\, e_k\, e_m \, dr = 0, \quad k \neq m.$$

To evaluate the normalization constants N_k we can use an analogous procedure as in § 5.2.1 to obtain

$$\int_0^R r\, e_k^2 \, dr = N_k^2 R^2 \mathrm{J}_0^2(\mu_k)\, \mathrm{I}_0^2(\mu_k).$$

Thus, choosing $N_k = (R|\mathrm{J}_0(\mu_k)|\, \mathrm{I}_0(\mu_k))^{-1}$ in (5.42) we obtain a system $\{e_k\}_{k\in\mathbb{N}}$ which is orthonormal and complete in the weight space $L_{2,r}(0, R)$ (see § 3.3.7).

On account of Eq. (5.40) we see that every function of the form

$$[A_k \cos(\omega_k t) + B_k \sin(\omega_k t)]\, e_k(r), \quad k \in \mathbb{N}, \tag{5.43$_1$}$$

where

$$\omega_k = a\lambda_k^{\frac{1}{2}} = \left(\frac{D}{\rho h}\right)^{\frac{1}{2}} \frac{\mu_k^2}{R^2} \tag{5.43$_2$}$$

and e_k are given by formula (5.42), is a solution of the problem (5.35), (5.38).

The solution of the problem with initial conditions has been always looked for as the superposition of the functions (5.43$_1$) with the orthonormality of the system of eigenfunctions being the basic tool. The fact that in our case the system $\{e_k\}_{k\in\mathbb{N}}$ is not orthonormal but orthonormal with a weight causes no essential changes and the usual procedure yields a formal solution in the form

$$u(r,t) = \sum_{k=1}^{\infty} \left[\varphi_k \cos(\omega_k t) + \frac{\psi_k}{\omega_k} \sin(\omega_k t)\right] e_k(r), \tag{5.44}$$

where

$$\varphi_k = \int_0^R r\, \varphi(r)\, e_k(r)\, dr,$$

$$\psi_k = \int_0^R r\, \psi(r)\, e_k(r)\, dr.$$

The series gives an "actual" solution of the problem (5.35) through (5.38) under suitable assumptions on φ and ψ. The detailed discussion which rests upon verifying the properties stated in § 5.3.3 is beyond the scope of this book.

5.3.3 A general remark on the Fourier method

To conclude we state a remark for readers with some knowledge of functional analysis. In an abstract setting, the applicability of the (time-"space") Fourier method for solving initial-"boundary" value problems for equations of the types $u_t + Lu = g$ and $u_{tt} + Lu + 2\alpha\, u_t = g$ $(\alpha \in \mathbf{R})$ is ensured under the following (relatively easily verifiable) assumptions on L. For details (such as in what sense the solutions are to be understood) we refer to Herrmann (1988). The pertinent conditions are:
1) L is a linear operator defined on a dense subspace D_L of a Hilbert space H [with a scalar product $(\cdot,\cdot)$];
2) L is symmetric: $(Lu, v) = (u, Lv), \quad u, v \in D_L$;
3) L is non-negative: $(Lu, u) \geq 0, \quad u \in D_L$;
4) the energy space of L is compactly embedded in H;
5) if $w \in H$ and $(Lu, w) = 0$ for all $u \in D_L$ then $w = 0$ (in other words, the null-space of the adjoint operator L^* is trivial).

5.3.4 Exercise

Investigate free transverse vibrations of a thin simply supported rectangular plate. Hint. Solve the problem given by the equation

$$u_{tt}(x,y,t) + a^2\Delta^2 u(x,y,t) = 0, \quad (x,y) \in \Omega,\ t > 0$$

where

$$\Omega = \{\,(x,y) \in \mathbf{R}^2 \mid 0 < x < c,\ 0 < y < d\},$$

by the initial conditions (5.9), (5.10) and by the boundary conditions

$$u(x,y,t) = \Delta\, u(x,y,t) = 0, \quad (x,y) \in \partial\Omega,\ t > 0.$$

Use the Fourier method.
Answer.

$$u(x,y,t) = \sum_{k=1}^{\infty}\sum_{m=1}^{\infty} [a_{km}\cos(\omega_{km}t) + b_{km}\sin(\omega_{km}t)]\sin\frac{k\pi x}{c}\sin\frac{m\pi y}{d},$$

$$\omega_{km}^2 = \frac{D}{\rho h}\lambda_{km} = \frac{D\pi^4}{\rho h}\left(\frac{k^2}{c^2} + \frac{m^2}{d^2}\right),$$

$$a_{km} = \frac{4}{cd}\int_0^c\int_0^d \varphi(x,y)\sin\frac{k\pi x}{c}\sin\frac{m\pi y}{d}\,dxdy,$$

$$b_{km} = \frac{4}{cd\omega_{km}}\int_0^c\int_0^d \psi(x,y)\sin\frac{k\pi x}{c}\sin\frac{m\pi y}{d}\,dxdy.$$

This solution is a classical one if

$$\varphi \in C^6(\bar{\Omega}), \quad \varphi|_{\partial\Omega} = \Delta\varphi|_{\partial\Omega} = \Delta^2\varphi|_{\partial\Omega} = 0,$$
$$\psi \in C^4(\bar{\Omega}), \quad \psi|_{\partial\Omega} = \Delta\psi|_{\partial\Omega} = 0.$$

Bibliography

Adams, R.A. (1975) *Sobolev Spaces*, Academic Press, New York.

Baldock, G.R. and Bridgeman, T. (1987) *The Mathematical Theory of Wave Motion*, Ellis Horwood, Chichester.

Bers, J., John, F. and Schechter, M. (1979) *Partial Differential Equations*, 5th ed., Lectures in Applied Mathematics, AMS, Providence – Rhode Island.

Budak, B.M., Samarskiĭ, A.A. and Tikhonov, A.N. (1964) *A Collection of Problems on Mathematical Physics*, International series of monographs in pure and applied mathematics 52, Pergamon Press, Oxford.

Cannon, J.R. (1984) *The One-dimensional Heat Equation*, Encyklopedia of Mathematics, Vol. 23, Addison–Wesley Inc., Advanced book program, Menlo Park – California.

Chorin, A.J. and Marsden, J.E. (1984) *A Mathematical Introduction to Fluid Mechanics*, Springer Verlag, New York.

Churchill, R.V. (1958) *Operational Mathematics*, 2nd ed., McGraw Hill, New York.

Courant, R. (1962) *Partial Differential Equations*, Interscience Publishers, John Wiley and Sons, New York.

Crank, J. (1984) *Free and Moving Boundary Problems*, University Press, Oxford.

Doetsch, G. (1967) *Guide to the Applications of Laplace Transforms*, 3rd ed., D. van Nostrand, London.

Erdélyi, A. *et al.* (1954) *Tables of Integral Transforms*, Vol. I, McGraw Hill, New York.

Friedman, A. (1964) *Partial Differential Equations of Parabolic Type*, Prentice–Hall Inc., Englewood Cliffs – New Jersey.

Friedman, B. (1956) *Principles and Techniques of Applied Mathematics*, John Wiley and Sons, New York.

Godunov, S.K. (1979) *Equations of Mathematical Physics*, (Russian), 2nd ed., Nauka, Moscow.

Herrmann, L. (1988) 'A study of an operator arising in the theory of circular plates', *Aplikace matematiky*, **33**, 337-353.

Hochstadt, H. (1964) *Differential Equations: A Modern Approach*, Holt, Rinehart and Winston, New York.

Jahnke, E. and Emde, F. (1945) *Tables of Functions*, 4th ed., Dover Publ. Inc., New York.

Jeffrey, A. (1976) *Quasilinear Hyperbolic Systems and Waves*, Pitman Publishing, London.

Kneschke, A. (1960) *Differentialgleichungen und Randwertprobleme, B. 2 Partielle Differentialgleichungen*, VEB Verlag Technik, Berlin.

Ladyzhenskaya, O.A., Solonnikov, V.A. and Ural'ceva, N.N. (1968) *Linear and Quasilinear Equations of Parabolic Type*, Translations of Mathematical Monographs, Vol. 23, AMS, Providence - Rhode Island.

McFarland, D.E., Smith, B.L. and Bernhart, W.D. (1972) *Analysis of Plates*, Spartan books, New York - Washington.

Mizohata, S. (1973) *The Theory of Partial Differential Equations*, University Press, Cambridge.

Özisik-Necati, M. (1968) *Boundary Value Problems of Heat Conduction*, International Textbook Company, Scranton - Pennsylvania.

Protter, M.H. and Weinberger, H.F. (1967) *Maximum Principles in Differential Equations*, Prentice–Hall Inc., Englewood Cliffs - New Jersey.

Roach, G.F. (1982) *Green's Functions*, 2nd ed., University Press, Cambridge.

Roberts, G.E. and Kaufman H. (1966) *Table of Laplace Transforms*, Saunders, Philadelphia.

Romanovskiĭ, R.K. (1985) 'On the Riemann matrices of the first and second kinds', (Russian), *Matem. Sbornik*, **127 (169)**, 494-501.

Rozhdestvenskiĭ, B.L. and Yanenko, N.N. (1983) *Systems of Quasilinear Equations and Their Applications to Gas Dynamics*, AMS, Providence - Rhode Island.

Smirnov, V.I. (1981) *A Course of Higher Mathematics*, Vol. IV, Pergamon Press, London.

Smoller, J. (1983) *Shock Waves and Reaction-Diffusion Equations*, Springer Verlag, New York.

Sneddon, I.N. (1951) *Fourier Transforms*, McGraw Hill, New York.

Stakgold, I. (1967, 1968) *Boundary Value Problem of Mathematical Physics*, Vol. I,II, The Macmillan Company, New York - London.

Timoshenko, S. and Woinowski-Krieger S. (1959) *Theory of Plates and Shells*, McGraw Hill, New York.

Timoshenko, S., Young, D.H. and Weaver Jr., W. (1974) *Vibrations Problems in Engineering*, 4th ed., John Wiley and Sons, New York.

Vejvoda, O. *et al.* (1982) *Partial Differential Equations: Time-Periodic Solutions*, Martinus Nijhoff, The Hague – Boston – London.

Watson, G.N. (1958) *A Treatise on the Theory of Bessel Functions*, 2nd ed., University Press, Cambridge.

Weidmann, J. (1980) *Linear Operators in Hilbert Spaces*, Graduate Texts in Mathematics 68, Springer Verlag, New York.

Whitham, G.B. (1974) *Linear and Nonlinear Waves*, A Wiley – Interscience Publication, John Wiley and Sons, New York.

Widder, D.W. (1975) *The Heat Equation*, Academic Press, New York.

Index

Mathematics and its Applications

Series Editor: G. M. BELL,

Emeritus Professor of Mathematics, King's College, University of London

Ball, M. A. **Mathematics in the Social and Life Sciences: Theories, Models and Methods**
de Barra, G. **Measure Theory and Integration**
Barták, J., Herrmann, L., Lovicar, V. & Vejvoda, O. **Partial Differential Equations of Evolution**
Bejancu, A. **Finsler Geometry and Applications**
Bell, G. M. and Lavis, D. A. **Statistical Mechanics of Lattice Models, Vols. 1 & 2**
Berry, J. S., Burghes, D. N., Huntley, I. D., James, D. J. G. & Moscardini, A. O. **Mathematical Modelling Courses**
Berry, J. S., Burghes, D. N., Huntley, I. D., James, D. J. G. & Moscardini, A. O. **Mathematical Modelling Methodology, Models and Micros**
Berry, J. S., Burghes, D. N., Huntley, I. D., James, D. J. G. & Moscardini, A. O. **Teaching and Applying Mathematical Modelling**
Blum, W. **Applications and Modelling in Learning and Teaching Mathematics**
Brown, R. **Topology: A Geometric Account of General Topology, Homotopy Types and the Fundamental Groupoid**
Burghes, D. N. & Borrie, M. **Modelling with Differential Equations**
Burghes, D. N. & Downs, A. M. **Modern Introduction to Classical Mechanics and Control**
Burghes, D. N. & Graham, A. **Introduction to Control Theory, Including Optimal Control**
Burghes, D. N., Huntley, I. & McDonald, J. **Applying Mathematics**
Burghes, D. N. & Wood, A. D. **Mathematical Models in the Social, Management and Life Sciences**
Butkovskiy, A. G. **Green's Functions and Transfer Functions Handbook**
Cartwright, M. **Fourier Methods: for Mathematicians, Scientists and Engineers**
Cerny, I. **Complex Domain Analysis**
Chorlton, F. **Textbook of Dynamics, 2nd Edition**
Chorlton, F. **Vector and Tensor Methods**
Cohen, D. E. **Computability and Logic**
Cordier, J.-M. & Porter, T. **Shape Theory: Categorical Methods of Approximation**
Crapper, G. D. **Introduction to Water Waves**
Cross, M. & Moscardini, A. O. **Learning the Art of Mathematical Modelling**
Cullen, M. R. **Linear Models in Biology**
Dunning-Davies, J. **Mathematical Methods for Mathematicians, Physical Scientists and Engineers**
Eason, G., Coles, C. W. & Gettinby, G. **Mathematics and Statistics for the Biosciences**
El Jai, A. & Pritchard, A. J. **Sensors and Controls in the Analysis of Distributed Systems**
Exton, H. **Multiple Hypergeometric Functions and Applications**
Exton, H. **Handbook of Hypergeometric Integrals**
Exton, H. ***q*-Hypergeometric Functions and Applications**
Faux, I. D. & Pratt, M. J. **Computational Geometry for Design and Manufacture**
Firby, P. A. & Gardiner, C. F. **Surface Topology**
Gardiner, C. F. **Modern Algebra**
Gardiner, C. F. **Algebraic Structures**
Gasson, P. C. **Geometry of Spatial Forms**
Goodbody, A. M. **Cartesian Tensors**
Goult, R. J. **Applied Linear Algebra**
Graham, A. **Kronecker Products and Matrix Calculus: with Applications**
Graham, A. **Matrix Theory and Applications for Engineers and Mathematicians**
Graham, A. **Nonnegative Matrices and Applicable Topics in Linear Algebra**
Griffel, D. H. **Applied Functional Analysis**
Griffel, D. H. **Linear Algebra and its Applications: Vol. 1, A First Course; Vol. 2, More Advanced**
Guest, P. B. **The Laplace Transform and Applications**
Hanyga, A. **Mathematical Theory of Non-linear Elasticity**
Harris, D. J. **Mathematics for Business, Management and Economics**
Hart, D. & Croft, A. **Modelling with Projectiles**
Hoskins, R. F. **Generalised Functions**
Hoskins, R. F. **Standard and Nonstandard Analysis**
Hunter, S. C. **Mechanics of Continuous Media, 2nd (Revised) Edition**
Huntley, I. & Johnson, R. M. **Linear and Nonlinear Differential Equations**
Irons, B. M. & Shrive, N. G. **Numerical Methods in Engineering and Applied Science**
Ivanov, L. L. **Algebraic Recursion Theory**
Johnson, R. M. **Theory and Applications of Linear Differential and Difference Equations**
Johnson, R. M. **Calculus: Theory and Applications in Technology and the Physical and Life Sciences**
Jones, R. H. & Steele, N. C. **Mathematics in Communication Theory**
Jordan, D. **Geometric Topology**
Kelly, J. C. **Abstract Algebra**
Kim, K. H. & Roush, F. W. **Applied Abstract Algebra**
Kim, K. H. & Roush, F. W. **Team Theory**
Kosinski, W. **Field Singularities and Wave Analysis in Continuum Mechanics**
Krishnamurthy, V. **Combinatorics: Theory and Applications**
Lindfield, G. & Penny, J. E. T. **Microcomputers in Numerical Analysis**
Livesley, K. **Mathematical Methods for Engineers**
Lord, E. A. & Wilson, C. B. **The Mathematical Description of Shape and Form**
Malik, M., Riznichenko, G. Y. & Rubin, A. B. **Biological Electron Transport Processes and their Computer Simulation**

Mathematics and its Applications

Series Editor: G. M. BELL,

Emeritus Professor of Mathematics, King's College, University of London

Martin, D.	**Manifold Theory: An Introduction for Mathematical Physicists**
Massey, B. S.	**Measures in Science and Engineering**
Meek, B. L. & Fairthorne, S.	**Using Computers**
Menell, A. & Bazin, M.	**Mathematics for the Biosciences**
Mikolas, M.	**Real Functions and Orthogonal Series**
Moore, R.	**Computational Functional Analysis**
Moshier, S. L. B.	**Methods and Programs for Mathematical Functions**
Murphy, J. A., Ridout, D. & McShane, B.	**Numerical Analysis, Algorithms and Computation**
Nonweiler, T. R. F.	**Computational Mathematics: An Introduction to Numerical Approximation**
Norcliffe, A. & Slater, G.	**Mathematics of Software Construction**
Ogden, R. W.	**Non-linear Elastic Deformations**
Oldknow, A.	**Microcomputers in Geometry**
Oldknow, A. & Smith, D.	**Learning Mathematics with Micros**
O'Neill, M. E. & Chorlton, F.	**Ideal and Incompressible Fluid Dynamics**
O'Neill, M. E. & Chorlton, F.	**Viscous and Compressible Fluid Dynamics**
Page, S. G.	**Mathematics: A Second Start**
Prior, D. & Moscardini, A. O.	**Model Formulation Analysis**
Rankin, R. A.	**Modular Forms**
Scorer, R. S.	**Environmental Aerodynamics**
Shivamoggi, B. K.	**Stability of Parallel Gas Flows**
Smith, D. K.	**Network Optimisation Practice: A Computational Guide**
Srivastava, H. M. & Manocha, L.	**A Treatise on Generating Functions**
Stirling, D. S. G.	**Mathematical Analysis**
Sweet, M. V.	**Algebra, Geometry and Trigonometry in Science, Engineering and Mathematics**
Temperley, H. N. V.	**Graph Theory and Applications**
Temperley, H. N. V.	**Liquids and Their Properties**
Thom, R.	**Mathematical Models of Morphogenesis**
Toth, G.	**Harmonic and Minimal Maps and Applications in Geometry and Physics**
Townend, M. S.	**Mathematics in Sport**
Townend, M. S. & Pountney, D. C.	**Computer-aided Engineering Mathematics**
Trinajstic, N.	**Mathematical and Computational Concepts in Chemistry**
Twizell, E. H.	**Computational Methods for Partial Differential Equations**
Vince, A. and Morris, C.	**Discrete Mathematics for Computing**
Walton, K., Marshall, J., Gorecki, H. & Korytowski, A.	**Control Theory for Time Delay Systems**
Warren, M. D.	**Flow Modelling in Industrial Processes**
Wheeler, R. F.	**Rethinking Mathematical Concepts**
Willmore, T. J.	**Total Curvature in Riemannian Geometry**
Willmore, T. J. & Hitchin, N.	**Global Riemannian Geometry**

Statistics, Operational Research and Computational Mathematics

Editor: B. W. CONOLLY,

Emeritus Professor of Mathematics (Operational Research), Queen Mary College, University of London

Abaffy, J. & Spedicato, E.	**ABS Projection Algorithms: Mathematical Techniques for Linear and Nonlinear Equations**
Beaumont, G. P.	**Introductory Applied Probability**
Beaumont, G. P.	**Probability and Random Variables**
Conolly, B. W.	**Techniques in Operational Research: Vol. 1, Queueing Systems**
Conolly, B. W.	**Techniques in Operational Research: Vol. 2, Models, Search, Randomization**
Conolly, B. W.	**Lecture Notes in Queueing Systems**
Conolly, B. W. & Pierce, J. G.	**Information Mechanics: Transformation of Information in Management, Command, Control and Communication**
French, S.	**Sequencing and Scheduling: Mathematics of the Job Shop**
French, S.	**Decision Theory: An Introduction to the Mathematics of Rationality**
Griffiths, P. & Hill, I. D.	**Applied Statistics Algorithms**
Hartley, R.	**Linear and Non-linear Programming**
Jolliffe, F. R.	**Survey Design and Analysis**
Jones, A. J.	**Game Theory**
Kapadia, R. & Andersson, G.	**Statistics Explained: Basic Concepts and Methods**
Lootsma, F.	**Operational Research in Long Term Planning**
Moscardini, A. O. & Robson, E. H.	**Mathematical Modelling for Information Technology**
Moshier, S. L. B.	**Mathematical Functions for Computers**
Oliveira-Pinto, F.	**Simulation Concepts in Mathematical Modelling**
Ratschek, J. & Rokne, J.	**New Computer Methods for Global Optimization**
Schendel, U.	**Introduction to Numerical Methods for Parallel Computers**

Schendel, U.	**Sparse Matrices**
Sehmi, N. S.	**Large Order Structural Eigenanalysis Techniques: Algorithms for Finite Element Systems**
Sewell, G.	**Computational Methods of Linear Algebra**
Sharma, O. P.	**Markovian Queues**
Smith, D. K.	**Dynamic Programming: A Practical Introduction**
Späth, H.	**Mathematical Software for Linear Regression**
Stoodley, K. D. C.	**Applied and Computational Statistics: A First Course**
Stoodley, K. D. C., Lewis, T. & Stainton, C. L. S.	**Applied Statistical Techniques**
Thomas, L. C.	**Games, Theory and Applications**
Vajda, S.	**Fibonacci & Lucas Numbers, and the Golden Section**
Whitehead, J. R.	**The Design and Analysis of Sequential Clinical Trials**